Contributions in Mathematical and Computational Sciences • Volume 4

Editors
Hans Georg Bock
Willi Jäger
Otmar Venjakob

For other titles published in this series, go to
http://www.springer.com/series/8861

Hans Georg Bock
Thomas Carraro
Willi Jäger
Stefan Körkel
Rolf Rannacher
Johannes P. Schlöder
Editors

Model Based Parameter Estimation

Theory and Applications

Editors
Hans Georg Bock
Willi Jäger
Stefan Körkel
Johannes P. Schlöder
Interdisciplinary Center
   for Scientific Computing (IWR)
Heidelberg University
Heidelberg, Germany

Thomas Carraro
Rolf Rannacher
Institute for Applied Mathematics
Heidelberg University
Heidelberg, Germany

ISBN 978-3-642-30366-1   ISBN 978-3-642-30367-8 (eBook)
DOI 10.1007/978-3-642-30367-8
Springer Heidelberg New York Dordrecht London

Library of Congress Control Number: 2012954944

Math. Subj. Classification (2010): 34-XX, 35-XX, 49-XX, 62-XX, 65-XX, 68-XX, 90-XX, 92-XX

© Springer-Verlag Berlin Heidelberg 2013
This work is subject to copyright. All rights are reserved by the Publisher, whether the whole or part of
the material is concerned, specifically the rights of translation, reprinting, reuse of illustrations, recitation,
broadcasting, reproduction on microfilms or in any other physical way, and transmission or information
storage and retrieval, electronic adaptation, computer software, or by similar or dissimilar methodology
now known or hereafter developed. Exempted from this legal reservation are brief excerpts in connection
with reviews or scholarly analysis or material supplied specifically for the purpose of being entered
and executed on a computer system, for exclusive use by the purchaser of the work. Duplication of
this publication or parts thereof is permitted only under the provisions of the Copyright Law of the
Publisher's location, in its current version, and permission for use must always be obtained from Springer.
Permissions for use may be obtained through RightsLink at the Copyright Clearance Center. Violations
are liable to prosecution under the respective Copyright Law.
The use of general descriptive names, registered names, trademarks, service marks, etc. in this publication
does not imply, even in the absence of a specific statement, that such names are exempt from the relevant
protective laws and regulations and therefore free for general use.
While the advice and information in this book are believed to be true and accurate at the date of
publication, neither the authors nor the editors nor the publisher can accept any legal responsibility for
any errors or omissions that may be made. The publisher makes no warranty, express or implied, with
respect to the material contained herein.

Printed on acid-free paper

Springer is part of Springer Science+Business Media (www.springer.com)

Preface to the Series

Contributions to Mathematical and Computational Sciences

Mathematical theories and methods and effective computational algorithms are crucial in coping with the challenges arising in the sciences and in many areas of their application. New concepts and approaches are necessary in order to overcome the complexity barriers particularly created by nonlinearity, high-dimensionality, multiple scales and uncertainty. Combining advanced mathematical and computational methods and computer technology is an essential key to achieving progress, often even in purely theoretical research.

The term mathematical sciences refers to mathematics and its genuine subfields, as well as to scientific disciplines that are based on mathematical concepts and methods, including sub-fields of the natural and life sciences, the engineering and social sciences and recently also of the humanities. It is a major aim of this series to integrate the different sub-fields within mathematics and the computational sciences, and to build bridges to all academic disciplines, to industry and other fields of society, where mathematical and computational methods are necessary tools for progress. Fundamental and application-oriented research will be covered in proper balance.

The series will further offer contributions on areas at the frontier of research, providing both detailed information on topical research, as well as surveys of the state-of-the-art in a manner not usually possible in standard journal publications. Its volumes are intended to cover themes involving more than just a single "spectral line" of the rich spectrum of mathematical and computational research.

The Mathematics Center Heidelberg (MATCH) and the Interdisciplinary Center for Scientific Computing (IWR) with its Heidelberg Graduate School of Mathematical and Computational Methods for the Sciences (HGS) are in charge of providing and preparing the material for publication. A substantial part of the material will be acquired in workshops and symposia organized by these institutions in topical areas of research. The resulting volumes should be more than just proceedings collecting papers submitted in advance. The exchange of information and the discussions during the meetings should also have a substantial influence on the contributions.

This series is a venture posing challenges to all partners involved. A unique style attracting a larger audience beyond the group of experts in the subject areas of specific volumes will have to be developed.

Springer Verlag deserves our special appreciation for its most efficient support in structuring and initiating this series.

Heidelberg University
Germany

Hans Georg Bock
Willi Jäger
Otmar Venjakob

Preface

The articles presented in this book are focused on mathematical and numerical methods for parameter estimation on differential equations-based models. Applications are taken from chemistry, electro-chemistry, environmental physics, biology, medicine, image processing and computer vision. This collection results from a workshop held in summer 2009 at Heidelberg University, supported jointly by the MAThematics Center Heidelberg (MATCH) and the Heidelberg Graduate School of Mathematical and Computational Methods for the Sciences (HGS MathComp). This activity is part of a larger educational and research program promoted by the German "Excellence Initiative." These two institutions are central in providing an interdisciplinary environment for students, researchers and external cooperators, including industrial partners. Special emphasis is given to the creation of research links between different areas of mathematics, computer science and several applied disciplines. In particular, HGS MathComp is located at the Interdisciplinary Center for Scientific Computing (IWR), where different scientific disciplines are interconnected with a special focus on the application of mathematical methods. Within this program the aim of the workshop "Model Based Parameter Estimation: Theory and Applications" was to create a platform for doctoral students and researchers from different disciplines to start interdisciplinary collaborations within institutes in Heidelberg and with external collaboration partners. The main topic of the workshop was the application of mathematical models and numerical methods for the solution of parameter estimation problems in various scientific areas. Well-founded methods of applied mathematics and computer-based modeling are essential to improving the quality of interdisciplinary research. As a result, scientific computing for the simulation of complex phenomena has become a standard in physics, chemistry, the life sciences and engineering, in research and industrial applications. Nevertheless, the use of optimization techniques in several research areas, though continuously increasing over the last years, still needs to be improved with the most advanced methods within interdisciplinary projects. To this end the workshop was mainly conceived for the use of model-based optimization for parameter estimation. The

vii

invited articles in this volume give an overview of related advanced computational methods and of problems occurring in prototypical applications. It is our hope that this book can serve as starting point for even more challenging interdisciplinary projects.

The contributions are ordered in two main groups according to the following organization:

- The first part of the book introduces state-of-the-art mathematical and numerical methods for solving parameter estimation problems. More concretely, mathematical and numerical methods for parameter estimation and optimum experimental design in the framework of differential equations are presented. Adaptive methods for the finite element solution of parameter estimation problems with models based on partial differential equations are shown, and robust methods for the parameter identification for data with outliers are introduced. Furthermore, a fast numerical method for the optimal sensor activation is presented.
- The second part of the book is focused on applications of parameter estimation in the following fields: the life sciences, which comprise systems biology and applications to biomedical sciences, chemistry, physics, image processing and computer vision. More specifically, the integration of experimental data and mathematical models is presented on quantitative analysis of signaling pathways, fluorescence recovery after photo-bleaching (FRAP), gene expression profiling, treatment of infectious diseases and stem cell transplantation. In addition, applications in combustion chemistry, catalytic reactors, heat transport in tubular reactors and solid oxide fuel cells (SOFC) are shown. Application of robust parameter estimation techniques in environmental physics is presented. Examples of parameter estimation in image processing and computer vision complete the second part of the book.

The editors would like to thank all the participants of the workshop and the contributors to this volume. Finally, the support by MATCH and the HGS Math-Comp is gratefully acknowledged.

Contents

Parameter Estimation and Optimum Experimental Design for Differential Equation Models 1
Hans Georg Bock, Stefan Körkel, and Johannes P. Schlöder

Adaptive Finite Element Methods for Parameter Identification Problems 31
Boris Vexler

Gauss–Newton Methods for Robust Parameter Estimation 55
Tanja Binder and Ekaterina Kostina

An Optimal Scanning Sensor Activation Policy for Parameter Estimation of Distributed Systems 89
Dariusz Uciński

Interaction Between Experiment, Modeling and Simulation of Spatial Aspects in the JAK2/STAT5 Signaling Pathway 125
Elfriede Friedmann, Andrea C. Pfeifer, Rebecca Neumann,
Ursula Klingmüller, and Rolf Rannacher

The Importance and Challenges of Bayesian Parameter Learning in Systems Biology 145
Johanna Mazur and Lars Kaderali

Experiment Setups and Parameter Estimation in Fluorescence Recovery After Photobleaching Experiments: A Review of Current Practice 157
J. Beaudouin, Mario S. Mommer, Hans Georg Bock,
and Roland Eils

Drug Resistance in Infectious Diseases: Modeling, Parameter Estimation and Numerical Simulation 171
Le Thi Thanh An and Willi Jäger

Mathematical Models of Hematopoietic Reconstitution After Stem Cell Transplantation 191
Anna Marciniak-Czochra and Thomas Stiehl

Combustion Chemistry and Parameter Estimation 207
Marc Fischer and Uwe Riedel

Numerical Simulation of Catalytic Reactors by Molecular-Based Models 227
Olaf Deutschmann and Steffen Tischer

Model-Based Design of Experiments for Estimating Heat-Transport Parameters in Tubular Reactors 251
Alexander Badinski and Daniel Corbett

Parameter Estimation for a Reconstructed SOFC Mixed-Conducting LSCF-Cathode 267
Thomas Carraro and Jochen Joos

An Application of Robust Parameter Estimation in Environmental Physics 287
Alexandra G. Herzog and Felix R. Vogel

Parameter Estimation in Image Processing and Computer Vision 311
Christoph S. Garbe and Björn Ommer

Parameter Estimation and Optimum Experimental Design for Differential Equation Models

Hans Georg Bock, Stefan Körkel, and Johannes P. Schlöder

Abstract This article reviews state-of-the-art methods for parameter estimation and optimum experimental design in optimization based modeling. For the calibration of differential equation models for nonlinear processes, constrained parameter estimation problems are considered. For their solution, numerical methods based on the boundary value problem method optimization approach consisting of multiple shooting and a generalized Gauß–Newton method are discussed. To suggest experiments that deliver data to minimize the statistical uncertainty of parameter estimates, optimum experimental design problems are formulated, an intricate class of non-standard optimal control problems, and derivative-based methods for their solution are presented.

Keywords Gauß–Newton method • Multiple shooting • Nonlinear differential-algebraic equations • Optimum experimental design • Parameter estimation • Variance–covariance matrix

1 Introduction

Dynamic processes in science, engineering or medicine often can be described by differential equation models. Solutions of the model equations, usually obtained by numerical methods, give simulations of the process behavior under various conditions.

To obtain realistic model output results, the model has to be validated. An important task for this is the calibration of the model, i.e. to explain experimental data by appropriate simulation results. In order to fit the model to the data, unknown

---

H.G. Bock · S. Körkel (✉) · J.P. Schlöder
Interdisciplinary Center for Scientific Computing, Heidelberg University,
Im Neuenheimer Feld 368, 69120 Heidelberg, Germany
e-mail: stefan.koerkel@iwr.uni-heidelberg.de

H.G. Bock et al. (eds.), *Model Based Parameter Estimation*, Contributions
in Mathematical and Computational Sciences 4, DOI 10.1007/978-3-642-30367-8_1,
© Springer-Verlag Berlin Heidelberg 2013

quantities in the model, called parameters, have to be estimated by minimizing a norm of the residuals between data and model response.

A fit of given data yet does not mean a validated model. Due to the randomness of experimental errors, the parameter estimate is also a random variable. Only if the statistical uncertainty of the parameters is small, simulations can also predict the outcome of future experiments and explain the behavior of the real process in a qualitatively and quantitatively correct way.

The statistical uncertainty of a parameter estimate depends on layout, setup, control and sampling of the experiments. Experimental design optimization problems minimize a function of the variance–covariance matrix of the parameter estimation problem. Optimal experiments can reduce the experimental effort for model validation drastically.

In this article we review the formulation of parameter estimation and optimum experimental design problems and discuss methods for their numerical solution.

The article is structured as follows. After this introduction, Sect. 2 introduces the class of differential equation models we consider in this article. In Sect. 3, parameter estimation problems are formulated. Section 4 deals with parameterization of the model equations by multiple shooting and Sect. 5 with the evaluation of derivatives. In Sect. 6, the generalized Gauß–Newton method is reviewed, and in Sect. 7, the local convergence behavior of this method is analyzed. Section 8 treats optimum experimental design, the problem formulation, variants and a sketch of numerical solution approaches.

2 Differential Equation Models

In this article we consider models of dynamic processes described by differential algebraic equations (DAE):

$$\dot{y}(t) = f(t, y(t), z(t), p) \tag{1}$$

$$0 = g(t, y(t), z(t), p) \tag{2}$$

with $t \in I := [t_0; t_f]$. $y : I \to \mathbb{R}^{n_y}$ are the differential states, $z : I \to \mathbb{R}^{n_z}$, the algebraic states. We subsume all states $x := (y, z) : I \to \mathbb{R}^{n_x} = \mathbb{R}^{n_y + n_z}$. Further the model depends on variables $p \in \mathbb{R}^{n_p}$ the values of which are known only imprecisely. We call these variables *parameters*.

We assume that $f : \mathbb{R} \times \mathbb{R}^{n_y} \times \mathbb{R}^{n_z} \times \mathbb{R}^{n_p} \to \mathbb{R}^{n_y}$ and $g : \mathbb{R} \times \mathbb{R}^{n_y} \times \mathbb{R}^{n_z} \times \mathbb{R}^{n_p} \to \mathbb{R}^{n_z}$ are sufficiently piecewise smooth for the solution of DAE initial value problems by numerical integration schemes.

In applications from, e.g., chemical reaction kinetics, chemical engineering, systems biology or epidemiology, the DAE systems are usually highly nonlinear in states and parameters and often stiff. For spatially distributed systems described by partial differential equations we apply a semi-discretization by a method of lines end obtain high-dimensional and sparse systems where the Courant–Friedrichs–Lewy condition also causes stiffness.

Further we assume that the DAE is of index 1, i.e.

$$\text{rank } \frac{\partial g}{\partial z} = n_z. \tag{3}$$

Then the existence and uniqueness of solutions can be established based on theory of ordinary differential equations.

We consider initial value problems with initial conditions

$$y(t_0) = y_0(p). \tag{4}$$

Initial values are required only for differential states, initial algebraic states are determined by the *consistency* of the algebraic equations

$$g(t_0, y(t_0), z(t_0), p) = 0. \tag{5}$$

For the solution by numerical methods it is required that the initial values for y and z are consistent. Then the numerical solutions stay consistent on the whole integration interval. To meet this requirement without always solving (5), we use modifications of the DAE which are automatically consistent, so-called relaxed formulations:

$$\dot{y}(t) = f(t, y(t), z(t), p) \tag{6}$$

$$0 = g(t, y(t), z(t), p) - \beta \left(\frac{t - t_0}{t_f - t_0} \right) \cdot g(t_0, y(t_0), z(t_0), p) \tag{7}$$

$$y(t_0) = y_0(p) \tag{8}$$

$$z(t_0) = s_0^z \tag{9}$$

with non-increasing $\beta : [0; 1] \to [0; 1]$, $\beta(0) = 1$. The modified DAE is consistent for every choice of s_0^z and can be solved by numerical integration. In optimization problems with DAE constraints, we then treat the consistency conditions (5) as additional constraints of the optimization problem, and s_0^z as additional variables.

Further, multiple-point boundary conditions may be given

$$r(x(t_0), \ldots, x(t_N), p) = 0 \tag{10}$$

respectively in a separable formulation

$$\sum_{i=1}^{N} r_i(x(t_i), p) = 0$$

with the time points $t_0 \leq t_1 \leq \ldots \leq t_N \leq t_f$ and the functions $r : \mathbb{R}^{n_x} \times \ldots \times \mathbb{R}^{n_x} \times \mathbb{R}^{n_p} \to \mathbb{R}^{n_r}$ respectively $r_i : \mathbb{R}^{n_x} \times \mathbb{R}^{n_p} \to \mathbb{R}^{n_r}$ sufficiently smooth. In the following we assume that the initial conditions (4) are part of the boundary conditions (10).

3 Parameter Estimation Problems

Let experimental data be given with measurement values $\eta_i \in \mathbb{R}$, $i = 1, \ldots, M$, measured at times $t_0 \le t_1 \le \ldots \le t_M \le t_f$ (not necessarily the same as the times in the boundary conditions above but for the sake of simplicity not distinguished in the notation here). We assume that the data values correspond to model responses $h_i(\bar{x}(t_i), \bar{p})$ for the true values $\bar{x}$ and $\bar{p}$ with $h_i : \mathbb{R}^{n_x} \times \mathbb{R}^{n_p} \to \mathbb{R}$ sufficiently smooth, $i = 1, \ldots, M$. Experimental data is always afflicted with measurement errors originating from various sources in the measurement process. Modeling these errors may be an intricate issue. In this article, we make the assumption that the measurement errors are additive, independent and normally distributed

$$\epsilon_i := \eta_i - h_i(\bar{x}(t_i), \bar{p}) \sim \mathcal{N}(0, \sigma_i^2)$$

with known variances σ_i^2, $i = 1, \ldots, M$ and $\mathrm{cov}(\epsilon_i, \epsilon_j) = 0$, $i \ne j$.

For the estimation of the model parameters, we minimize the sum of squares of the residuals $\eta_i - h_i(x(t_i), p)$ weighted by the inverse of the variances σ_i^2. This least squares estimator can be interpreted as a maximum likelihood estimator for the parameters, see e.g. [26].

Writing the model equations and boundary conditions as constraints, we obtain a constrained parameter estimation problem.

$$\min_{p,x} \frac{1}{2} \sum_{i=1}^{M} \left(\frac{\eta_i - h_i(x(t_i), p)}{\sigma_i} \right)^2 \tag{11}$$

$$\text{s.t. } \dot{y}(t) = f(t, y(t), z(t), p)$$

$$0 = g(t, y(t), z(t), p)$$

$$0 = r(x(t_0), \ldots, x(t_N), p) \tag{12}$$

With the vectors $h(x, p) := (h_1(x(t_1), p), \ldots, h_M(x(t_M), p))^T$ and $\eta := (\eta_1, \ldots, \eta_M)^T$ and the matrix $\Sigma = \mathrm{diag}(\sigma_1, \ldots, \sigma_M)$, the error model can be written shortly as

$$\eta - h(x, p) \sim \mathcal{N}(0, \Sigma^2) \tag{13}$$

and the objective (11) as

$$\min \frac{1}{2} (\eta - h(x, p))^T \Sigma^{-2} (\eta - h(x, p)). \tag{14}$$

In a more general setting, a non-diagonal positive-definite matrix $\Sigma^2 \in \mathbb{R}^{M \times M}$ can also be considered for correlated measurement errors. Different error distributions require different estimators, see [26], e.g. the l_1 estimation, see [7, 10]. For an overview article on models, criteria and algorithms for nonlinear parameter estimation problems see [25] a comparison of derivative-free methods for nonlinear differential-algebraic models can be found in [21].

4 Parameterization of the DAE by Multiple Shooting

To solve the parameter estimation problem (11)–(12), we transform it into a finite-dimensional problem.

A simple and obvious approach is *single shooting* also known as the black box approach. It consists in solving the full model equations, i.e. integrating the DAE on the full interval $[t_0, t_f]$. In order to apply a numerical integrator, the relaxed formulation (6)–(9) has to be used and the consistency conditions (5) have to be added as additional constraints to the problem with additional variables s_0^z. With the solution obtained from this single shooting integration, the least squares terms and the boundary conditions can be evaluated. Then we obtain a finite dimensional nonlinear least squares problem in the variables $v := (p, s_0^z)$:

$$\min_v \frac{1}{2} \| F_1(v) \|_2^2 \tag{15}$$

$$\text{s.t.} \ F_2(v) = 0 \tag{16}$$

where

$$F_1(v) = \Sigma^{-1} (\eta - h(x(s_0^z, p), p))$$

$$F_2(v) = \begin{pmatrix} r(x(t_0; s_0^z, p), \dots, x(t_N; s_0^z, p), p) \\ g(t_0, y_0(p), s_0^z, p) \end{pmatrix}.$$

$x(t; s_0^z, p)$ is the DAE solution $x =: x(s_0^z, p)$ evaluated at t. Problem (15)–(16) can be solved by a generalized Gauß–Newton method, see Section 6. To this end, derivatives $J_1 := \frac{d F_1}{d v}$ and $J_2 := \frac{d F_2}{d v}$ are needed, methods for their evaluation are discussed in Sect. 5.

The single shooting approach is easy to implement, but it is difficult to get good initial guesses for the variables v and hence difficult to start close enough to the solution in order to achieve convergence of the generalized Gauß–Newton method. This holds because of the high nonlinearity which arises due to the propagation of perturbations of the model as described by the following Lemma.

Lemma 1. *Let $f \in C^0(D)$ and Lipschitz continuous on $\bar{D} = [a, b] \times \{ \| y - y_0 \| \leq K \} \subset D$ with Lipschitz constant $L < \infty$. The solutions y, w of*

$$\dot{y}(t) = f(t, y(t)), \qquad\qquad y(t_0) = y_0$$

$$\dot{w}(t) = f(t, w(t)) + \delta f(t, w(t)), \qquad w(t_0) = w_0 = y_0 + \delta y_0$$

both exist on $[t_0, t_f] \subset [a, b]$ in $\bar{D}$. Under the two assumptions $\| \delta y_0 \| \leq \varepsilon_1$ and $\| \delta f(t, w(t)) \| \leq \varepsilon_2$ for all $(t, w(t))$, the deviation $\delta y(t) := w(t) - y(t)$ satisfies

$$\| \delta y(t) \| \leq \varepsilon_1 \exp(L \cdot (t - t_0)) + \varepsilon_2 \exp(L \cdot (t - t_0))(t - t_0)$$

The proof uses Gronwall's Lemma.

In other words, perturbations of initial values and the right-hand-side function, in particular by uncertain parameter values, may propagate exponentially in time.

As remedy we suggest to use the boundary value problem method optimization approach with a *multiple shooting* parameterization of the differential equation. We define a time grid with multiple shooting nodes

$$t_0 = \tau_0 < \tau_1 < \ldots < \tau_K < \tau_{K+1} = t_f$$

and multiple shooting variables $s_k := (s_k^y, s_k^z)$, $k = 0, \ldots, K$ where s_0^y satisfies $s_0^y = y_0(p)$. For $k = 0, \ldots, K$, we solve initial value problems on the multiple shooting intervals, i.e. for $t \in [\tau_k; \tau_{k+1}[$ we solve

$$\dot{y}(t) = f(t, y(t), z(t), p)$$
$$0 = g(t, y(t), z(t), p) - \beta \left(\frac{t - \tau_k}{\tau_{k+1} - \tau_k} \right) \cdot g(\tau_k, s_k^y, s_k^z, p)$$
$$y(\tau_k) = s_k^y$$
$$z(\tau_k) = s_k^z.$$

Here we apply the relaxed formulation of the DAE to start consistent in τ_k and stay consistent in $[\tau_k; \tau_{k+1}[$. The solutions $x(t) =: x(t; \tau_k, s_k, p)$ for $t \in [\tau_k; \tau_{k+1}[$ form a piecewise solution $x(t; s, p)$ for $t \in [\tau_0; \tau_f]$.

We need to satisfy constraints, namely the boundary conditions

$$0 = r(x(t_0; s, p), \ldots, x(t_N; s, p), p),$$

including the initial conditions for the differential states

$$s_0^y = y_0(p),$$

and additionally, at the multiple shooting nodes, continuity conditions for the differential states

$$0 = y(\tau_{k+1}; \tau_k, s_k, p) - s_{k+1}^y, \quad k = 0, \ldots, K - 1,$$

and consistency conditions for the algebraic states

$$0 = g(\tau_k, s_k^y, s_k^z, p), \quad k = 0, \ldots, K$$

to eventually ensure continuity and consistency of the solution.

Substituting $x = x(s, p)$ in the parameter estimation problem (11)–(12), we obtain a finite-dimensional constrained nonlinear least squares problem in the variables $v = (s_0, s_1, s_2, \ldots, s_K, p) = (s, p)$:

Parameter Estimation and Optimum Experimental Design

$$\min_v \frac{1}{2} \| F_1(v) \|_2^2 \tag{17}$$

$$\text{s.t. } F_2(v) = 0 \tag{18}$$

where

$$F_1(v) = \Sigma^{-1}(\eta - h(x(s, p), p))$$

$$F_2(v) = \begin{pmatrix} r(x(t_0; s, p), \ldots, x(t_N; s, p), p) \\ \left(y(\tau_{k+1}; \tau_k, s_k, p) - s_{k+1}^y \right)_{k=0,\ldots,K-1} \\ \left(g(\tau_k, s_k^y, s_k^z, p) \right)_{k=0,\ldots,K} \end{pmatrix}$$

To solve problem (17)–(18) by a generalized Gauß–Newton method, see Sect. 6, we need the Jacobian which consists of the blocks

$$J = \begin{pmatrix} J_1 \\ J_2 \end{pmatrix} = \begin{pmatrix} \dfrac{\mathrm{d} F_1}{\mathrm{d} v} \\[2mm] \dfrac{\mathrm{d} F_2}{\mathrm{d} v} \end{pmatrix} = \left(\begin{array}{cccccc|c} D_0^1 & D_1^1 & D_2^1 & \cdots & D_K^1 & & D_p^1 \\ \hline D_0^2 & D_1^2 & D_2^2 & \cdots & D_K^2 & & D_p^2 \\ G_{y_0} & (-I\ 0) & & & & & G_{y\,0}^p \\ & G_{y_1} & (-I\ 0) & & & & G_{y\,1}^p \\ & & \ddots & \ddots & & & \vdots \\ & & & G_{y_{K-1}} & (-I\ 0) & & G_{y\,K-1}^p \\ H_0 & & & & & & H_0^p \\ & H_1 & & & & & H_1^p \\ & & H_2 & & & & H_2^p \\ & & & \ddots & & & \vdots \\ & & & & & H_K & H_K^p \end{array} \right),$$

where

$$D_j^1 = \frac{\mathrm{d} F_1}{\mathrm{d} s_k}, \quad k = 0, \ldots, K, \qquad\qquad D_p^1 = \frac{\mathrm{d} F_1}{\mathrm{d} p},$$

$$D_j^2 = \frac{\mathrm{d} r}{\mathrm{d} s_k}, \quad k = 0, \ldots, K, \qquad\qquad D_p^2 = \frac{\mathrm{d} r}{\mathrm{d} p},$$

$$G_{y_k} = \frac{\partial y}{\partial s_k}(\tau_{k+1}; \tau_k, s_k, p), \quad k = 0, \ldots, K-1,$$

$$G_{y\,k}^p = \frac{\partial y}{\partial p}(\tau_{k+1}; \tau_k, s_k, p), \quad k = 0, \ldots, K-1,$$

$$H_k = \frac{\partial g}{\partial s_j}(\tau_k, s_k^y, s_k^z, p), \quad k = 0, \ldots, K,$$

$$H_k^p = \frac{\partial g}{\partial p}(\tau_k, s_k^y, s_k^z, p), \quad k = 0, \ldots, K.$$

Most of the variables can easily be eliminated from the structured equality constraints:

1. s_k^z from the H_k-row, because of the index-1-assumption for the DAE (3), each matrix H_k is regular for $k = 0, \ldots, K$.
2. s_k^y, from the G_k-row, $j = 1, \ldots, K$.

This elimination is called *condensing*, see [8, 9]. Thereafter a system with the remaining variables s_0^y and p has to be solved, the size of which is essentially the same as the size of the linear system in the single shooting approach.

An advantage of multiple shooting over single shooting is that perturbations of the problem caused by uncertain parameter values are propagated only over the multiple shooting intervals instead of the whole integration interval. This reduces the nonlinearity of the constrained least squares problem significantly and leads to better local convergence of the Gauß–Newton method. Moreover, the additional multiple shooting variables may be initialized by the experimental data, giving a starting point for the v-variables close to the solution, whereas in single shooting an uncertain starting p may be far away from the solution. The integration of the DAE on the intervals is decoupled and may be computed in parallel. Further algorithmic strategies to improve efficiency are described in [11].

5 Derivative Evaluation

For the computation of the Jacobians, derivatives of the states x w.r.t. initial values s_k

$$G = \frac{\partial x}{\partial s} = \frac{\partial(y, z)}{\partial s} = (G_y, G_z)$$

and w.r.t. parameters p

$$G^p = \frac{\partial x}{\partial p} = \frac{\partial(y, z)}{\partial p} = (G_y^p, G_z^p)$$

have to be evaluated at the multiple shooting nodes τ_k. They are solutions of the *variational differential equations* w.r.t. initial values

Parameter Estimation and Optimum Experimental Design

$$\dot{G}_y(t) = \frac{\partial f}{\partial x}(t, y(t), z(t), p) \, G(t)$$

$$0 = \frac{\partial g}{\partial x}(t, y(t), z(t), p) \, G(t)$$

$$G_y(t_k) = I$$

and w.r.t. parameters

$$\dot{G}_y^p(t) = \frac{\partial f}{\partial x}(t, y(t), z(t), p) \, G^p(t) + \frac{\partial f}{\partial p}(t, y(t), z(t), p)$$

$$0 = \frac{\partial g}{\partial x}(t, y(t), z(t), p) \, G^p(t) + \frac{\partial g}{\partial p}(t, y(t), z(t), p)$$

$$G_y^p(\tau_k) = 0$$

which have to be solved as a system together with the nominal DAE (1)–(2). For inconsistent initial values of the algebraic variational states, a relaxed formulation has to be applied here as well.

For numerical solution, we suggest the approach of Internal Numerical Differentiation, i.e. of incorporating the derivative computation into the integrator scheme, see e.g. [1, 5]. The required derivatives of the model functions are evaluated by automated Algorithmic Differentiation [15].

6 Generalized Gauß–Newton Method

To solve nonlinear constrained least squares problems like (17)–(18), we apply the generalized Gauß–Newton method. In this section we formulate the algorithm. In the next section, we review its local convergence properties and discuss, why Gauß–Newton methods should be used instead of Newton/SQP methods for the solution of nonlinear least squares problems.

The general outline of the algorithm is as follows:

Algorithm 1 (Generalized Gauß–Newton Method)

0. Set $j := 0$. Choose an initial guess v^0.
1. Solve the linear constrained least squares problem

$$\min_{\Delta v^j} \frac{1}{2} \| F_1(v^j) + J_1(v^j) \Delta v^j \|_2^2$$

$$s.t. \ F_2(v^j) + J_2(v^j) \Delta v^j = 0.$$

2. Determine a step-length $\alpha^j \in (0; 1]$ by a globalization method.

3. *Compute the new iterate*

$$v^{j+1} := v^j + \alpha^j \Delta v^j.$$

4. *If* $\|\Delta v^j\|_2 \leq TOL$, *TOL prescribed tolerance, terminate, otherwise*
5. *Set* $j := j + 1$. *Go to 1.*

We assume regularity of the problem: Let $J_1 \in \mathbb{R}^{M \times n}$, $J_2 \in \mathbb{R}^{N_2 \times n}$, $J = \begin{pmatrix} J_1 \\ J_2 \end{pmatrix}$ with $M + N_2 \geq n \geq N_2$. In all points v where J has to be evaluated, we assume

- **Constraint Qualification (CQ)**

$$\text{rank } J_2(v) = N_2. \tag{19}$$

- **Positive Definiteness (PD)**

$$\text{rank } J(v) = n \tag{20}$$

which is equivalent to

$$\Delta v^T J_1(v)^T J_1(v) \Delta v > 0 \text{ for all } \Delta v \neq 0 \text{ with } J_2 \Delta v = 0.$$

Lemma 2. *Let (CQ) and (PD) hold. The solution of the constrained linear least squares problem*

$$\min \frac{1}{2} \|F_1 + J_1 \Delta v\|_2^2 \tag{21}$$

$$\text{s.t. } F_2 + J_2 \Delta v = 0 \tag{22}$$

can be represented as

$$\Delta v^* = -\begin{pmatrix} I & 0 \end{pmatrix} \begin{pmatrix} J_1^T J_1 & J_2^T \\ J_2 & 0 \end{pmatrix}^{-1} \begin{pmatrix} J_1^T & 0 \\ 0 & I \end{pmatrix} \begin{pmatrix} F_1 \\ F_2 \end{pmatrix}$$

and the Lagrange multiplier is

$$\lambda^* = \begin{pmatrix} 0 & I \end{pmatrix} \begin{pmatrix} J_1^T J_1 & J_2^T \\ J_2 & 0 \end{pmatrix}^{-1} \begin{pmatrix} J_1^T & 0 \\ 0 & I \end{pmatrix} \begin{pmatrix} F_1 \\ F_2 \end{pmatrix}.$$

Proof. Because of (CQ) and (PD), $\begin{pmatrix} J_1^T J_1 & J_2^T \\ J_2 & 0 \end{pmatrix}$ is regular. The Lagrangian of the constrained linear least squares problem is

$$\mathcal{L}(\Delta v, \lambda) = \frac{1}{2} F_1^T F_1 + F_1^T J_1 \Delta v + \frac{1}{2} \Delta v^T J_1^T J_1 \Delta v - \lambda^T (F_2 + J_2 \Delta v).$$

Parameter Estimation and Optimum Experimental Design

Setting the gradient of the Lagrangian to zero

$$0 = \nabla \mathcal{L}(\Delta v^*, \lambda^*) = \begin{pmatrix} J_1^T F_1 + J_1^T J_1 \Delta v^* - J_2^T \lambda^* \\ F_2 + J_2 \Delta v^* \end{pmatrix},$$

which here is necessary and sufficient for optimality, results in

$$\begin{pmatrix} \Delta v^* \\ -\lambda^* \end{pmatrix} = - \begin{pmatrix} J_1^T J_1 & J_2^T \\ J_2 & 0 \end{pmatrix}^{-1} \begin{pmatrix} J_1^T & 0 \\ 0 & I \end{pmatrix} \begin{pmatrix} F_1 \\ F_2 \end{pmatrix}.$$

We define the *generalized inverse* of J by

$$J^+ = \begin{pmatrix} I & 0 \end{pmatrix} \begin{pmatrix} J_1^T J_1 & J_2^T \\ J_2 & 0 \end{pmatrix}^{-1} \begin{pmatrix} J_1^T & 0 \\ 0 & I \end{pmatrix} \tag{23}$$

and obtain the representation

$$\Delta v = -J^+ F$$

for the solution of the linear constrained least squares problem (21)–(22).

J^+ satisfies the Moore–Penrose property $J^+ J J^+ = J^+$, but only in the unconstrained case it is a Moore–Penrose pseudo inverse satisfying all four Moore–Penrose properties.

A stable numerical realization of the solution of the linearized problems is described in [11], based on the approach of Stoer [27].

7 Local Convergence of Newton Type Methods

In this section we write x instead of v for the variables of the finite dimensional problem. We consider constrained nonlinear least squares problems

$$\min \frac{1}{2} \| F_1(x) \|_2^2 \tag{24}$$

$$\text{s.t. } F_2(x) = 0 \tag{25}$$

as discussed above, unconstrained nonlinear least squares problems

$$\min \frac{1}{2} \| F(x) \|_2^2 \tag{26}$$

and nonlinear equations systems

$$F(x) = 0. \tag{27}$$

Newton type methods solve these problems iteratively, starting from x_0 and computing iterates $x_{k+1} = x_k + \alpha^k \Delta x_k$ where Δx_k is the solution of some in x_k linearized problem and can be written as $\Delta x_k = -M(x_k)F(x_k)$.

First we formulate the local contraction theorem proved in [8] to state the local convergence behavior of Newton type methods.

Theorem 3 (Local Contraction Theorem). *Let* $F : \mathbb{R}^m \supseteq D \to \mathbb{R}^n$, $F \in C^1(D, \mathbb{R}^n)$, $J := \frac{\mathrm{d}F}{\mathrm{d}x}$. *For all* $x, y \in D$, $\theta \in [0, 1]$ *with* $y - x = -M(x)F(x)$:

1. There exists $\omega < \infty$, *such that*

$$\|M(y)\,(J(x + \theta(y - x)) - J(x))\,(y - x)\| \le \omega \cdot \theta \cdot \|y - x\|^2.$$

2. There exists $\kappa(x) \le \kappa < 1$, *such that*

$$\|M(y)R(x)\| \le \kappa(x)\,\|y - x\|$$

for the residual $R(x) := F(x) - J(x)M(x)F(x)$.

Let $x_0 \in D$ *be given with*

$$\Delta x_k := -M(x_k)F(x_k)$$

$$\delta_0 := \kappa + \frac{\omega}{2}\,\|\Delta x_0\| < 1, \quad \delta_k := \kappa + \frac{\omega}{2}\,\|\Delta x_k\|,$$

$$D_0 := \left\{ x : \|x - x_0\| \le \frac{\|\Delta x_0\|}{1 - \delta_0} \right\} \subseteq D.$$

Then:

1. The iteration $x_{k+1} = x_k + \Delta x_k$ *is well defined and stays in* D_0.
2. There exists an $x_* \in D_0$, *such that* $x_k \to x_*$ *converges for* $k \to \infty$.
3. The following a priori estimate holds:

$$\left\| x_{k+j} - x_* \right\| \le \frac{\delta_k^j}{1 - \delta_k}\,\|\Delta x_k\|.$$

4. The following a posteriori estimate holds:

$$\|\Delta x_{k+1}\| \le \delta_k\,\|\Delta x_k\| = \kappa\,\|\Delta x_k\| + \frac{\omega}{2}\,\|\Delta x_k\|^2.$$

From this Theorem one can directly conclude that Newton's method for nonlinear equation systems (27) with $M(x) = J(x)^{-1}$ converges locally quadratically because $R(x) = 0$ and therefore $\kappa = 0$. Applied to Gauß–Newton methods for constrained nonlinear least squares problems, we can draw the following conclusions:

Remark 4 (Local convergence of Gauß–Newton methods). Let $F = \begin{pmatrix} F_1 \\ F_2 \end{pmatrix} : \mathbb{R}^m \supseteq D \to \mathbb{R}^n$, $F \in C^1(D, \mathbb{R}^n)$, $J := \frac{\mathrm{d}F}{\mathrm{d}x}$. Let (CQ) and (PD) hold and

$$M(x) = \begin{pmatrix} I & 0 \end{pmatrix} \begin{pmatrix} J_1(x)^T J_1(x) & J_2(x)^T \\ J_2(x) & 0 \end{pmatrix}^{-1} \begin{pmatrix} J_1(x)^T & 0 \\ 0 & I \end{pmatrix}$$

be the solution operator of the linearized constrained least squares problem

$$\min_{\Delta x} \| F_1(x) + J_1(x)\Delta x \|_2^2$$

$$\text{s.t. } F_2(x) + J_2(x)\Delta x = 0.$$

Let in a neighborhood of the solution x_* of the nonlinear problem, $M(y)$ be bounded, $\| M(y) \| \leq \beta$, and let J satisfy the following Lipschitz condition, $\| (J(x + \theta(y - x)) - J(x))(y - x) \| \leq \gamma \cdot \theta \cdot \| y - x \|^2$. M is continuously differentiable, thus $\| M(y) - M(x) \| \leq L \| y - x \|$. Let $\| R(x) \| \leq \rho$. Then the following holds:

1. $\omega = \beta \cdot \gamma$ is large if

   - $\| J' \|$ respectively $\| F'' \|$ is large, i.e. the problem is very nonlinear.
   - $\| M(y) \|$ is large, i.e. $\begin{pmatrix} J_1^T J_1 & J_2^T \\ J_2 & 0 \end{pmatrix}$ is almost singular.

2. Due to the transformation $M(x)R(x) = M(x)(F(x) - J(x)M(x)F(x)) = (M(x) - M(x)J(x)M(x))F(x) = 0$ holds

$$\| M(y)R(x) \| = \| (M(y) - M(x))R(x) \| \leq \rho L \| y - x \|.$$

   Hence $\kappa = \rho L < 1$ if

   - the residual R is small,
   - M satisfies a Lipschitz condition respectively the first derivative of M is small.

3. If the starting value is close to the solution, where "close" is determined by

$$\kappa + \frac{\omega}{2} \| \Delta x_0 \| < 1,$$

   then the Gauß–Newton method converges, and the convergence is linear with convergence rate $\kappa < 1$.

7.1 Small and Large Residual Problems

In this subsection we want to discuss what happens if we treat parameter estimation problems by an SQP method which is essentially Newton's method for the first order optimality conditions. For simplicity of notation, in this subsection we consider the unconstrained case: minimize a nonlinear least squares functional

$$\min \frac{1}{2} \| F(x) \|_2^2 = \frac{1}{2} F(x)^T F(x) =: \phi(x). \tag{28}$$

Gradient and Hessian of the functional are

$$g := \nabla \phi(x) = J(x)^T F(x),$$

$$H := \nabla^2 \phi(x) = J(x)^T J(x) + \sum_{i=1}^{m} F_i(x) \frac{\partial J_i(x)}{\partial x} =: B(x) + E(x).$$

The Hessian H has a random part E depending on F which depends on the random measurement data, and a deterministic part B where the dependency on the random data vanishes due to differentiation. We assume regularity (PD), i.e. rank $J = n$, then $B = J^T J$ is a regular matrix.

7.1.1 Gauß–Newton and SQP Method

For the solution of (28), we apply a Newton-type method started from x_0 and the iterates are computed by $x_{k+1} = x_k + \alpha_k \Delta x_k$. For simplicity of notation we now omit x_k: $F := F(x_k)$, $J := J(x_k)$, $\phi := \phi(x_k)$.

In a Gauß–Newton method Δx_k solves

$$\min \frac{1}{2} \| F + J \Delta x \|_2^2 = \frac{1}{2} F^T F + F^T J \Delta x + \frac{1}{2} \Delta x^T J^T J \Delta x$$

respectively

$$J^T J \Delta x^k + J^T F \Delta x_k = 0$$

and hence

$$\Delta x_k = -(J^T J)^{-1} J^T F.$$

In an SQP method which is equivalent to Newton's method for $\nabla \phi(x) = 0$, Δx_k solves

$$\min \nabla \phi^T \Delta x + \frac{1}{2} \Delta x^T \nabla^2 \phi \Delta x$$

respectively

$$\nabla \phi + \nabla^2 \phi \Delta x_k = 0$$

Parameter Estimation and Optimum Experimental Design

and hence
$$\Delta x_k = -(\nabla^2 \phi)^{-1} \nabla \phi = -(\nabla^2 \phi)^{-1} J^T F.$$

Generically, the method can be described by $\Delta x_k = -A^{-1} g = -MF$. Applying the Gauß–Newton method results in $A = J^T J$ and $M = J^\dagger = (J^T J)^{-1} J^T$ and the local convergence is linear with rate κ if $\kappa < 1$. For Newton's method, $A = \nabla^2 \phi$, $M = (\nabla^2 \phi)^{-1} J^T$ and the local convergence is quadratic.

7.1.2 Small and Large Residuals

Remember that the condition $\kappa < 1$ is equivalent to the fact that M satisfies a Lipschitz condition and R is small. We address such problems as *small residual problems*.

Theorem 5 (Small residual problems). *Equivalent to $\kappa < 1$ is the following condition: the spectral radius*

$$\varrho(B(x_*)^{-1} E(x_*)) < 1$$

in any stationary point x_* of the Gauß–Newton method.

Proof. $M = J^\dagger = (J^T J)^{-1} J^T$, $R = F - JJ^\dagger F$.

$$J^\dagger(y) R(x) = \underbrace{J^\dagger(x) R(x)}_{=0} + \left(\left. \frac{\partial J^\dagger(y)}{\partial y} \right|_{y=x} (y-x) \right) R(x) + O(\|y-x\|^2)$$

$$= (J^T(x) J(x))^{-1} \left(\left. \frac{\partial J^T(y)}{\partial y} \right|_{y=x} (y-x) \right) R(x)$$

$$+ \left(\left. \frac{\partial J^T(y) J(y)}{\partial y} \right|_{y=x} (y-x) \right) \underbrace{J^T(x) R(x)}_{=0} + O(\|y-x\|^2)$$

$$= B(x)^{-1} \left(\left. \frac{\partial J^T(y)}{\partial y} \right|_{y=x} (y-x) \right) F(x)$$

$$- B(x)^{-1} \left(\left. \frac{\partial J^T(y)}{\partial y} \right|_{y=x} (y-x) \right) J(x) \underbrace{J^\dagger(x) F(x)}_{=-(y-x)} + O(\|y-x\|^2)$$

$$= B(x)^{-1} E(x)(y-x) + O(\|y-x\|^2).$$

- Assume $\rho(B(x_*)^{-1} E(x_*)) =: \kappa_1 < 1$. Choose a neighborhood of x_* and a norm such that $\|B(x)^{-1} E(x)\| \le \kappa_2 < 1$ for all x in this neighborhood. Reduce the neighborhood, such that also $O(\|y - x\|^2) \leq \frac{1-\kappa_2}{2}\|y - x\|$ for all x, y in this neighborhood. Then

$$\|J(y)^\dagger R(x)\| \leq \|B(x)^{-1} E(x)\|\|y - x\| + O(\|y - x\|^2)$$

$$\leq \kappa_2 \|y - x\| + \frac{1 - \kappa_2}{2}\|y - x\|$$

$$= \frac{1 + \kappa_2}{2}\|y - x\| =: \kappa_3\|y - x\|, \quad \kappa_3 < 1.$$

- Assume vice versa that the condition for the local contraction theorem is satisfied for a norm $\|.\|$ and all x in a neighborhood of x_*:

$$\|J^\dagger(y) R(x)\| = \|B(x)^{-1} E(x)(y - x) + O(\|y - x\|^2)\| \leq \kappa\|y - x\|, \quad \kappa < 1.$$

Then
$$\|B(x)^{-1} E(x)(y - x)\| - O(\|y - x\|^2) \leq \kappa\|y - x\|.$$

Make the neighborhood so small that $O(\|y - x\|^2) \leq \frac{1-\kappa}{2}\|y - x\|$. Then

$$\|B(x)^{-1} E(x)(y - x)\| \leq \frac{1 + \kappa}{2}\|y - x\| =: \kappa_1\|y - x\|, \quad \kappa_1 < 1.$$

Hence $\|B(x)^{-1} E(x)\| \leq \kappa_1 < 1$ and $\rho(B(x)^{-1} E(x)) < 1$.

In other words: for small residual problems, the random part E of the Hessian is relatively bounded by 1 w.r.t. the deterministic part B, i.e. B dominates E.

Theorem 6. *If x_* is a stationary point of the Gauß–Newton method and*

$$\varrho(B(x_*)^{-1} E(x_*)) < 1$$

then $H(x_*)$ is positive definite.

Proof. $J(x_*)$ has full rank, hence $J(x_*)^T J(x_*) = B(x_*) \succ 0$ is positive definite, hence $B(x_*)^{\frac{1}{2}} \succ 0$ exists. Then

$$H(x_*) = B(x_*) + E(x_*) = B(x_*)^{\frac{1}{2}} \left(I + B(x_*)^{-\frac{1}{2}} E(x_*) B(x_*)^{-\frac{1}{2}} \right) B(x_*)^{\frac{1}{2}}$$

and
$$H(x_*) \succ 0 \iff I + B(x_*)^{-\frac{1}{2}} E(x_*) B(x_*)^{-\frac{1}{2}} \succ 0. \tag{29}$$

$B(x_*)^{-1} E(x_*) = B(x_*)^{-\frac{1}{2}} \left(B(x_*)^{-\frac{1}{2}} E(x_*) B(x_*)^{-\frac{1}{2}} \right) B(x_*)^{\frac{1}{2}}$ is a similarity transformation, hence $B(x_*)^{-\frac{1}{2}} E(x_*) B(x_*)^{-\frac{1}{2}}$ has the same eigenvalues as $B(x_*)^{-1} E(x_*)$. Thus $\rho(B(x_*)^{-\frac{1}{2}} E(x_*) B(x_*)^{-\frac{1}{2}}) < 1$ and all the eigenvalues of

Parameter Estimation and Optimum Experimental Design

$$I + B(x_*)^{-\frac{1}{2}} E(x_*) B(x_*)^{-\frac{1}{2}}$$

are within $]0; 2[$. Consequently, $H(x_*)$ is positive definite.

Corollary 7. *In Theorem 6, x_* is not only a stationary point, but also a local minimizer and stable against perturbations.*

For *large residual problems*, i.e. if $\kappa > 1$ respectively $\varrho(B(x_*)^{-1})E(x_*)) > 1$, the Gauß–Newton method does not converge. Now we will discuss what happens if we apply Newton's method and compute a minimizer x_* of a large residual problem.

7.1.3 Statistical Perturbation of the Problem

Let us consider a parameter estimation problem with random data $\eta \in \mathbb{R}^m$, model $h(x) \in \mathbb{R}^m$, true parameter values $\bar{x} \in \mathbb{R}^n$ and $\eta - h(\bar{x}) \sim \mathcal{N}(0, I)$. Let $F(x) := \eta - h(x) \in \mathbb{R}^m$. Let x_* be the minimizer of a large residual problem. Then

$$F(x_*) = \eta - h(x_*), \quad \eta = h(x_*) + F(x_*).$$

Create perturbed data $\hat{\eta}$ by reflecting the errors at the estimated model trajectory, see Fig. 1:
$$\hat{\eta} := h(x_*) - F(x_*), \quad F(x_*) = h(x_*) - \hat{\eta}.$$

Then $\|\hat{\eta} - h(x_*)\| = \|F(x_*)\| = \|\eta - h(x_*)\|$. We now construct a homotopy of perturbed problems

$$\tilde{F}(x, \tau) := F(x) + (\tau - 1)F(x_*).$$

For $\tau = 1$, $\tilde{F}(x, \tau) = F(x) = \eta - h(x)$, we have the original problem with data η. For $\tau = -1$, $\tilde{F}(x, \tau) = \hat{\eta} - h(x)$, we have the problem with the reflected data $\hat{\eta}$.

Theorem 8. *Let x_* be a minimizer with $\tilde{\kappa} := \varrho(B(x_*)^{-1})E(x_*)) > 1$. Then*

1. *for all τ, x_* is a stationary point of $\min \frac{1}{2}\|\tilde{F}(x, \tau)\|_2^2$,*
2. *the Hessian in x_* is $\tilde{H}(x_*, \tau) = B(x_*) + \tau E(x_*)$,*
3. *$\tilde{H}(x_*, \tau)$ is not positive definite for all $\tau < -\frac{1}{\kappa} \geq -1$.*

Proof. x_* is a minimizer, hence $J(x_*)^T F(x_*) = 0$. Because of $\tilde{J}(x, \tau) = J(x)$,

$$\tilde{J}(x_*, \tau)^T \tilde{F}(x_*, \tau) = J(x_*)^T (F(x_*) + (\tau - 1)F(x_*)) = 0,$$

$$\tilde{H}(x, \tau) = \tilde{J}(x, \tau)^T \tilde{J}(x, \tau) + \tilde{F}(x, \tau)\frac{\partial \tilde{J}}{\partial x}(x, \tau)$$

$$= B(x) + (F(x) + (\tau - 1)F(x_*))\frac{\partial J}{\partial x}(x),$$

$$\tilde{H}(x_*, \tau) = B(x_*) + \tau F(x_*)\frac{\partial J}{\partial x}(x_*) = B(x_*) + \tau E(x_*).$$

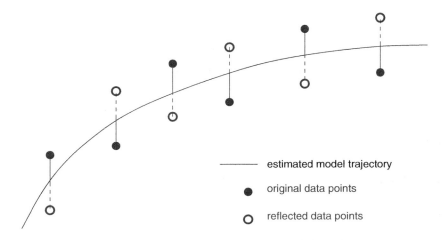

Fig. 1 Create perturbed data by reflecting the errors at the estimated model trajectory

Hence $\tilde{H}(x_*, \tau) \succ 0$ if and only if

$$I + B(x_*)^{-\frac{1}{2}} \tau E(x_*) B(x_*)^{-\frac{1}{2}} =: I + \tau \hat{E}(x_*) \succ 0,$$

cf. (29).

$$\rho(B(x_*)^{-1} E(x_*)) = \rho(B(x_*)^{-\frac{1}{2}} E(x_*) B(x_*)^{-\frac{1}{2}}) = \rho(\hat{E}(x_*)) = \tilde{\kappa} > 1,$$

hence $\hat{E}(x_*)$ has an eigenvalue $\tilde{\kappa} > 1$. By assumption, $\tilde{H}(x_*, 0) = B(x_*) + E(x_*) \succ 0$. $I + \tau \hat{E}(x_*)$ has an eigenvalue $1 + \tau \tilde{\kappa}$ which becomes negative for $\tau < -\frac{1}{\tilde{\kappa}} \geq -1$. Then $\tilde{H}(x_*, \tau)$ is not positive definite for all $\tau < -\frac{1}{\tilde{\kappa}} \geq -1$.

Corollary 9.

1. For large residual problem, the minimizer of $\min \frac{1}{2} \|\tilde{F}(x, \tau)\|_2^2$ is not stable against perturbations. It jumps away from x_* which may become a saddle point or maximizer.
2. Large residual minimizers are not statistically stable.
3. They are minimizers but not estimators.
4. The Gauß–Newton method does not converge to them.

8 Optimum Experimental Design

So far we have discussed how to fit a model to experimental data and to estimate the parameters. Now we want to analyze the reliability of the parameter estimate w.r.t. the randomness of the data due to measurement errors.

Parameter Estimation and Optimum Experimental Design

We now assume that we know the true model and the true parameters and want to design optimal experiments to verify the parameters with minimal statistical uncertainty. For the discrimination between rivaling model variants, methods for experimental design for model discrimination have been developed, see e.g. [3, 12, 13].

An introduction to nonlinear experimental design can be found in the textbooks by Atkinson and Donev [2] and Pukelsheim [22]. Methods for experimental design for differential equation models are developed and applied by several groups, for recent publications see, e.g., Uciński [29, 30], Banga and Balsa-Canto [4] or Telen et al. [28]. A survey article with many applications in chemical engineering has been published by Franceschini and Macchietto [14].

In this article, we describe the statistical uncertainty by the variance–covariance matrix of the parameter estimation. The task of *optimum experimental design* consists in finding experiments the data of which yields maximal reliability of the parameter estimate.

To achieve this, we can optimize the layout, setup and processing of experiments. As an *experiment* we define a run of the process under certain conditions described by control variables and time-dependent control functions entering the model equations and additionally by the choice of sampling points and corresponding measurement methods. Usually data is not only acquired from one experiment, but from series of *multiple experiments*.

8.1 Multiple Experiment Parameter Estimation Problem

We now consider series of N_{ex} multiple experiments. We allow that the processes in different experiments are described by different models

$$\dot{y}^j(t) = f^j(t, y^j(t), z^j(t), p, q^j, u^j(t))$$

$$0 = g^j(t, y^j(t), z^j(t), p, q^j(t))$$

$$0 = r^j(x^j(t_0^j), \ldots, x^j(t_{N^j}^j), p, q^j)$$

with states $x^j = (y^j, z^j)$, $j = 1, \ldots, N_{ex}$. We additionally introduce control variables $q^j \in \mathbb{R}^{n_q^j}$ and control functions $u^j(t) : [t_0^j; t_f^j] \to \mathbb{R}^{n_u^j}$ which describe the layout, setup and processing of the jth experiment. We assume that the parameters p are common in all models.

For the data $\eta^j \in \mathbb{R}^{M^j}$ measured in experiment j we again assume normal distribution

$$\epsilon_i^j = \eta_i^j - h_i^j(x^j(t_i^j), p, q^j) \sim \mathcal{N}\left(0, \frac{\sigma_i^{j\,2}}{w_i^j}\right), \tag{30}$$

$i = 1, \ldots, M^j$, $j = 1, \ldots, N_{ex}$, and independence

$$\mathrm{cov}(\epsilon_{i_1}^{j_1}, \epsilon_{i_2}^{j_2}) = 0, \quad (j_1, i_1) \neq (j_2, i_2),$$

of the measurement errors.

Notice that we have modified the variances, namely divided by weights w_i^j. If data is available for already processed experiments, the corresponding w_i^j are 1. For new experiments which are being designed, the variables $w_i^j \in \{0, 1\}$ describe if the ith measurement in experiment j shall be actually carried out or not. Usually constraints on the number of measurements are given.

The modified multiple experiment parameter estimation problem can then be written as

$$\min_{p,x} \frac{1}{2} \sum_{j=1}^{N_{ex}} \sum_{i=1}^{M^j} w_i^j \left(\frac{\eta_i^j - h_i^j(x^j(t_i^j), p, q^j)}{\sigma_i^j} \right)^2$$

$$\text{s.t. } \dot{y}^j(t) = f^j(t, y^j(t), z^j(t), p, q^j, u^j(t))$$

$$0 = g^j(t, y^j(t), z^j(t), p, q^j(t))$$

$$0 = r^j(x^j(t_0^j), \ldots, x^j(t_{N^j}^j), p, q^j), \quad j = 1, \ldots, N_{ex}$$

As described above, we apply a specific experiment-wise parameterization of the state variables. For the numerical solution, an experiment-wise multiple-shooting with exploitation of matrix structures in the solution of the linear systems within a Gauß–Newton method provides an efficient numerical approach, see [23, 24].

For the following uncertainty discussion, the parameterization of the states may be chosen in a way that the additional variables s describe quantities the uncertainty of which is desired to be reduced, so called *key-performance-indicators*. For a case study, see [18].

After parameterization, we obtain a finite-dimensional multiple experiment constrained nonlinear least squares problem with variables $v = (p, s^1, \ldots, s^{N_{ex}})$:

$$\min_v \frac{1}{2} \sum_{j=1}^{N_{ex}} \| F_1^j(p, s^j) \|_2^2$$

$$\text{s.t. } F_2^j(p, s^j) = 0, \quad j = 1, \ldots, N_{ex}$$

with

$$F_1^j(p, s^j) = \left(\sqrt{w_i^j} \, \frac{\eta_i^j - h_i^j(x^j(t_i^j, s^j, p, q^j), p, q^j)}{\sigma_i^j} \right)_{i=1,\ldots,M^j}$$

$$F_2^j(p, s^j) = d^j(s^j, p)$$

where d^j comprises all the constraints belonging to the jth experiment after parameterization of the states by the variables s^j. Define

$$F_1(v) := \left(F_1^j(p, s^j) \right)_{j=1,\ldots,N_{ex}}, \qquad F_2(v) = \left(F_2^j(p, s^j) \right)_{j=1,\ldots,N_{ex}}.$$

For numerical solution and statistical analysis, the Jacobians of least squares terms and constraints are needed. They have experiment-wise block structure

$$J_1 = \frac{dF_1}{dv} = \begin{pmatrix} J_1^{1,p} & J_1^{1,s^1} & 0 & \cdots & 0 \\ \vdots & & \ddots & & \\ J_1^{N_{ex},p} & 0 & \cdots & 0 & J_1^{N_{ex},s^{N_{ex}}} \end{pmatrix},$$

$$J_2 = \frac{dF_2}{dv} = \begin{pmatrix} J_2^{1,p} & J_2^{1,s^1} & 0 & \cdots & 0 \\ \vdots & & \ddots & & \\ J_2^{N_{ex},p} & 0 & \cdots & 0 & J_2^{N_{ex},s^{N_{ex}}} \end{pmatrix}$$

with

$$J_1^{j,p} = \frac{\partial F_1^j}{\partial p} = \left(-\frac{\sqrt{w_i^j}}{\sigma_i^j} \left(\frac{\partial h_i^j}{\partial x} \frac{\partial x^j}{\partial p}(t_i) + \frac{\partial h_i^j}{\partial p} \right) \right)_{i=1,\ldots,M^j}$$

$$J_1^{j,s^j} = \frac{\partial F_1^j}{\partial s^j} = \left(-\frac{\sqrt{w_i^j}}{\sigma_i^j} \left(\frac{\partial h_i^j}{\partial x} \frac{\partial x^j}{\partial s^j}(t_i) \right) \right)_{i=1,\ldots,M^j}$$

and

$$J_2^{j,p} = \frac{\partial F_2^j}{\partial p} = \frac{\partial d^j}{\partial x^j} \frac{\partial x^j}{\partial p} + \frac{\partial d^j}{\partial p}$$

$$J_2^{j,s^j} = \frac{\partial F_2^j}{\partial s^j} = \frac{\partial d^j}{\partial x^j} \frac{\partial x^j}{\partial s^j} + \frac{\partial d^j}{\partial s^j}.$$

The derivatives $G^j = (\frac{\partial x^j}{\partial p}, \frac{\partial x^j}{\partial s^j})$ are solutions of variational differential equations, see Sect. 5:

$$\dot{G}_y^j(t) = \frac{\partial f^j}{\partial x}(t, y^j(t), z^j(t), p, q^j)\, G^j(t) + \frac{\partial f^j}{\partial(p, s^j)}(t, y^j(t), z^j(t), p, q^j)$$

$$0 = \frac{\partial g^j}{\partial x}(t, y^j(t), z^j(t), p, q^j)\, G^j(t) + \frac{\partial g^j}{\partial(p, s^j)}(t, y^j(t), z^j(t), p, q^j)$$

$$G_y^j(t_0) = \frac{\partial y_0^j}{\partial(p, s^j)}$$

where $\frac{\partial f^j}{\partial s^j} = 0$ and $\frac{\partial g^j}{\partial s^j} = 0$. We write

$$J_1^j = -\sqrt{W^j}\,\Sigma^{j-1}\left(\frac{\partial h_i^j}{\partial x}G^j(t_i) + \frac{\partial h_i^j}{\partial(p,s^j)}\right)_{i=1,\ldots,M^j},$$

where $W^j := \mathrm{diag}(w_1^j,\ldots,w_{M^j}^j)$ and $\frac{\partial h_i^j}{\partial s^j} = 0$, and

$$J_2^j = \frac{\partial d^j}{\partial x^j}G^j + \frac{\partial d^j}{\partial(p,s^j)}.$$

8.2 Statistical Uncertainty of the Parameter Estimate

We want to describe the propagation of the uncertainty of the experimental data given by the random distribution of the measurement errors to the uncertainty of the parameter estimate.

Let $\hat{v}$ be the solution of the nonlinear constrained parameter estimation problem. We apply a local analysis and therefore linearize the problem in $\hat{v}$ as is done in a generalized Gauß–Newton method obtaining

$$\min \frac{1}{2}\|F_1(\hat{v}) + J_1(\hat{v})\Delta v\|_2^2$$

$$\text{s.t. } F_2(\hat{v}) + J_2(\hat{v})\Delta v = 0,$$

We again assume regularity (CQ) and (PD), see (19) and (20), and can write the solution as $\Delta v = -J^+(\hat{v})\begin{pmatrix}F_1(\hat{v})\\F_2(\hat{v})\end{pmatrix}$ where J^+ is the generalized inverse of $\begin{pmatrix}J_1(\hat{v})\\J_2(\hat{v})\end{pmatrix}$, see (23). The expected value

$$E(\Delta v) = E\left(-J^+(\hat{v})\begin{pmatrix}F_1(\hat{v})\\F_2(\hat{v})\end{pmatrix}\right) = -J^+(\hat{v})E\left(\begin{pmatrix}F_1(\hat{v})\\F_2(\hat{v})\end{pmatrix}\right) = 0$$

of the increment is zero in the solution point $\hat{v}$. The uncertainty is expressed by the *variance–covariance matrix*

$$C(\hat{v}) := E(\Delta v \Delta v^T) = E\left(J^+(\hat{v})\begin{pmatrix}F_1(\hat{v})\\F_2(\hat{v})\end{pmatrix}\begin{pmatrix}F_1(\hat{v})\\F_2(\hat{v})\end{pmatrix}^T J^+(\hat{v})^T\right)$$

$$= J^+(\hat{v})\begin{pmatrix}E(F_1(\hat{v})F_1(\hat{v})^T) & E(F_1(\hat{v})F_2(\hat{v})^T)\\E(F_2(\hat{v})F_1(\hat{v})^T) & E(F_2(\hat{v})F_2(\hat{v})^T)\end{pmatrix}J^+(\hat{v})^T$$

$$= (I\ 0) \begin{pmatrix} J_1^T J_1 & J_2^T \\ J_2 & 0 \end{pmatrix}^{-1} \begin{pmatrix} J_1^T & 0 \\ 0 & I \end{pmatrix} \begin{pmatrix} I & 0 \\ 0 & 0 \end{pmatrix} \begin{pmatrix} J_1 & 0 \\ 0 & I \end{pmatrix} \begin{pmatrix} J_1^T J_1 & J_2^T \\ J_2 & 0 \end{pmatrix}^{-1} \begin{pmatrix} I \\ 0 \end{pmatrix}$$

$$= (I\ 0) \begin{pmatrix} J_1^T J_1 & J_2^T \\ J_2 & 0 \end{pmatrix}^{-1} \begin{pmatrix} J_1^T J_1 & 0 \\ 0 & 0 \end{pmatrix} \begin{pmatrix} J_1^T J_1 & J_2^T \\ J_2 & 0 \end{pmatrix}^{-1} \begin{pmatrix} I \\ 0 \end{pmatrix} \tag{31}$$

$$= (I\ 0) \begin{pmatrix} J_1^T J_1 & J_2^T \\ J_2 & 0 \end{pmatrix}^{-1} \begin{pmatrix} I \\ 0 \end{pmatrix}. \tag{32}$$

Here $E(F_1(\hat{v}) F_2(\hat{v})^T)$, $E(F_2(\hat{v}) F_1(\hat{v})^T)$ and $E(F_2(\hat{v}) F_2(\hat{v})^T)$ are zero because F_2 is not random and $E(F_1(\hat{v}) F_2(\hat{v})^T) = I$ because of the construction of the weighted least squares functional. The identity from (31) to (32) can be concluded from the following Lemma [10].

Lemma 10. *Let*

$$\begin{pmatrix} X & Y^T \\ Y & Z \end{pmatrix} := \begin{pmatrix} J_1^T J_1 & J_2^T \\ J_2 & 0 \end{pmatrix}^{-1} =: M^{-1}.$$

Then the variance–covariance matrix C is equal to X and satisfies

$$I = J_1^T J_1 C + J_2^T Y$$
$$0 = J_2 C.$$

Proof. Multiplying M by M^{-1}

$$I = J_1^T J_1 X + J_2^T Y \qquad\qquad 0 = J_1^T J_1 Y^T + J_2^T Z$$
$$0 = J_2 X \qquad\qquad I = J_2 Y^T$$

and substituting terms in the representation (31)

$$C = (I\ 0)\, M^{-1} \begin{pmatrix} J_1^T J_1 & 0 \\ 0 & 0 \end{pmatrix} M^{-1} \begin{pmatrix} I \\ 0 \end{pmatrix} = (X\ Y^T) \begin{pmatrix} J_1^T J_1 & 0 \\ 0 & 0 \end{pmatrix} \begin{pmatrix} X \\ Y \end{pmatrix}$$

$$= X J_1^T J_1 X = X(I - J_2^T Y) = X - (J_2 X)^T Y = X$$

yields the desired result.

For multiple experiments, the variance–covariance matrix is

$$
C = \begin{pmatrix} I & 0 \end{pmatrix} \left(\begin{array}{ccccc|ccc} \sum\limits_{j=1}^{N_{ex}} J_1^{j,p\,T} J_1^{j,p} & J_1^{1,p\,T} J_1^{1,s^1} & \cdots & J_1^{N_{ex},p\,T} J_1^{N_{ex},s^{N_{ex}}} & J_2^{1,p\,T} & \cdots & J_2^{N_{ex},p\,T} \\ \hline J_1^{1,s^1\,T} J_1^{1,p} & J_1^{1,s^1\,T} J_1^{1,s^1} & & & J_2^{1,s^1\,T} & & \\ \vdots & & \ddots & & & \ddots & \\ J_1^{N_{ex},s^{N_{ex}}\,T} J_1^{N_{ex},p} & & & J_1^{N_{ex},s^{N_{ex}}\,T} J_1^{N_{ex},s^{N_{ex}}} & & & J_2^{N_{ex},s^{N_{ex}}\,T} \\ \hline J_2^{1,p} & J_2^{1,s^1} & & & & & \\ \vdots & & & \ddots & & & \\ J_2^{N_{ez},p} & & & J_2^{N_{ex},s^{N_{ex}}} & & & \end{array} \right)^{-1} \begin{pmatrix} I \\ 0 \end{pmatrix}
$$

The variance–covariance matrix can be used to describe confidence regions of the parameters. For details, see [8].

8.3   The Experimental Design Optimization Problem

For nonlinear regression models the variance–covariance matrix depends on the values of the parameters. Moreover, it depends on the controls q and $u(t)$ describing experimental layout, setup and processing and on the weights for the choice of sample points. The variance–covariance matrix depends on J and hence on derivatives $\frac{\partial x}{\partial v}$ of the states x w.r.t. to the variables v. But it does not depend on the experimental data η, hence it can be computed before the experiments are actually carried out.

For now, we keep v fixed and want to find experiments which allow to estimate v with minimal statistical uncertainty. This task of *optimum experimental design* means the calculation of *experimental design variables* $\xi = (q, u(t), w)$ which minimize a suitable objective functional $\phi(C)$ on the variance–covariance matrix C, e.g. the A-criterion trace(C), the D-criterion det(C), the E-criterion $\|C\|_2$ or the M-criterion max C_{ii}.

Usually, experiment-wise (vector-valued) constraints to controls and states are given:

$$
L_1^j \le c^j(q^j) \le U_1^j
$$
$$
L_2^j \le b^j(x^j(t), u^j(t)) \le U_2^j,
$$

$j = 1, \ldots, N_{ex}$. Constraints to the choice of measurements can be expressed by

$$
L_{I_k}^j \le \sum_{i \in I_k} w_i^j \le U_{I_k}^j
$$

for some subsets $I_k \subseteq \{1, \ldots, M^j\}$, $k = 1, \ldots, K^j$ $j = 1, \ldots, N_{ex}$ where the w_i^j are 0–1-variables

$$w_i^j \in \{0, 1\}, \; i = 1, \ldots, M^j.$$

Furthermore, the model equations including boundary conditions and the variational model equations, see Sect. 5, for the derivatives of the states w.r.t. v have to be satisfied. Altogether, the experimental design optimization problem reads as follows.

$$\min_{q,u,w} \phi(C)$$

$$\text{s.t. } C = \begin{pmatrix} I & 0 \end{pmatrix} \begin{pmatrix} J_1^T J_1 & J_2^T \\ J_2 & 0 \end{pmatrix}^{-1} \begin{pmatrix} I \\ 0 \end{pmatrix}$$

$$\text{for } j = 1, \ldots, N_{ex}:$$

$$J_1^j = -\sqrt{W^j} \Sigma^{j\,-1} \left(\frac{\partial h_i^j}{\partial x} G^j(t_i) + \frac{\partial h_i^j}{\partial(p, s^j)} \right)_{i=1,\ldots,M^j},$$

$$J_2^j = \frac{\partial d^j}{\partial x^j} G^j + \frac{\partial d^j}{\partial(p, s^j)}$$

$$\dot{y}^j(t) = f^j(t, y^j(t), z^j(t), p, q^j, u^j(t))$$

$$0 = g^j(t, y^j(t), z^j(t), p, q^j(t))$$

$$0 = r^j(x^j(t_0^j), \ldots, x^j(t_{N^j}^j), p, q^j)$$

$$\dot{G}_y^j(t) = \frac{\partial f^j}{\partial x}(t, y^j(t), z^j(t), p, q^j) G^j(t) + \frac{\partial f^j}{\partial(p, s^j)}(t, y^j(t), z^j(t), p, q^j)$$

$$0 = \frac{\partial g^j}{\partial x}(t, y^j(t), z^j(t), p, q^j) G^j(t) + \frac{\partial g^j}{\partial(p, s^j)}(t, y^j(t), z^j(t), p, q^j)$$

$$G_y^j(t_0) = \frac{\partial y_0^j}{\partial(p, s^j)}$$

$$L_1^j \leq c^j(q^j) \leq U_1^j$$

$$L_2^j \leq b^j(x^j(t), u^j(t)) \leq U_2^j$$

$$L_{I_k}^j \leq \sum_{i \in I_k} w_i^j \leq U_{I_k}^j, \; I_k \subseteq \{1, \ldots, M^j\}, \; k = 1, \ldots K^j$$

$$w_i^j \in \{0, 1\}, \; i = 1, \ldots, M^j.$$

8.4 Modifications and Enhancements

8.4.1 Experimental Design for Maximal Information Gain

To take a priori information from previous experiments into account, the Jacobians of already processed ("old") experiments may be included in the computation of the variance–covariance matrix. It then describes the uncertainty of the parameter estimation from a multiple experiment consisting of the old experiments and some "new" experiments which are being designed. In the optimization problem, then only the controls and sampling weights for the new experiments are optimization variables whereas the settings for the old experiments are constants. Experimental designs computed from this modified problem maximize the information gain by the new experiments under consideration of the old experiments.

8.4.2 Modification for Robust Experimental Design w.r.t. Parameter Uncertainty

For regression models which are nonlinear in the parameters, the variance-covariance matrix depends on the values of the parameters. We now use the short notation $C = C(\xi, p)$. We assume that the parameter values are only known roughly and that we can describe this by a multi-normal distribution

$$p \sim \mathcal{N}(p_0, C_0).$$

For experimental design we consider a worst-case approach [20]

$$\min_{\xi} \max_{\|p-p_0\|_{2,C_0^{-1}} \leq \gamma} \phi(C(\xi, p))$$

where we minimize the experimental design over the worst case within a confidence region of the parameters. The norm is defined by $\|p\|_{2,C_0^{-1}} := \sqrt{p^T C_0^{-1} p}$ and γ is a confidence quantile. To treat this semi-infinite optimization problem, we apply a Taylor expansion w.r.t. p around p_0:

$$\min_{\xi} \max_{\|p-p_0\|_{2,C_0^{-1}} \leq \gamma} \phi(C(\xi, p_0)) + \frac{\mathrm{d}}{\mathrm{d}p}\phi(C(\xi, p_0))(p - p_0)$$

and can state the solution of the inner problem explicitly:

$$\max_{\|p-p_0\|_{2,C_0^{-1}} \leq \gamma} \phi(C(\xi, p_0)) + \frac{\mathrm{d}}{\mathrm{d}p}\phi(C(\xi, p_0))(p - p_0)$$

$$= \phi(C(\xi, p_0)) + \gamma \left\| \frac{\mathrm{d}}{\mathrm{d}p}\phi(C(\xi, p_0)) \right\|_{2,C_0},$$

see [20]. The formulation

$$\min_{\xi} \phi(C(\xi, p_0)) + \gamma \left\| \frac{d}{dp} \phi(C(\xi, p_0)) \right\|_{2, C_0}$$

hence yields a robustification of the experimental design optimization problem. Notice that we end up with a multi-objective optimization problem where the two goals, the minimization of the variance–covariance matrix and the minimization of the sensitivity w.r.t. the parameters, are weighted by the confidence quantile γ.

The robustified objective depends on the derivative of ϕ w.r.t. p and thus on second derivatives of the system states w.r.t. the parameters.

8.4.3 Sequential Strategy

Because experimental design problems depend on the current parameter values and parameter estimation problems depend on the data evaluated so far in previously designed experiments, we propose to solve both problems in a sequential way, [19]:

Algorithm 2 (Sequential experimental design and parameter estimation)

0. $k := 0$. Start with an in initial guess p_0 for the parameter values.
1. With $p = p_k$, design N_k new optimal experiments maximizing the information gain taking all previous experiments into account.
2. Process the new experiments to obtain experimental data.
3. Compute a parameter estimate p_{k+1} using the data from all previous and the N_k new experiments.
4. Compute the variance–covariance matrix C_{k+1} of the new parameter estimate p_{k+1}. If the confidence region is sufficiently small or the experimental budget has been spent, STOP.
5. Set $k := k + 1$ and go to 1.

The experimental design optimization problems in step 2 may be robustified using the current variance–covariance matrix C_k of the current parameter estimate p_k. This sequential approach may be applied offline by designing one single or multiple parallel experiments in every loop cycle. Or it may be applied online by computing new controls for the continuation of experiments after each or some new data has been measured. For a case study of the online approach, see [17].

8.5 Methods for the Numerical Solution

We want to give a short overview of our methods for the solution of experimental design optimization problems. These are nonlinear optimal control problems.

Specific difficulties are the non-standard objective and the dependence of the variance–covariance matrix on derivatives of the state variables.

We apply the direct approach for optimal control, i.e. we parameterize the control functions u by finite-dimensional variables. Commonly, piecewise polynomials are chosen, but any other finite-dimensional parameterization is also applicable. The state constraints are evaluated on a finite grid. The DAE system is evaluated by single shooting or, in a tailored reformulation of the problem, by multiple shooting.

We relax the 0–1-conditions by $w_i^j \in [0, 1]$, $i = 1, \ldots, M^j$. In the end, non-integer solutions are rounded by appropriate heuristics [16].

After these transformations we obtain a finite dimensional constrained nonlinear optimization problem which we solve by an structured SQP method. Within this algorithm, we need to evaluate derivatives of objective and constraints. The Hessian of the Lagrangian is approximated by low-rank update-formulas. Rather intricate is the numerical computation of the gradient of the objective, i.e. the gradient of a functional of the variance–covariance matrix depending on derivatives $\frac{\partial x}{\partial p}$ of the states w.r.t. the parameters. For the evaluation we apply matrix differentiation calculus [31], Internal Numerical Differentiation for mixed second derivatives $\frac{\partial^2 x}{\partial q \partial p}$ of the states w.r.t. parameters and controls [5, 6], and Automatic Differentiation to evaluate the required first and second derivatives of the model functions. For the robustified problem, one more order of derivatives has to be provided. Structures in the Jacobians from the multiple experiment formulation are exploited by a tailored derivative evaluation and linear algebra.

These methods are implemented in our software package VPLAN, see [16] for the fundamental design.

9 Summary

In this paper we have reviewed parameter estimation problems and optimum experimental design problem, two important steps in optimization based modeling. We apply the boundary problem method optimization approach for the solution of parameter estimation problems with multiple shooting and the generalized Gauß–Newton method. We have discussed convergence properties of this approach and shown that parameter estimation problems should be solved by Gauß–Newton rather than SQP methods. To not only fit models to data, but also validate the parameters, we suggest to minimize the statistical uncertainty of parameter estimates by optimal model based design of experiments. For this intricate optimal control problem, we provide numerical solution methods.

Acknowledgements The authors want to thank DFG for providing excellent research conditions within the Heidelberg Graduate School of Mathematical and Computational Methods for the Sciences. S. Körkel wants to thank BASF SE for funding his position and parts of his research group. Additional funding is granted by the German Federal Ministry for Education and Research within the initiative *Mathematik für Innovationen in Industrie und Dienstleistungen*.

References

1. Albersmeyer, J., Bock, H.: Sensitivity Generation in an Adaptive BDF-Method. In: H.G. Bock, E. Kostina, X. Phu, R. Rannacher (eds.) Modeling, Simulation and Optimization of Complex Processes: Proceedings of the International Conference on High Performance Scientific Computing, March 6–10, 2006, Hanoi, Vietnam, pp. 15–24. Springer Verlag Berlin Heidelberg New York (2008)
2. Atkinson, A.C., Donev, A.N.: Optimum Experimental Designs. Oxford University Press (1992)
3. Atkinson, A.C., Fedorov, V.V.: Optimal design: Experiments for discriminating between several models. Biometrika **62**(2), 289 (1975)
4. Banga, J., Balsa-Canto, E.: Parameter estimation and optimal experimental design. Essays Biochem. **45**, 195–209 (2008)
5. Bauer, I.: Numerische Verfahren zur Lösung von Anfangswertaufgaben und zur Generierung von ersten und zweiten Ableitungen mit Anwendungen bei Optimierungsaufgaben in Chemie und Verfahrenstechnik. Ph.D. thesis, Universität Heidelberg (1999). URL http://www.ub.uni-heidelberg.de/archiv/1513
6. Bauer, I., Bock, H.G., Körkel, S., Schlöder, J.P.: Numerical methods for initial value problems and derivative generation for DAE models with application to optimum experimental design of chemical processes. In: F. Keil, W. Mackens, H. Voss, J. Werther (eds.) Scientific Computing in Chemical Engineering II, vol. 2, pp. 282–289. Springer-Verlag, Berlin Heidelberg (1999)
7. Binder, T., Kostina, E.: Robust parameter estimation in differential equations. In: H.G. Bock, T. Carraro, W. Jäger, S. Körkel, R. Rannacher, J.P. Schlöder (eds.) Model Based Parameter Estimation: Theory and Applications, Contributions in Mathematical and Computational Sciences. Springer (2012)
8. Bock, H.: Randwertproblemmethoden zur Parameteridentifizierung in Systemen nichtlinearer Differentialgleichungen, *Bonner Mathematische Schriften*, vol. 183. Universität Bonn, Bonn (1987). URL http://www.iwr.uni-heidelberg.de/groups/agbock/FILES/Bock1987.pdf
9. Bock, H.G., Eich, E., Schlöder, J.P.: Numerical Solution of Constrained Least Squares Boundary Value Problems in Differential-Algebraic Equations. In: K. Strehmel (ed.) Differential-Algebraic Equations. Numerical Treatment of Differential Equations. BG Teubner (1988)
10. Bock, H.G., Körkel, S., Kostina, E., Schlöder, J.P.: Robustness aspects in parameter estimation, optimal design of experiments and optimal control. In: W. Jäger, R. Rannacher, J. Warnatz (eds.) Reactive Flows, Diffusion and Transport, pp. 117–146. Springer-Verlag, Berlin Heidelberg (2007)
11. Bock, H.G., Kostina, E.A., Schlöder, J.P.: Numerical methods for parameter estimation in nonlinear differential algebraic equations. GAMM-Mitteilungen **30**(2), 352–375 (2007)
12. Chen, B.H., Asprey, S.P.: On the design of optimally informative dynamic experiments for model discrimination in multiresponse nonlinear situations. Ind. Eng. Chem. Res. **42**(7), 13791390 (2003)
13. Dieses, A.: Numerische Verfahren zur Diskriminierung nichtlinearer Modelle für dynamische chemische Prozesse. Diplomarbeit, Interdisziplinäres Zentrum für Wissenschaftliches Rechnen der Universität Heidelberg (1997)
14. Franceschini, G., Macchietto, S.: Model-based design of experiments for parameter precision: State of the art. Chemical Engineering Science **63**, 4846–4872 (2008)
15. Griewank, A., Walther, A.: Evaluating Derivatives: Principles and Techniques of Algorithmic Differentiation, 2nd edn. No. 105 in Other Titles in Applied Mathematics. SIAM, Philadelphia, PA (2008). URL http://www.ec-securehost.com/SIAM/OT105.html
16. Körkel, S.: Numerische Methoden für optimale Versuchsplanungsprobleme bei nichtlinearen DAE-Modellen. Ph.D. thesis, Universität Heidelberg, Heidelberg (2002). URL http://www.koerkel.de

17. Körkel, S., Arellano-Garcia, H.: Online experimental design for model validation. In: R.M. de Brito Alves, C.A.O. do Nascimento, E.C.B. Jr. (eds.) Proceedings of 10th International Symposium on Process Systems Engineering — PSE2009 (2009)
18. Körkel, S., Arellano-Garcia, H., Schöneberger, J., Wozny, G.: Optimum experimental design for key performance indicators. In: B. Braunschweig, X. Joulia (eds.) Proceedings of 18th European Symposium on Computer Aided Process Engineering - ESCAPE 18 (2008)
19. Körkel, S., Bauer, I., Bock, H., Schlöder, J.: A sequential approach for nonlinear optimum experimental design in dae systems. In: Scientific Computing in Chemical Engineering II, pp. 338–345. Springer (1999)
20. Körkel, S., Kostina, E., Bock, H., Schlöder, J.: Numerical methods for optimal control problems in design of robust optimal experiments for nonlinear dynamic processes. Optimization Methods and Software **19**, 327–338 (2004)
21. Moles, C.G., Mendes, P., Banga, J.R.: Parameter estimation in biochemical pathways: A comparison of global optimization methods. Genome Res. **13**(11), 2467–2474 (2003)
22. Pukelsheim, F.: Optimal Design of Experiments. Wiley (1993)
23. Schlöder, J.: Numerische Methoden zur Behandlung hochdimensionaler Aufgaben der Parameteridentifizierung, *Bonner Mathematische Schriften*, vol. 187. Universität Bonn, Bonn (1988)
24. Schlöder, J.P., Bock, H.G.: Identification of rate constants in bistable chemical reaction systems. In: P. Deuflhard, E. Hairer (eds.) Inverse Problems, *Progress in Scientific Computing*, vol. 2, pp. 27–47. Birkhäuser Boston (1983)
25. Schwetlick, H.: Nonlinear parameter estimation: Models, criteria and algorithms. In: D.F. Griffiths, G.A. Watson (eds.) Numerical Analysis 1991, Proceedings of the 14th Dundee Conference on Numerical Analysis, *Pitman Research Notes in Mathematics Series*, vol. 260, pp. 164–193. Longman Scientific & Technical (1992)
26. Seber, G.A.F., Wild, C.J.: Nonlinear Regression. Wiley (1989)
27. Stoer, J.: On the numerical solution of constrained least-squares problems. SIAM J. Numer. Anal. **8**(2), 382–411 (1971)
28. Telen, D., Logist, F., Derlinden, E.V., Tack, I., Impe, J.V.: Optimal experiment design for dynamic bioprocesses: a multi-objective approach. Chemical Engineering Science **78**, 82–97 (2012)
29. Uciński, D.: Optimal Measurement Methods for Distributed Parameter System Identification. CRC Press (2005)
30. Uciński, D.: An optimal scanning sensor activation policy for parameter estimation of distributed systems. In: H.G. Bock, T. Carraro, W. Jäger, S. Körkel, R. Rannacher, J.P. Schlöder (eds.) Model Based Parameter Estimation: Theory and Applications, Contributions in Mathematical and Computational Sciences. Springer (2012)
31. Walter, S.: Structured higher-order algorithmic differentiation in the forward and reverse mode with application in optimum experimental design. Ph.D. thesis, Humboldt-Universität zu Berlin (2011)

Adaptive Finite Element Methods for Parameter Identification Problems

Boris Vexler

Abstract This chapter provides an overview on the use of adaptive finite element methods for parameter identification problems governed by partial differential equations. We discuss a posteriori error estimates for the finite element discretization error with respect to a given quantity of interest for both stationary (elliptic) and nonstationary (parabolic) problems. These error estimates guide adaptive algorithms for mesh refinement, which are tailored to the parameter identification problem. The capability of the presented methods is demonstrated on two model examples of (stationary and nonstationary) combustion problems. Moreover we present a recently developed technique for efficient computation of the Tikhonov regularization parameter in the context of distributed parameter estimation using adaptive finite element methods.

1 Introduction

Many application problems coming from physics, chemistry, biology, or engineering sciences are described by mathematical models involving partial differential equations (PDEs). All these models involve parameters, which are often either unknown or only imprecisely known. In the majority of cases the unknown parameters can not be measured in a direct way. This fact gives the motivation for systematic comparison between experiments and numerical simulations leading to parameter identification problems. As described in other chapters of this volume, parameter identification problems are usually reformulated as optimization problems, where the mathematical model plays the role of the constraint. This leads to optimization problems with partial differential equations.

B. Vexler (✉)
Lehrstuhl für Mathematische Optimierung, Technische Universität München, Fakultät für
Mathematik, Boltzmannstraße 3, 85748 Garching b. München, Germany
e-mail: vexler@ma.tum.de

H.G. Bock et al. (eds.), *Model Based Parameter Estimation*, Contributions
in Mathematical and Computational Sciences 4, DOI 10.1007/978-3-642-30367-8_2,
© Springer-Verlag Berlin Heidelberg 2013

For numerical solution the resulting optimization problems have to be discretized. This leads inevitably to *discretization errors* between the solution of the continuous (infinite dimensional) and the discretized (finite dimensional) problems. For the discretization of problems involving partial differential equations we employ *finite element methods* which are constructed using triangulations (meshes) of the computational domain. For efficient solution of parameter identification problems the computational meshes have to be chosen such that on the one hand the numerical effort to solve the discretized problem is as small as possible, and on the other hand the discretization error is under a given tolerance. These considerations lead to construction of adaptive algorithms for mesh refinement.

The use of adaptive techniques based on a posteriori error estimation is well accepted in the context of finite element discretization of partial differential equations, see, e.g., [5, 13, 31].

In the last years application of these techniques to optimization problems with PDEs is actively investigated. In the articles [17, 18, 22, 24], the authors provide a posteriori error estimates for elliptic optimal control problems with distributed or Neumann control subject to box constraints on the control variable. These estimates assess the error in the control, state, and adjoint variables with respect to the natural norms of the corresponding spaces.

However, in many applications, the error in global norms does not provide a useful error bound for the error in the quantity of physical interest. In [3,5] a general concept for a posteriori estimation of the discretization error with respect to the cost functional in the context of optimal control problems is presented. In the papers [6, 7, 32], this approach is extended to the estimation of the discretization error with respect to an arbitrary functional depending on both the control (parameter) and the state variable, i.e., with respect to a quantity of interest. This allows, among other things, an efficient treatment of parameter identification problems. This approach is summarized in Sect. 3.

The extension of these techniques to optimal control and parameter identification problems governed by parabolic equations was done in recent publications [4, 25, 26] and is summarized in Sect. 4.

In several situations, the unknown parameter variable is infinite dimensional, the parameter identification problem is ill-posed, and one has to employ regularization methods, see, e. g., [12, 19]. In Sect. 5 we present the techniques developed in [16] for efficient computation of Tikhonov regularization parameter using adaptive finite element methods.

2 Basic Concept

In this section we present our basic concept for a posteriori error estimation and adaptivity. We consider a (system of) partial differential equations in the abstract setting

$$A(q, u) = 0, \tag{1}$$

Adaptive Finite Element Methods for Parameter Identification Problems 33

where u denotes the state variable in a state space V and may describe such physical quantities like temperature, velocity, pressure, concentration, etc. The variable q in the space Q describes the (set of) unknown parameters. The goal of the parameter identification problem is to estimate the parameter variable q from measurements $\hat{C}$ in some measurement space Z corresponding to the observation operator $C : V \to Z$. As described in other chapters of this volume, this problem is usually reformulated as an optimization problem using the least squares approach:

$$\text{minimize } J(q,u) = \frac{1}{2}\|C(u) - \hat{C}\|_Z^2 + \frac{\alpha}{2}\|q - \hat{q}\|_Q^2 \quad \text{subject to (1)} \quad (2)$$

Here, $\hat{q} \in Q$ denotes the reference parameter and $\alpha \geq 0$ is the Tikhonov regularization parameter. For a finite dimensional parameter space $Q = \mathbb{R}^l$ the regularization parameter is often chosen as $\alpha = 0$. In the case of an ill posed parameter identification problem the choice of α is a delicate issue. In Sect. 5 we discuss a strategy for choosing α using adaptive finite elements.

For their numerical solution the infinite-dimensional optimization problem (2) has to be discretized. Usually one uses finite dimensional spaces $Q_h \subset Q$ and $V_h \subset V$ for the discretization of the parameter and the state variable respectively. The differential operator A is approximated by its discrete analogue A_h resulting in the following discretized optimization problem:

$$\text{minimize } J(q_h, u_h) \quad \text{subject to}$$
$$q_h \in Q_h, u_h \in V_h, \quad (3)$$
$$A_h(q_h, u_h) = 0.$$

In several publications the asymptotic behavior of the error between the solution (q,u) of the continuous problem (2) and the solution (q_h, u_h) of the discrete problem (3) is studied and a priori error estimates are derived, see, e.g., [20, 27] for the a priori error analysis in the context of parameter identification problems. However, a priori error estimates do not provide information how to construct an efficient discretization which allows to bring the discretization error below a given tolerance with minimum of numerical effort.

For assessing the quality of a discrete model we develop a posteriori error estimates for the error between the solution of the continuous optimization problem and the solution of the discrete one. A crucial ingredient of the error analysis is the choice of a quantity, which describes the goal of the computation. We call this quantity the *quantity of interest* $I : Q \times V \to \mathbb{R}$ and derive an a posteriori error estimate η_h for the error with respect to it, i.e.

$$I(q,u) - I(q_h, u_h) \approx \eta_h .$$

It is the task of an adaptive algorithm to automatically construct an appropriate discretization. One starts with a coarse mesh which is successively enriched (refined) using information from the error estimator η_h computed on the present mesh. Thereby, the aim is to generate economical meshes in the sense that the number of unknowns $N_h = \dim V_h$ is as small as possible for achieving a given tolerance.

In the following, we sketch a typical mesh refinement algorithm.

Algorithm: Basic Adaptive Mesh Refinement

1. Choose starting discretization (Q_{h_0}, V_{h_0}) and set $l = 0$
2. Compute $u_{h_l} \in V_{h_l}, q_{h_l} \in Q_{h_l}$ solving discrete optimization problem
3. Compute error estimator $\eta_{h_l} = \eta(u_{h_l}, q_{h_l})$
4. If $\eta_{h_l} \leq TOL$ stop
5. Refine: $(Q_{h_l}, V_{h_l}) \to (Q_{h_{l+1}}, V_{h_{l+1}})$ using information from η_{h_l}
6. Increment l and goto 2.

In this algorithm, the a posteriori error estimator is used for two goals: first, in step (4) η_{h_l} is used as a *stopping criterion* for a given tolerance TOL. Second, in step (5) the error estimator should provide information for construction of new discretizations. In the context of finite element methods, this step is based on cell-wise (or node-wise) representation of the error estimator η_h, e.g.

$$\eta_h = \sum_K \eta_{h,K}$$

with cell contributions $\eta_{h,K}$. The value of $\eta_{h,K}$ is used to decide how the whole error can be reduced, if the cell K is refined. The common feature of different strategies is, that the local mesh refinement should be done for those K with "large" $\eta_{h,K}$. Different marking strategies are described, e. g., in [5].

3   Adaptivity for Parameter Estimation in Stationary Problems

In this section we describe our approach for parameter identification problems governed by stationary partial differential equations. We start with an abstract formulation of the problem (2) in a weak form and discuss necessary optimality conditions for it. In Sect. 3.1 we describe the finite element discretization for the problem under consideration. A posteriori error estimates are presented in Sect. 3.2. Thereafter, we illustrate the efficiency of our method for estimation of Arrhenius parameters in a simple stationary combustion model.

Throughout this section we assume the state equation (1) to be formulated in a weak form:

$$a(q, u)(v) = f(v) \quad \text{for all } v \in V. \tag{4}$$

Here, $a \colon Q \times V \times V$ denotes a semi-linear form which is linear in the third argument and f in the dual space V^* represents given data. The optimization problem is then formulated as follows:

Adaptive Finite Element Methods for Parameter Identification Problems

$$\text{minimize } J(q, u) \quad \text{subject to} \tag{5}$$

$$q \in Q, \ u \in V \text{ and } (4), \tag{6}$$

where the cost functional $J(q, u)$ is given as in (2).

Remark 1. Often additional inequality constraints, for instance box constraints $q_a \leq q \leq q_b$, are included in the above optimization problem. For extension of adaptive techniques presented in this chapter to problems with inequality constrains we refer to [9, 33].

For the discussion of existence of solutions for different classes of optimization problems governed by PDEs we refer, e. g., to [15, 23, 30]. Throughout we assume that the optimization problem (4)–(6) possesses a solution.

The majority of optimization algorithms for the solution of (4)–(6) are based on necessary optimality conditions derived with the Lagrange approach. For problem (4)–(6) the Lagrange functional is introduced by

$$L(q, u, z) = J(q, u) + f(z) - a(q, u)(z). \tag{7}$$

In several situations one can prove, see, e. g., [30], that for a local minimum (q, u) of (4)–(6) there exists an adjoint solution $z \in V$ such that the triple $x = (q, u, z) \in X = Q \times V \times V$ is a stationary point of L, i. e.

$$L'(x)(\delta x) = 0 \quad \forall \delta x \in X.$$

This condition results in the optimality system consisting of three equations:

$$
\begin{aligned}
a(q, u)(v) &= f(v) & \text{for all } v \in V & \quad \text{(state equation)}, \\
a'_u(q, u)(v, z) &= J'_u(q, u)(v) & \text{for all } v \in V & \quad \text{(adjoint equation)}, \\
a'_q(q, u)(p, z) &= J'_q(q, u)(p) & \text{for all } p \in Q & \quad \text{(optimality condition)}.
\end{aligned}
\tag{8}
$$

In the presence of additional inequality constraints on the parameter variable q, the optimality condition becomes a variational inequality, see, e. g, [30].

3.1 Finite Element Discretization

We consider conforming finite element discretizations. Starting from a non-overlapping partition $\mathcal{T}_h$ of the computational domain Ω into cells K (quadrilaterals or triangles in two space dimensions, hexahedra or tetrahedra in three space dimensions), we construct the finite-dimensional spaces

$$V_h \subset V, \qquad Q_h \subset Q, \tag{9}$$

Fig. 1 Quadrilateral mesh in two dimensions with two hanging nodes

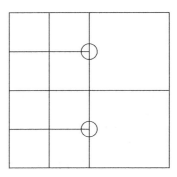

see, e.g., [10, 11]. Here, h denotes a generic discretization parameter, describing essentially the local fineness of the mesh. It is defined as a cell-wise constant function by setting $h|_K = h_K$ with the diameter h_K of the cell K. Due to the fact that we allow local mesh refinement, we do not assume that the cells are of comparable size.

Remark 2. In order to ease mesh refinement, we allow for cells to have nodes which lie on midpoints of faces of the neighboring cells. But at most one such *hanging node* (see Fig. 1) is permitted for each face of a cell. There are no degrees of freedom corresponding to these irregular nodes and the value of a finite element function is determined by pointwise interpolation.

The optimization problem on the discrete level is given through the discretization of the state equation (4):

$$\text{Minimize} \quad J(q_h, u_h), \quad \text{subject to} \tag{10}$$

$$q_h \in Q_h, \ u_h \in V_h, \tag{11}$$

$$a(q_h, u_h)(v_h) = f(v_h) \quad \text{for all } v_h \in V_h. \tag{12}$$

Due to the Galerkin-type discretization, the necessary optimality conditions for the discrete optimization problem (10)–(12) can be directly translated from the continuous level. The triple $x_h = (q_h, u_h, z_h) \in X_h = Q_h \times V_h \times V_h$ is a stationary point for the Lagrangian L on the discrete space $Q_h \times V_h \times V_h$, i.e.

$$L'(x_h)(\delta x_h) = 0 \quad \text{for all } \delta x_h \in X_h.$$

This results in the discrete optimality system:

$$\begin{aligned}
a(q_h, u_h)(v_h) &= f(v_h) & &\text{for all } v_h \in V_h, \\
a'_u(q_h, u_h)(v_h, z_h) &= J'_u(q_h, u_h)(v_h) & &\text{for all } v_h \in V_h, \\
a'_q(q_h, u_h)(p_h, z_h) &= J'_q(q_h, u_h)(p_h) & &\text{for all } p_h \in Q_h.
\end{aligned} \tag{13}$$

3.2 A Posteriori Error Estimators

The aim of this section is to present an error estimator η_h for the error between the solution (q, u) of the continuous optimization problem (4)–(6) and the solution (q_h, u_h) of the discrete analog (10)–(12) with respect to a given quantity of interest $I: Q \times V \to \mathbb{R}$, i. e.

$$I(q, u) - I(q_h, u_h) \approx \eta_h.$$

For parameter identification problems the quantity of interest is often chosen as a component of a finite dimensional parameter $I(q, u) = q_i$ (one can consider several quantities of interest) or as some weighted mean value in the case of a distributed parameter q.

We start with a special case, where the quantity of interest coincide with the cost functional, i. e. $I(q, u) = J(q, u)$. This case is in general not of importance for parameter identification problems, but the error estimates for a general quantity of interest are derived based on this special case. The following error representation is shown in [5]:

Proposition 1. *Let* $x = (q, u, z) \in X$ *fulfill the optimality system* (8) *and let* $x_h = (q_h, u_h, z_h) \in X_h$ *fulfill the discrete counterpart* (13). *Then there holds:*

$$J(q, u) - J(q_h, u_h) = \frac{1}{2} L'(x_h)(x - \tilde{x}_h) + \mathcal{R},$$

where $\tilde{x}_h \in X_h$ *is arbitrary and* $\mathcal{R}$ *is a third order remainder term given by*

$$\mathcal{R} = \frac{1}{2} \int_0^1 L'''(x + se_x)(e_x, e_x, e_x) \cdot s \cdot (s - 1) \, ds$$

with $e_x = x - x_h$.

In order to turn the above error representation into a computable error estimator, we proceed as follows. First we choose $\tilde{x}_h = i_h x$ with a suitable interpolation operator $i_h: X \to X_h$, then the interpolation error is approximated using an operator $\pi: X_h \to \widetilde{X}_h$, with $\widetilde{X}_h \neq X_h$, such that $x - \pi x_h$ has a better local asymptotical behavior as $x - i_h x$. Then we approximate:

$$J(q, u) - J(q_h, u_h) \approx \eta^J = \frac{1}{2} L'(x_h)(\pi_h x_h - x_h).$$

Such an operator can be constructed for example by the interpolation of the computed bilinear finite element solution in the space of biquadratic finite elements on patches of cells. For this operator the improved approximation property relies on local smoothness of the solution and super-convergence properties of the approximation x_h. The use of such "local higher-order approximations" is observed to work very successfully in the context of a posteriori error estimation, see, e.g., [5, 6].

For extension of this result to the case of a general quantity of interest we proceed as in [6, 7] and introduce an auxiliary Lagrange functional

$$M: X \times X \to \mathbb{R}, \quad M(x, x_1) = I(q, u) + L'(x)(x_1).$$

We abbreviate $y = (x, x_1)$ and $x_1 = (q_1, u_1, z_1)$. Then similarly to Proposition 1 an error representation for the error in I can be formulated using continuous and discrete stationary points of M, cf. [6, 7]:

Proposition 2. *Let* $y = (x, x_1) \in X \times X$ *be a stationary point of* M, *i.e.,*

$$M'_y(y)(\delta y) = 0 \quad \forall \delta y \in X \times X,$$

and let $y_h = (x_h, x_{1,h}) \in X_h \times X_h$ *be a discrete stationary point, i.e,*

$$M'_y(y_h)(\delta y_h) = 0 \quad \forall \delta y_h \in X_h \times X_h,$$

then there holds

$$I(q, u) - I(q_h, u_h) = \frac{1}{2} M'(y_h)(y - \tilde{y}_h) + \mathcal{R}_1,$$

where $\tilde{y}_h \in X_h \times X_h$ *is arbitrary and the remainder term is given as*

$$\mathcal{R}_1 = \frac{1}{2} \int_0^1 M'''(y + s e_y)(e_y, e_y, e_y) \cdot s \cdot (s - 1) \, ds$$

with $e_y = y - y_h$.

Again, in order to turn this error identity into a computable error estimator, we neglect the remainder term $\mathcal{R}_1$ and approximate the interpolation error using a suitable approximation of the interpolation error leading to

$$I(q, u) - I(q_h, u_h) \approx \eta^I = \frac{1}{2} M'_y(y_h)(\pi_h y_h - y_h).$$

For a concrete form of this error estimator consisting of some residuals we refer to [6, 7].

A crucial question is of course how to compute the discrete stationary point y_h of M required for this error estimator. At the first glance it seems that the solution of the stationarity equation for M leads to coupled system of double size compared with the optimality system (8). However, solving this stationarity equation can be easily done using the already computed stationary point $x_h = (q_h, u_h, z_h)$ of L and exploiting existing structures. The following proposition shows that the computation

Adaptive Finite Element Methods for Parameter Identification Problems

of auxiliary variables $x_{1,h} = (q_{1,h}, u_{1,h}, z_{1,h})$ is equivalent to one step of an SQP method, which is often applied for solving (8). The corresponding equation can be also solved by a Schur complement technique reducing the problem to the control space, cf., e.g., [26].

Proposition 3. Let $x = (q, u, z)$ and $x_h = (q_h, u_h, z_h)$ be continuous and discrete stationary points of L. Then $y = (x, x_1)$ is a stationary point of M if and only if

$$L''(x)(\delta x, x_1) = -I_q'(q, u)(\delta q) - I_u'(q, u)(\delta u) \quad \forall \delta x = (\delta q, \delta u, \delta z) \in X.$$

Moreover, $y_h = (x_h, x_{1,h})$ is a discrete stationary point of M if and only if

$$L''(x_h)(\delta x_h, x_{1,h}) = -I_q'(q_h, u_h)(\delta q_h) - I_u'(u_h)(\delta u_h) \quad \forall \delta x_h = (\delta q_h, \delta u_h, \delta z_h) \in X_h.$$

3.3 Estimation of Arrhenius Parameters for a Stationary Combustion Model

In this section we demonstrate the behavior of the presented adaptive strategy for the problem of estimating Arrhenius parameters in a simple combustion model. The results are taken from [1, 32]. For the application to parameter estimation in multidimensional reactive flows we refer to [1, 2].

The state equation is given by scalar stationary convection-diffusion-reaction equation for the variable u in a domain $\Omega \subset \mathbb{R}^2$ with a divergence-free vector field β and a diffusion coefficient D:

$$\beta \cdot \nabla u - \mathrm{div}(D\nabla u) + s(u, q) = f, \tag{14}$$

provided with Dirichlet boundary conditions $u = \widehat{u}$ at the inflow boundary $\Gamma_{\mathrm{in}} \subset \partial\Omega$ and Neumann conditions $\partial_n u = 0$ on $\partial\Omega \backslash \Gamma_{\mathrm{in}}$. As usual in combustion problems, the reaction term is of Arrhenius type

$$s(u, q) := A \exp\{-E/(d - u)\}u(c - u). \tag{15}$$

The variable u stands for the mole fraction of a fuel, while the mole fraction of the oxidizer is $0.2 - u$. Since the Arrhenius law is a heuristic law and can not be derived by physical laws, the involved parameters are a priori unknown and have to be calibrated. This parameter fitting is usually done by comparison of experimental data and simulation results. We assume the parameters d, c to be fixed and the parameters A, E are considered as unknown and form the vector-valued parameter $q = (\ln(A), E) \in Q = \mathbb{R}^2$.

We consider the following configuration: Fuel (F) and oxidizer (Ox) are injected in different pipes and diffuse in a reaction chamber with overall length 35 mm and high 7 mm, see Fig. 2. At the center tube, the Dirichlet condition for the

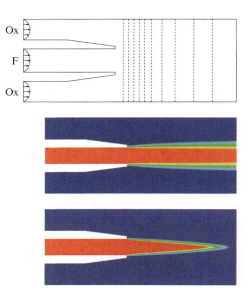

Fig. 2 Configuration of the reaction chamber for estimating Arrhenius coefficients. *Dashed vertical lines* show the lines where the measurements are modeled

Fig. 3 Mole fraction of the fuel (u^0) for the initial parameters q^0. *Blue*: $u = 0$, *red*: $u = 0.2$

Fig. 4 Mole fraction of the fuel (u) for the optimal parameters q

fuel is $u = u_{in} := 0.2$, and at the upper and lower tube, $u = 0$. On all other parts of the boundary, homogeneous Neumann conditions are opposed.

The fixed parameters in the Arrhenius law (15) are $c = u_{in}$ and $d = 0.24$. The convection direction $\beta(x, y)$ is a velocity field obtained by solving the incompressible Navier–Stokes equations with parabolic inflow profile at the tubes with peak flow $\beta_{max} = 0.2$ m/s. The diffusion coefficient D is chosen $D = 2.e - 6$.

The measurements $C(u) \in Z = \mathbb{R}^{n_m}$ are modeled by mean values along $n_m = 10$ straight lines Γ_i at different positions in the reaction chamber, see dashed lines in Fig. 2:

$$C_i(v) = \int_{\Gamma_i} v\, dx, \quad i = 1, \ldots, n_m.$$

For the measurement data $\hat{C} \in Z$ we use artificial measurements obtained by choosing optimal parameters $q = (6.9, 0.07)$ and solving the state equation on a very fine mesh. The initial parameters are set to $q^0 = (\log(A^0), E^0) = (4, 0.15)$, leading to low reaction rates and a diffusion dominated solution. In Fig. 3 the corresponding state variable (fuel) u^0 is shown. For the optimal q, in contrast to the initial guess q^0, a sharp reaction front occurs, see Fig. 4. Obviously, the difference in the parameters has a substantial impact on the state variable u.

The state equation is discretized by finite elements using local projection stabilization for the convection term. This symmetric stabilization allows for a natural formulation of the discrete adjoint equation, see [8] for details.

For the quantity of interest we choose $I(q) = q_2$ to control the discretization error in the second parameter. Comparing the error in the parameters with a more conventional strategy on globally refined meshes, our proposed algorithm is much more efficient. In Fig. 5, the relative error in the second parameter is plotted in

Adaptive Finite Element Methods for Parameter Identification Problems

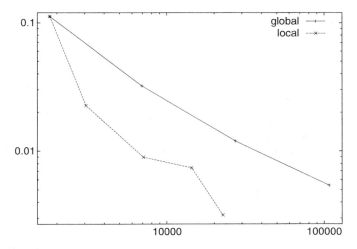

Fig. 5 Relative error in the second Arrhenius parameter in dependence of the number of mesh points. *Solid line*: globally refined meshes, *dashed line*: locally refined meshes on the basis of a posteriori error estimation

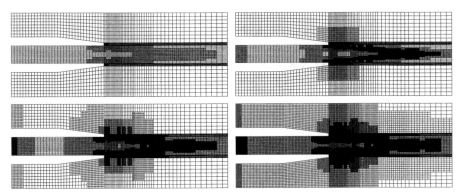

Fig. 6 Obtained meshes for estimating Arrhenius parameters with 2,825, 6,704, 13,676, and 21,752 nodes (from *upper left* to *lower right*)

dependence of the number of mesh points. The dashed line results from our method on locally refined meshes. The solid line stands for parameter estimation with the same optimization loop but on uniformly refined meshes. For a relative error of less than 1%, only 6,704 nodes are necessary with a locally refined mesh, whereas more than 100,000 nodes are necessary on a uniformly refined mesh.

In Fig. 6, a sequence of locally refined meshes produced by the refinement algorithm is shown. The highest amount of mesh points is located near the flame front and close to the measurement lines.

4 Adaptivity for Parameter Estimation in Nonstationary Problems

In this section we discuss an extension of our techniques to parameter identification problems with state equations described by parabolic partial differential equations following [4, 25, 26]. In what follows we partially use the same notation for the semilinear form a, the Lagrange functional L, etc. as in Sect. 3 with slightly different meaning due to the time dependency of involved equations.

Throughout this section we assume the state equation (1) to be formulated in a weak form as follows:

$$(\partial_t u, \phi) + a(q, u)(\phi) = (f, \phi) \quad \forall \phi \in X,$$
$$u(0) = u_0(q). \tag{16}$$

Here, we use the following notation: Q is again the parameter (control) space, the Hilbert spaces V, H and the dual V^* build a Gelfand triple, $I = (0, T)$ is the time interval, $(\cdot, \cdot)$ denotes the inner product in $L^2(0, T; L^2(\Omega))$, and the space X is given as

$$X = W(0, T) = \left\{ v \mid v \in L^2((0, T), V) \text{ and } \partial_t v \in L^2((0, T), V^*) \right\}. \tag{17}$$

The semilinear form $a: Q \times X \times X \to \mathbb{R}$ as well as the initial condition u_0 may depend on unknown parameters q. The form a results from a stationary semilinear form $\bar{a}: Q \times V \times V \to \mathbb{R}$ by

$$a(q, u)(\phi) := \int_0^T \bar{a}(q, u(t))(\phi(t)) \, dt.$$

The optimization problem is then formulated as follows:

$$\text{minimize } J(q, u) \quad \text{subject to} \tag{18}$$

$$q \in Q, \ u \in X \text{ and (16)}, \tag{19}$$

where the cost functional $J(q, u)$ is given as in (2).

As for stationary problems, the optimality conditions can be formulated using a Lagrange functional $L: Q \times X \times X \to \mathbb{R}$ defined as

$$L(q, u, z) = J(q, u) + (f - \partial_t u, z) - a(q, u)(z) - (u(0) - u_0(q), z(0))_H. \tag{20}$$

The necessary optimality condition for $x = (q, u, z) \in X = Q \times X \times X$

$$L'(x)(\delta x) = 0 \quad \forall \delta x \in X$$

Adaptive Finite Element Methods for Parameter Identification Problems | 43

results in the state equation (16), the adjoint equation, which is formulated backwards in time and the gradient equation, see, e. g., [23] or [30]. For details of the optimality system in the above notation we refer to [26].

4.1 Space-Time Finite Element Discretization

For discretization of the state equation (16) we employ Galerkin finite element methods in space and time. This allows for natural computable representation of the discrete gradient and Hessian due to the fact that optimization (building up the optimality system) and discretization commute in this case, see [4, 26] for details. Moreover, our systematic approach to a posteriori error estimation relies on using the Galerkin-type discretizations.

For temporal discretization we employ *discontinuous Galerkin (dG)* methods, see, e. g., [14] or [29] and the discretization in space is done with usual conforming finite elements.

4.1.1 Temporal Discretization

To define a semi-discretization in time, let us partition the time interval $[0, T]$ as

$$[0, T] = \{0\} \cup I_1 \cup I_2 \cup \cdots \cup I_M$$

with subintervals $I_m = (t_{m-1}, t_m]$ of size k_m and time points

$$0 = t_0 < t_1 < \cdots < t_{M-1} < t_M = T.$$

We define the discretization parameter k as a piecewise constant function by setting $k\big|_{I_m} = k_m$ for $m = 1, \ldots, M$. By means of the subintervals I_m, we define for $r \in \mathbb{N}_0$ the semi-discrete space $\widetilde{X}_k^r$ by

$$\widetilde{X}_k^r = \left\{ v_k \in L^2((0, T), V) \,\middle|\, v_k\big|_{I_m} \in \mathcal{P}^r(I_m, V) \text{ and } v_k(0) \in H \right\}$$

Here, $\mathcal{P}^r(I_m, V)$ denotes the space of polynomials up to order r defined on I_m with values in V. Thus, the functions in $\widetilde{X}_k^r$ may have discontinuities at the ends of the subintervals I_m. This space will be used in the sequel as trial and test space in the discontinuous Galerkin method.

To define the dG(r) discretization we employ the following definition for functions $v_k \in \widetilde{X}_k^r$:

$$v_{k,m}^+ := \lim_{t \to 0+} v_k(t_m + t), \quad v_{k,m}^- := \lim_{t \to 0+} v_k(t_m - t) = v_k(t_m), \quad [v_k]_m := v_{k,m}^+ - v_{k,m}^-.$$

Then, the dG(r) semi-discretization of the state equation (16) reads: Find for given parameter $q_k \in Q$ a state $u_k \in \widetilde{X}_k^r$ such that

$$\sum_{m=1}^{M} \int_{I_m} (\partial_t u_k, \phi)_H \, dt + a(q_k, u_k)(\phi) + \sum_{m=0}^{M-1} ([u_k]_m, \phi_m^+)_H = (f, \phi), \ \forall \phi \in \widetilde{X}_k^r,$$

$$u_{k,0}^- = u_0(q_k).$$

(21)

The semi-discrete optimization problem for the dG(r) time discretization has the form:

Minimize $J(q_k, u_k)$ subject to the state equation (21), $(q_k, u_k) \in Q \times \widetilde{X}_k^r$. (22)

The corresponding Lagrangian $\widetilde{L}: Q \times \widetilde{X}_k^r \times \widetilde{X}_k^r \to \mathbb{R}$ is defined by

$$\widetilde{L}(q_k, u_k, z_k) = J(q_k, u_k) + (f, z_k) - \sum_{m=1}^{M} \int_{I_m} (\partial_t u_k, z_k)_H \, dt$$

$$-a(q_k, u_k)(z_k) - \sum_{m=0}^{M-1} ([u_k]_m, z_{k,m}^+)_H - (u_{k,0}^- - u_0(q_k), z_{k,0}^-)_H$$

and the optimality condition for the triple $x_k = (q_k, u_k, z_k) \in X_k = Q \times \widetilde{X}_k^r \times \widetilde{X}_k^r$ on this semidiscrete level is given as

$$\widetilde{L}(x_k)(\delta x_k) = 0 \quad \forall \delta x_k \in X_k.$$

4.1.2 Spatial Discretization

The discontinuous Galerkin discretization in time naturally allows for spatial finite element meshes, which are different in time. The use of such dynamic meshes leads to very efficient numerical schemes, see [28] for details. For each time interval I_m we construct (conforming) finite element spaces $V_h^{s,m}$ (of order s) in a usual way. The corresponding state space $\widetilde{X}_{k,h}^{r,s} \subset \widetilde{X}_k^r$ is defined by

$$\widetilde{X}_{k,h}^{r,s} = \left\{ v_{kh} \in L^2((0.T), H) \ \middle| \ v_{kh}\big|_{I_m} \in \mathcal{P}^r(I_m, V_h^{s,m}) \text{ and } v_{kh}(0) \in V_h^{s,0} \right\}.$$

Then, the space-time finite element discretization of the state equation reads: Find for given parameter $q_{kh} \in Q$ a state $u_{kh} \in \widetilde{X}_{k,h}^{r,s}$ such that

$$\sum_{m=1}^{M} \int_{I_m} (\partial_t u_{kh}, \phi)_H \, dt + a(q_{kh}, u_{kh})(\phi) + \sum_{m=0}^{M-1} ([u_{kh}]_m, \phi_m^+)_H + (u_{kh,0}^-, \phi_0^-)_H$$

$$= (f, \phi) + (u_0(q_{kh}), \phi_0^-)_H \quad \forall \phi \in \widetilde{X}_{k,h}^{r,s}. \quad (23)$$

Adaptive Finite Element Methods for Parameter Identification Problems 45

Thus, the optimization problem with fully discretized state is given by

$$\text{Minimize } J(q_{kh}, u_{kh}) \text{ subject to (23)}, \quad (q_{kh}, u_{kh}) \in Q \times \widetilde{X}_{k,h}^{r,s}. \tag{24}$$

The optimality condition for the triple $x_{kh} = (q_{kh}, u_{kh}, z_{kh}) \in X_{kh} = Q \times \widetilde{X}_{k,h}^{r,s} \times \widetilde{X}_{k,h}^{r,s}$ on this discrete level is given as

$$\widetilde{L}(x_{kh})(\delta x_{kh}) = 0 \quad \forall \delta x_{kh} \in X_{kh}.$$

4.1.3 Full Discretization

As in the stationary case (cf. Sect. 3) we choose a finite dimensional subspace $Q_d \subset Q$ for the discretization of the parameter (control) space. In many situations the control space Q is finite dimensional and we set $Q_d = Q$. If the parameter q is given as a function (in space and/or in time) the subspace $Q_d \subset Q$ is typically constructed in a similar way as the discrete spaces for the state variable.

Then, the formulation of the state equation, the optimization problems and the Lagrangian defined on the fully discretized state space can directly be transferred to the level with fully discretized control and state spaces by replacing Q by Q_d. The fully discrete solutions is indicated by the subscript σ which collects the discretization indices k, h, and d.

The corresponding optimality condition for the triple $x_\sigma = (q_\sigma, u_\sigma, z_\sigma) \in X_\sigma = Q_d \times \widetilde{X}_{k,h}^{r,s} \times \widetilde{X}_{k,h}^{r,s}$ reads

$$\widetilde{L}(x_\sigma)(\delta x_\sigma) = 0 \quad \forall \delta x_\sigma \in X_\sigma.$$

4.2 A Posteriori Error Estimators

In the case of parameter estimation problems governed by parabolic equation, the construction of efficient adaptive algorithms requires separation of different error sources. Our aim is to estimates the discretization error due to temporal discretization, due to spatial discretization and due to control discretization separately. This allows to balance the influence of different parts of the discretization, see the discussion of such error balancing strategies in [26, 28]. Therefore, we split

$$
\begin{aligned}
J(q, u) - J(q_\sigma, u_\sigma) &= J(q, u) - J(q_k, u_k) \\
&\quad + J(q_k, u_k) - J(q_{kh}, u_{kh}) \\
&\quad + J(q_{kh}, u_{kh}) - J(q_\sigma, u_\sigma),
\end{aligned}
$$

where $(q_k, u_k) \in Q \times \widetilde{X}_k^r$ is the solution of the time discretized problem (22) and $(q_{kh}, u_{kh}) \in Q \times \widetilde{X}_{k,h}^{r,s}$ is the solution of the time and space discretized problem (24) with still undiscretized control space Q. Each term in this sum is then represented in a similar way as described in Sect. 3 using the residual of the equation in the optimality system. For the technical issues concerning evaluation of the resulting estimate we again refer to [26, 28].

As already discussed in Sect. 3 one is usually interested not in the error with respect to the cost functional but with respect to a quantity of interest, which can for example be a component of the parameter vector. To this end one consider the splitting

$$I(q, u) - I(q_\sigma, u_\sigma) = I(q, u) - I(q_k, u_k)$$
$$+ I(q_k, u_k) - I(q_{kh}, u_{kh})$$
$$+ I(q_{kh}, u_{kh}) - I(q_\sigma, u_\sigma),$$

and estimate each term in this sum using an additional Lagrangian M as in the stationary case, cf. Sect. 3.2 and see [26] for details.

4.3 Estimation of Model Parameters for Propagation of Laminar Flames

In this section we illustrate our method on parameter estimation for a model of propagation of laminar flames.

The governing equation for the considered problem is taken from an example given in [21]. It describes the major part of gaseous combustion under the low Mach number hypothesis. Introducing the dimensionless temperature θ, denoting by Y the species concentration, and assuming constant diffusion coefficients yields the system of equations

$$\begin{aligned}
\partial_t \theta - \Delta\theta &= \omega(Y, \theta) & \text{in } \Omega \times I, \\
\partial_t Y - \frac{1}{\text{Le}} \Delta Y &= -\omega(Y, \theta) & \text{in } \Omega \times I, \\
\theta &= \theta_0 & \text{on } \Omega \times \{0\}, \\
Y &= Y_0 & \text{on } \Omega \times \{0\},
\end{aligned} \tag{25}$$

where the Lewis number Le is the ratio of diffusivity of heat and diffusivity of mass. We use a simple one-species reaction mechanism governed by an Arrhenius law given by

$$\omega(Y, \theta) = \frac{\beta^2}{2\text{Le}} Y \, e^{\frac{\beta(\theta-1)}{1+\alpha(\theta-1)}}, \tag{26}$$

in which an approximation for large activation energy has been employed.

Fig. 7 Computational domain Ω and measurement points p_i

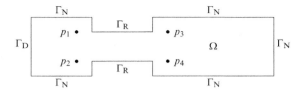

Here, we consider a freely propagating laminar flame described by (25) and its response to a heat absorbing obstacle, a set of cooled parallel rods with rectangular cross section (cf. Fig. 7). The computational domain has width $H = 16$ and length $L = 60$. The obstacle covers half of the width and has length $L/4$. The boundary conditions are chosen as

$$\theta = 1 \quad \text{on } \Gamma_\text{D} \times I, \qquad d_n \theta = 0 \quad \text{on } \Gamma_\text{N} \times I,$$
$$Y = 0 \quad \text{on } \Gamma_\text{D} \times I, \qquad d_n Y = 0 \quad \text{on } \Gamma_\text{N} \times I,$$

$$d_n \theta = -\kappa \theta \quad \text{on } \Gamma_\text{R} \times I,$$
$$d_n Y = 0 \quad \text{on } \Gamma_\text{R} \times I,$$

where the heat absorption is modeled by boundary conditions of Robin type on Γ_R.

The initial condition is the analytical solution of a one-dimensional right-traveling flame in the limit $\beta \to \infty$ located left of the obstacle:

$$\theta_0(x) = \begin{cases} 1 & \text{for } x_1 \leq \tilde{x}_1 \\ e^{\tilde{x}_1 - x_1} & \text{for } x_1 > \tilde{x}_1 \end{cases},$$

$$Y_0(x) = \begin{cases} 0 & \text{for } x_1 \leq \tilde{x}_1 \\ 1 - e^{\text{Le}(\tilde{x}_1 - x_1)} & \text{for } x_1 > \tilde{x}_1 \end{cases}.$$

For the computations, the occurring parameters are set as in [21] to

$$\text{Le} = 1, \qquad \beta = 10, \qquad \kappa = 0.1, \qquad \tilde{x}_1 = 9.$$

whereas the temperature ratio α, which determines the gas expansion in non-constant density flows, is the objective of the parameter estimation. The state variable u is given here as $u = (\theta, Y)$ and the parameter variable is $q = \alpha \in Q = \mathbb{R}$.

The unknown parameter α is estimated here using information from pointwise measurements of θ and Y at four points $p_i \in \Omega$, $i = 1, 2, 3, 4$, at final time $T = 60$. This parameter identification problem can be formulated by means of a cost functional of least-squares type, that is

$$J(q, u) = \frac{1}{2} \sum_{i=1}^{4} (\theta(p_i, T) - \hat{\theta}_i)^2 + \frac{1}{2} \sum_{i=1}^{4} (Y(p_i, T) - \hat{Y}_i)^2.$$

Table 1 Local refinement on a fixed mesh with equilibration

N	M	η_h^I	η_k^I	$\eta_h^I + \eta_k^I$	e^I	I_{eff}
269	512	$4.3 \cdot 10^{-02}$	$-8.4 \cdot 10^{-03}$	$3.551 \cdot 10^{-02}$	$-2.889 \cdot 10^{-02}$	-0.81
635	512	$5.5 \cdot 10^{-03}$	$-9.1 \cdot 10^{-03}$	$-3.533 \cdot 10^{-03}$	$-4.851 \cdot 10^{-02}$	13.72
1,847	722	$-1.5 \cdot 10^{-02}$	$-3.6 \cdot 10^{-03}$	$-1.889 \cdot 10^{-02}$	$-3.024 \cdot 10^{-02}$	1.60
5,549	1,048	$-6.5 \cdot 10^{-03}$	$-2.5 \cdot 10^{-03}$	$-9.074 \cdot 10^{-03}$	$-1.097 \cdot 10^{-02}$	1.20
14,419	1,088	$-2.4 \cdot 10^{-03}$	$-2.5 \cdot 10^{-03}$	$-5.064 \cdot 10^{-03}$	$-5.571 \cdot 10^{-03}$	1.10
43,343	1,102	$-8.5 \cdot 10^{-04}$	$-2.5 \cdot 10^{-03}$	$-3.453 \cdot 10^{-03}$	$-3.693 \cdot 10^{-03}$	1.06

Table 2 Local refinement on dynamic meshes with equilibration

N_{tot}	N_{max}	M	η_h^I	η_k^I	$\eta_h^I + \eta_k^I$	e^I	I_{eff}
137,997	269	512	$4.3 \cdot 10^{-02}$	$-8.4 \cdot 10^{-03}$	$3.551 \cdot 10^{-02}$	$-2.889 \cdot 10^{-02}$	-0.81
238,187	663	512	$3.5 \cdot 10^{-03}$	$-8.6 \cdot 10^{-03}$	$-5.192 \cdot 10^{-03}$	$-5.109 \cdot 10^{-02}$	9.84
633,941	1,677	724	$-1.6 \cdot 10^{-02}$	$-3.5 \cdot 10^{-03}$	$-2.015 \cdot 10^{-02}$	$-3.227 \cdot 10^{-02}$	1.60
1,741,185	2,909	1,048	$-7.3 \cdot 10^{-03}$	$-2.5 \cdot 10^{-03}$	$-9.869 \cdot 10^{-03}$	$-1.214 \cdot 10^{-02}$	1.23
3,875,029	4,785	1,098	$-2.2 \cdot 10^{-03}$	$-2.5 \cdot 10^{-03}$	$-4.792 \cdot 10^{-03}$	$-5.432 \cdot 10^{-03}$	1.13
9,382,027	10,587	1,140	$-7.9 \cdot 10^{-04}$	$-2.5 \cdot 10^{-03}$	$-3.301 \cdot 10^{-03}$	$-3.588 \cdot 10^{-03}$	1.08
23,702,227	25,571	1,160	$-2.8 \cdot 10^{-04}$	$-2.4 \cdot 10^{-03}$	$-2.756 \cdot 10^{-03}$	$-2.944 \cdot 10^{-03}$	1.06

The values of the artificial measurements $\hat{\theta}_i$ and $\hat{Y}_i$, $i = 1, 2, 3, 4$, are obtained from a reference solution computed on fine space and time discretizations. The quantity of interest is given here as

$$I(q, u) = q.$$

since the control space in this application is given by $Q = \mathbb{R}$, it is not necessary to discretize Q. Thus, there is no discretization error due to the control discretization and the a posteriori error estimator η^I consists of the error estimator for the temporal error η_k^I and the error estimator for the spatial error η_h^I.

For our numerical test we use piecewise constant polynomials in time and cell-wise bilinear polynomials in space, cf. Sect. 4.1. We apply our adaptive algorithm with simultaneous refinement of the space and time discretizations. The temporal and spatial discretization errors are equilibrated as described above, see [25, 26] for details.

Tables 1 and 2 demonstrate the effectivity of the error estimator $\eta_h^I + \eta_k^I$ on locally refined discretizations using fixed and dynamically changing spatial triangulations. Here, $e^I = I(q, u) - I(q_\sigma, u_\sigma)$ denotes the discretization error and the effectivity index I_{eff} is as usual defined by

$$I_{\text{eff}} = \frac{e_I}{\eta_h^I + \eta_k^I}.$$

In Fig. 8, we compare uniform refinement of the space and time discretizations with local refinement of both discretizations on a fixed spatial mesh and on dynamically changing triangulations. We gain a remarkable reduction of the required degrees of freedom for reaching a given tolerance. To meet for instance an error of $|e^I| \approx 10^{-2}$, the uniform refinement requires in total 15,056,225 degrees

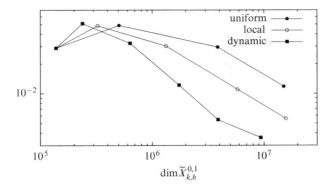

Fig. 8 Comparison of the error $|e^I|$ for different refinement strategies

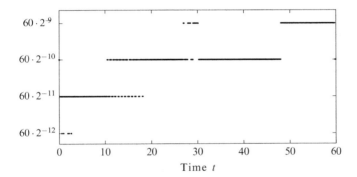

Fig. 9 Visualization of the adaptively determined time step size k

of freedom, the local refinement needs 5,820,901 degrees of freedom, and the dynamical refinement necessitates only 1,741,185 degrees of freedom. Thus, we gain a reduction by a factor of about 1:8.6.

Figure 9 depicts the distribution of the temporal step size k resulting from a fully adaptive computation on dynamic meshes. We observe a strong refinement of the time steps at the beginning of the time interval, whereas the time steps at the end are determined by the adaptation to be eight times larger.

Before presenting a sequence of dynamically changing meshes, we show in Fig. 10 a typical locally refined mesh obtained by computations on a fixed spatial mesh. We note, that the refinement is especially concentrated at the four reentrant corners and the two measurement points behind the obstacle. The interior of the region with restricted cross section is also strongly refined.

Finally, Fig. 11 shows the spatial triangulation and the reaction rate ω for certain selected time points. Thereby, ω is computed from the numerical solution by means of formula (26). We observe, that the refinement traces the front of the reaction rate ω until $t \approx 56$ (cf. Fig. 11d). Afterwards, the mesh around the front becomes coarser and the refinement is concentrated at the four measurement points p_i. Compared to the usage of one fixed triangulation, the usage of dynamically changing

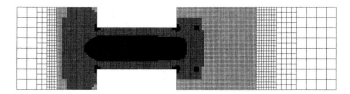

Fig. 10 Locally refined fixed mesh

meshes enables us here to reduce the discretization error in terms of the quantity of interest with lower number of degrees of freedom, cf. Fig. 8. Although computing on dynamically changing meshes requires some additional effort, see [28] for details, this technique pays off also in terms of CPU-time.

5 Adaptivity for Optimal Choice of Regularization Parameters

In this section we discuss an adaptive multilevel inexact Newton method for determining an optimal regularization parameter in Tikhonov regularization, see [16] for more details.

We again consider the state equation

$$a(q,u)(v) = f(v) \quad \forall v \in V, \qquad (27)$$

where the parameter q is now an infinite dimensional object, e. g., a function in an infinite dimensional Hilbert space Q. The measurement data usually posses noise and is denoted by $\hat{C}^\delta$ with

$$\|\hat{C} - \hat{C}^\delta\|_Z \leq \delta$$

and ideal measurements $\hat{C}$. In a variety of situations the problem of estimating the parameter q from $\hat{C}^\delta$ is ill-posed. Therefore, regularization methods are necessary for a stable numerical solution of the parameter identification problem. One of the well-known regularization techniques is Tikhonov regularization leading to the following optimal control problem which depends on the regularization parameter β:

$$\text{minimize } J(\beta, q, u) = \frac{1}{2}\|C(u) - \hat{C}^\delta\|_G^2 + \frac{1}{2\beta}\|q\|_Q^2, \quad \text{subject to (27).} \qquad (28)$$

Note, that for technical reasons we replaced the regularization parameter α, see (2), by $1/\beta$. The regularization parameter β should be chosen in such a way, that the solution of this optimal control problem denoted by (q^β, u^β) is close to the ideal solution $(q^\dagger, u^\dagger)$ for the problem without noise, see, e.g., [12] for precise definitions. A well-established strategy for choosing the Tikhonov parameter β is

Adaptive Finite Element Methods for Parameter Identification Problems 51

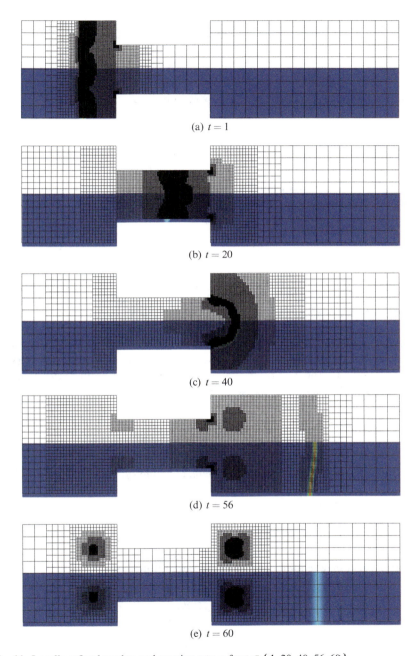

Fig. 11 Locally refined meshes and reaction rate ω for $t \in \{1, 20, 40, 56, 60\}$

the discrepancy principle: The parameter β^* should be chosen as the solution of the following one-dimensional equation

$$i(\beta^*) = \tau^2 \delta^2 \tag{29}$$

with some $\tau \geq 1$ and the function $i \colon \mathbb{R}_+ \to \mathbb{R}_+$ given as

$$i(\beta) = \|C(u^\beta) - \hat{C}^\delta\|_G^2.$$

In [12] Newton's method for solving (29) is analyzed. The corresponding algorithm would require evaluation of the function $i(\beta)$ and its derivative $i'(\beta)$ in each step. However, $i(\beta)$ is not available since it depends on the exact solution of the infinite dimensional optimal control problem (28). Therefore, one should replace the function $i(\beta)$ by its discrete analog $i_h(\beta)$ defined as

$$i_h(\beta) = \|C(u_h^\beta) - \hat{C}^\delta\|_G^2,$$

where $(q_h^\beta, u_h^\beta) \in Q_h \times V_h$ is the solution of the discretized version of the optimal control problem (28). In [16] we described and analyzed an inexact multilevel Newton's method for solution of (29), where $i(\beta)$ is replaced by $i_h(\beta)$ in each Newton step and the choice of the discrete finite element spaces Q_h and V_h is adaptively controlled using appropriate a posteriori error estimates. The finite element spaces Q_h and V_h should be chosen on the one hand as coarse as possible to save numerical effort and on the other hand fine enough to preserve quadratic convergence of the method to the solution β^* of (29). This algorithm is sketched below:

---

1. Choose initial guess $\beta^0 > 0$, initial discretization Q_{h_0}, V_{h_0}, set $k = 0$
2. Solve discrete optimal control problem, compute $(q_h^{\beta^k}, u_h^{\beta^k})$
3. Evaluate $i_h(\beta_k)$, $i_h'(\beta_k)$
4. Evaluate error estimators

$$|i(\beta_k) - i_{h_k}(\beta_k)| \leq \eta^I, \qquad |i'(\beta_k) - i_{h_k}'(\beta_k)| \leq \eta^K$$

5. If the accuracy requirements for η^I, η^K are fulfilled, set

$$\beta^{k+1} = \beta_k - \frac{i_{h_k}(\beta^k) - \tau^2 \delta^2}{i_{h_k}'(\beta^k)}$$

6. else: refine discretization $h_k \to h_{k+1}$ using local information from η^I, η^K
7. if stopping criterion is fulfilled: break
8. else: Set $k = k + 1$ and go to 2.

Adaptive Finite Element Methods for Parameter Identification Problems 53

In [16] error estimators are presented to be used in the step (4) of the algorithm, an efficient strategy for evaluation of $i'_h(\beta)$ is discussed, and accuracy requirements for the step (5) are provided allowing for quadratic convergence of the method. Moreover, we discussed the stopping criterion for the step (7) and proved convergence of the solution q_h^β with computed $\beta = \beta_{k*}$ to the ideal solution $q^\dagger$ as δ tends to 0.

References

1. Becker, R., Braack, M., Vexler, B.: Numerical parameter estimation in multidimensional reactive flows. Combustion Theory and Modelling **8**(4), 661 – 682 (2004)
2. Becker, R., Braack, M., Vexler, B.: Parameter identification for chemical models in combustion problems. Applied Numerical Mathematics **54**(3–4), 519–536 (2005)
3. Becker, R., Kapp, H., Rannacher, R.: Adaptive finite element methods for optimal control of partial differential equations: Basic concepts. SIAM J. Control Optim. **39**(1), 113–132 (2000)
4. Becker, R., Meidner, D., Vexler, B.: Efficient numerical solution of parabolic optimization problems by finite element methods. Optim. Methods Softw. **22**(5), 813–833 (2007)
5. Becker, R., Rannacher, R.: An optimal control approach to a-posteriori error estimation. In: A. Iserles (ed.) Acta Numerica 2001, pp. 1–102. Cambridge University Press (2001)
6. Becker, R., Vexler, B.: A posteriori error estimation for finite element discretizations of parameter identification problems. Numer. Math. **96**(3), 435–459 (2004)
7. Becker, R., Vexler, B.: Mesh refinement and numerical sensitivity analysis for parameter calibration of partial differential equations. J. Comp. Physics **206**(1), 95–110 (2005)
8. Becker, R., Vexler, B.: Optimal control of the convection-diffusion equation using stabilized finite element methods. Numer. Math. **106**(3), 349–367 (2007)
9. Benedix, O., Vexler, B.: A posteriori error estimation and adaptivity for elliptic optimal control problems with state constraints. Comput. Optim. Appl. **44**(1), 3–25 (2009)
10. Braess, D.: Finite Elements: Theory, Fast Solvers and Applications in Solid Mechanics. Cambridge University Press, Cambridge (2007)
11. Brenner, S., Scott, R.: The mathematical theory of finite element methods. Springer Verlag, Berlin Heidelberg New York (2002)
12. Engl, H., Hanke, M., Neubauer, A.: Regularization of Inverse Problems. Kluwer, Dordrecht (1996)
13. Eriksson, K., Estep, D., Hansbo, P., Johnson, C.: Introduction to adaptive methods for differential equations. In: A. Iserles (ed.) Acta Numerica 1995, pp. 105–158. Cambridge University Press (1995)
14. Eriksson, K., Estep, D., Hansbo, P., Johnson, C.: Computational differential equations. Cambridge University Press, Cambridge (1996)
15. Fursikov, A.V.: Optimal Control of Distributed Systems: Theory and Applications, *Transl. Math. Monogr.*, vol. 187. AMS, Providence (1999)
16. Griesbaum, A., Kaltenbacher, B., Vexler, B.: Efficient computation of the Tikhonov regularization parameter by goal oriented adaptive discretization. Inverse Problems **24**(2) (2008)
17. Hintermüller, M., Hoppe, R., Iliash, Y., Kieweg, M.: An a posteriori error analysis of adaptive finite element methods for distributed elliptic control problems with control constraints. ESIAM Control Optim. Calc. Var. **14**(3), 540–560 (2008)
18. Hoppe, R., Iliash, Y., Iyyunni, C., Sweilam, N.: A posteriori error estimates for adaptive finite element discretizations of boundary control problems. J. Numer. Math. **14**(1), 57–82 (2006)
19. Kaltenbacher, B., Neubauer, A., Scherzer, O.: Iterative Regularization Methods for Nonlinear Ill-Posed Problems. de Gruyter, Berlin, New York (2008)

20. Kröner, A., Vexler, B.: A priori error estimates for elliptic optimal control problems with bilinear state equation. J. Comput. Appl. Math. **230**(2), 251–284 (2009)
21. Lang, J.: Adaptive Multilevel Solution of Nonlinear Parabolic PDE Systems. Theory, Algorithm, and Applications, *Lecture Notes in Earth Sci.*, vol. 16. Springer-Verlag, Berlin (1999)
22. Li, R., Liu, W., Ma, H., Tang, T.: Adaptive finite element approximation for distributed elliptic optimal control problems. SIAM J. Control Optim. **41**(5), 1321–1349 (2002)
23. Lions, J.L.: Optimal Control of Systems Governed by Partial Differential Equations, *Grundlehren Math. Wiss.*, vol. 170. Springer-Verlag, Berlin (1971)
24. Liu, W., Yan, N.: A posteriori error estimates for distributed convex optimal control problems. Adv. Comput. Math **15**(1–4), 285–309 (2001)
25. Meidner, D.: Adaptive space-time finite element methods for optimization problems governed by nonlinear parabolic systems. PhD Thesis, Ruprecht-Karls-Universität Heidelberg (2008)
26. Meidner, D., Vexler, B.: Adaptive space-time finite element methods for parabolic optimization problems. SIAM J. Control Optim. **46**(1), 116–142 (2007)
27. Rannacher, R., Vexler, B.: A priori error estimates for the finite element discretization of elliptic parameter identification problems with pointwise measurements. SIAM J. Control Optim. **44**(5), 1844–1863 (2005)
28. Schmich, M., Vexler, B.: Adaptivity with dynamic meshes for space-time finite element discretizations of parabolic equations. SIAM J. Sci. Comput. **30**(1), 369–393 (2008)
29. Thomée, V.: Galerkin finite element methods for parabolic problems. Springer, Berlin (2006)
30. Tröltzsch, F.: Optimale Steuerung partieller Differentialgleichungen: Theorie, Verfahren und Anwendungen. Friedr. Vieweg & Sohn Verlag, Wiesbaden (2010)
31. Verfürth, R.: A Review of A Posteriori Error Estimation and Adaptive Mesh-Refinement Techniques. Wiley/Teubner, New York-Stuttgart (1996)
32. Vexler, B.: Adaptive finite elements for parameter identification problems. PhD Thesis, Institut für Angewandte Mathematik, Universität Heidelberg (2004)
33. Vexler, B., Wollner, W.: Adaptive finite elements for elliptic optimization problems with control constraints. SIAM J. Control Optim. **47**(1), 509–534 (2008)

Gauss–Newton Methods for Robust Parameter Estimation

Tanja Binder and Ekaterina Kostina

Abstract In this paper we treat robust parameter estimation procedures for problems constrained by differential equations. Our focus is on the l_1 norm estimator and Huber's M-estimator. Both of the estimators are briefly characterized and the corresponding optimality conditions are given. We describe the solution of the resulting minimization problems using the Gauss–Newton method and present local convergence results for both nonlinear constrained l_1 norm and Huber optimization. An approach for the efficient solution of the linearized problems of the Gauss–Newton iterations is also sketched as well as globalization strategies using line search methods. Two numerical examples are exercised to demonstrate the superiority of the two presented robust estimators over standard least squares estimation in case of outliers in the measurement data.

1 Introduction

All parameter estimation procedures are based on assumptions on the statistical distribution of the underlying data. For instance, least squares (LS) estimation was developed by Gauss [21] for normally distributed measurements. This procedure has been widely used since then, although it has been known from the end of the nineteenth century that even samples that consist of "good" measurements only are only approximately normal but tend to be rather longer-tailed (e.g. Hampel [23]). Already in 1852 and 1863, Peirce [39] and Chauvenet [16], respectively, independently developed rejection procedures for outlying measurements. Today it is commonly agreed that a correction of the data in advance of the parameter

---

T. Binder · E. Kostina (✉)
Department of Mathematics and Computer Science, University of Marburg,
Hans-Meerwein-Strasse, 35032 Marburg, Germany
e-mail: binder@mathematik.uni-marburg.de; kostina@mathematik.uni-marburg.de

H.G. Bock et al. (eds.), *Model Based Parameter Estimation*, Contributions
in Mathematical and Computational Sciences 4, DOI 10.1007/978-3-642-30367-8_3,
© Springer-Verlag Berlin Heidelberg 2013

estimation is not the way to proceed. Huber and Ronchetti [27] state two major difficulties (amongst others) with this approach: (a) outliers have to be determined correctly or the situation will even get worse (usually there will be both false rejections and false retentions), and (b) outliers may obscure each other so that they cannot be detected.

But nevertheless outliers in the data need to be treated efficiently. In any modeling process the implicit assumption is made that small errors of the model will cause only minor errors in the conclusions. Furthermore, all models make basic assumptions on the underlying situation, like randomness and independence of measurement errors and their statistical distribution. Neither of these assumptions need to be justified which poses a major problem as many common procedures are very sensitive to seemingly marginal deviations from the assumptions. A remedy for this problem are so-called "robust" methods. We use robust here in the sense of Huber and Ronchetti [27], meaning *insensitivity to small deviations from the assumptions*. Particularly we will be concerned with so-called distributional robustness, i.e. deviations from the assumed underlying distribution.

Data with some outliers generally follow a distribution which is longer-tailed than the normal distribution, i.e. there are more errors with a large absolute value than should be expected. Sensitivity to this longtailedness is typical for classical statistical methods, including least squares parameter estimation. One single outlier can completely spoil a least squares analysis if it is only located sufficiently far away from the rest of the data. Thus robust procedures have to be designed to formally spot these outliers and reduce their influence.

2 Robust Parameter Estimation Problem

The parameter estimation problem can be described as follows. Assume that we have a mathematical model of some real world process which is described by a system of differential equations and depends on a finite number of unknown parameters. Assume further that at times t_j, $j = 1, \ldots, N$, we have given measurements $\bar{S}_{ij}$, $i = 1, \ldots, K$, $j = 1, \ldots, N$, of observation functions S_{ij} depending on states u of the system and parameters q which are to be estimated and that these measurements are subject to measurement errors ε_{ij},

$$ \bar{S}_{ij} = S_{ij}(t_j, q^{\text{true}}, u^{\text{true}}) + \varepsilon_{ij}, i = 1, \ldots, K, \ j = 1, \ldots, N, $$

where true denotes the "true" values of the states and parameters.

The determination of the unknown parameters is commonly approached by solving an optimization problem in which a special functional is minimized under constraints that describe the specifics of the model. The known data are output signals from some output device and for the optimization functional a suitable norm of the deviation of the model response from these data is used. The choice which norm

Gauss–Newton Methods for Robust Parameter Estimation

is appropriate for a given problem should be based on the statistical properties of the measurement errors. A formal description of an abstract (simplified) parameter estimation problem may look as follows:

$$\min \quad J(q, u) := \rho(S(q, u(t_i, z_i)) - \bar{S}_i, i = 1, \ldots, M), \tag{1}$$

$$\text{s.t.} \quad \mathcal{F}\left(t, z, q, u, \frac{\partial u}{\partial t}, \frac{\partial u}{\partial z}, \frac{\partial^2 u}{\partial z^2}\right) = 0, t \in T, z \in Z;$$

some initial and boundary conditions,

where $t \in T \subset \mathbb{R}$ is time, $z \in Z \subset \mathbb{R}^d$ are coordinate variables of a d-dimensional space, T and Z bounded, the state variables $u = u(t, z)$ are determined in an appropriate space V, q are unknown parameters in an appropriate space Q, $\mathcal{F}$ is a vector function, $S(q, u)$ is an observation operator that maps the state variable u to the appropriate space of measurements, $\bar{S}_i, i = 1, \ldots, M$, is a given set of output measurements. How to choose an appropriate functional $\rho(\cdot)$ is discussed later.

There are several well-established numerical methods for the solution of parameter estimation problems governed by systems of differential-algebraic equations. Their common basis is the so-called "all-at-once" approach, see e.g. Ascher [2], Bock [9,10], Cervantes and Biegler [15], and Schulz [42]. This approach follows the principle of discretizing the dynamic model including possible boundary conditions as a boundary value problem and incorporating the resulting discretized model as a nonlinear constraint in the optimization problem. Possible discretization strategies are finite differences, collocation, see e.g. Ascher [2], Schulz [42], and Biegler [15], or multiple shooting methods, see e.g. Bock [9] and Bulirsch [14].

The discretization of the dynamic model in (1) yields a finite dimensional, large-scale, nonlinear constrained approximation problem which can be formally written as

$$\min_{x \in \mathbb{R}^{n_x}} \quad \rho(F_1(x)), \text{ s.t. } F_2(x) = 0. \tag{2}$$

Note that the equality conditions $F_2(x) = 0$ include the discretized dynamic model. We assume that the functions $F_i : D \subset R^{n_x} \to R^{m_i}, i = 1, 2$, are twice continuously differentiable. The vector $x \in \mathbb{R}^{n_x}$ combines the parameters and all variables resulting from discretization of the model dynamics.

The standard choice of the function $\rho(\cdot)$ is the least squares functional: $\rho(s) = \|s\|_2^2 = \sum_{i=1}^m s_i^2$, $s \in \mathbb{R}^m$. In contrast to standard least squares optimization problems, we consider robust objective functions which allow to get parameter estimates that are less sensitive to outliers in the measurement data. In this paper, we restrict our further considerations to the cases where the objective function is the l_1 norm,

$$\rho(s) = \|s\|_1 = \sum_{i=1}^m |s_i|, \tag{3}$$

or Huber's function,

$$\rho(s) = \sum_{i=1}^{m} \rho_\gamma(s_i) \quad \text{with} \quad \rho_\gamma(s) = \begin{cases} \frac{1}{2}s^2, & |s| \leq \gamma, \\ \gamma|s| - \frac{1}{2}\gamma^2, & |s| > \gamma, \end{cases} \quad (4)$$

for a prescribed threshold γ. For other objective functions that also yield robust parameter estimates, i.e. they are insensitive to outliers in the data and other deviations from the normal distribution, we refer to Andrews et al. [1], Beaton and Tukey [7], Hinich and Talwar [24], Dennis and Welsch [17], and Fair [19]. All these estimators share the common property that they can be computed using iteratively reweighted least squares (IRLS). Holland and Welsch [25] give a comparison of their features and performance.

2.1 The Least Absolute Deviations Estimator

Least absolute deviations (LAD) or minimum l_1 norm estimation, which we will use synonymously, is also known under a variety of other names including least absolute residuals, least lines, or minimum sum of absolute errors, as it minimizes the l_1 norm of the residual vector. Historically, this estimator dates back to 1755 when Boscovich investigated the figure of the earth from geodatic measurements in joint work with Maire (see Scales and Gersztenkorn [41]). Laplace [35] also used the methods developed by Boscovich and advanced the ideas on a more rigorous statistical basis. Thus LAD estimation dates back even longer than least squares estimation which was first used by Gauss [21] and Legendre [36] in the first decade of the nineteenth century. A brief review of the properties of the estimator as well as a bibliography with relevant literature until 1977 is given by Gentle [22].

A very simple example as given by Scales and Gersztenkorn [41] might serve to demonstrate the effect of robustness. Assume that we have m observations η_i, $i = 1, \ldots, m$, associated with a single parameter p. Then the least squares estimate is computed as the root of the first derivative with respect to p of the sum of the squared residuals,

$$\frac{d}{dp} \sum_{i=1}^{m} |p - \eta_i|^2 = 0.$$

The solution of this equation is $p = \frac{1}{m} \sum_{i=1}^{m} \eta_i$ and thus the mean of the observations. The minimum l_1 norm estimate, on the other hand, is given as the solution of

$$\sum_{i=1}^{m} \text{sgn}(p - \eta_i) = 0,$$

where $\text{sgn}(s)$ is some value between -1 and 1 for $s = 0$. This equation is solved by the median of the observations. While the first suffers from outliers in the data as the large deviations are averaged in the solution, the latter simply ignores such

Gauss–Newton Methods for Robust Parameter Estimation

"wild points" and only takes into account the "side" on which it lies from the bulk of the data.

If the residual components $F_{1i}(x)$, $i = 1,\ldots,m_1$, follow a Laplace distribution with zero mean and variances $2\sigma_i^2$, i.e. if their probability density is given by

$$f(F_{1i}(x)) = \frac{1}{2\sigma_i} \exp\left(-\frac{|F_{1i}(x)|}{\sigma_i}\right),$$

this estimator yields a maximum likelihood estimate of the unknown parameters (see Rice and White [40]).

Now let us have a closer look at the nonlinear l_1 optimization problem. The objective function in this problem is not smoothly differentiable. As this is a necessary requirement for most standard optimization procedures for nonlinear problems, we cannot apply them directly. Fortunately, the problem can be rewritten in form of a nonlinear optimization problem with a linear objective function. In order to eliminate the non-differentiability from the objective function, we introduce two new variables $w, u \in \mathbb{R}^{m_1}$ such that $F_1(x) = u - w$ and $u, w \geq 0$. The l_1 optimization problem (2), (3) can then be rewritten equivalently as

$$\begin{aligned} \min_{x \in \mathbb{R}^{n_x}, u, w \in \mathbb{R}^{m_1}} \quad & e^T(u + w), \\ \text{s.t.} \quad & F_1(x) = u - w, \qquad u \geq 0, w \geq 0, \\ & F_2(x) = 0, \qquad e^T = (1,\ldots,1) \in \mathbb{R}^{m_1}. \end{aligned} \tag{5}$$

For linear functions $F_1(\cdot)$ and $F_2(\cdot)$ the reformulation is a linear programming problem and there were several algorithms developed for the linear case, e.g. by Barrodale and Roberts [3,4].

The optimality conditions characterizing optimal solutions of the l_1-estimator may be proved as a consequence of the Fritz-John optimality conditions of nonlinear programming, see Ben-Tal [8]. First we define the set of "active" measurements at x as $I_A(x) := \{i = 1,\ldots,m_1 : F_{1i}(x) = 0\}$, and say that x satisfies a constraint qualification if the Jacobian of constraints and active measurements at x has full row rank,

$$\text{rank}\, \nabla F_A(x) = m_A + m_2, \, \nabla F_A(x) = \begin{pmatrix} \nabla F_{1A}(x) \\ \nabla F_2(x) \end{pmatrix}, \tag{6}$$

$$\nabla F_{1A}(x) := [\nabla F_1(x)]_{i \in I_A(x)}, \, m_A := |I_A(x)|.$$

Theorem 1. (Necessary conditions) *Let x^* be a regular solution of problem (2), (3). Then there exists a vector of Lagrange multipliers $\lambda^* = [\lambda_k^*]_{k=1,2}$, $\lambda_k^* \in \mathbb{R}^{m_k}$, $k = 1,2$, such that the following conditions (the first-order necessary optimality conditions) are true for the pair (x^*, λ^*):*

$$\nabla F_1^T(x^*)\lambda_1^* + \nabla F_2^T(x^*)\lambda_2^* = 0, \tag{7}$$

$$|\lambda_{1i}^*| \leq 1, \, i = 1,\ldots,m_1, \, \lambda_{1i}^* = \text{sgn}\, F_{1i}(x^*) \quad \text{if } F_{1i}(x^*) \neq 0, \, i = 1,\ldots,m_1.$$

Further, the pair (x^*, λ^*) satisfies the second-order necessary optimality conditions:

$$d^T H d \geq 0, \ H := \left(\sum_{i=1}^{m_1} \lambda_{1i}^* \nabla^2 F_{1i}(x^*) + \sum_{i=1}^{m_2} \lambda_{2i}^* \nabla^2 F_{2i}(x^*) \right), \quad (8)$$

for all critical directions d defined by

$$D_1(x^*, \lambda^*) = \{ d \in \mathbb{R}^{n_x} \mid \nabla F_{1i}(x^*)d = 0, \ if \ |\lambda_{1i}^*| < 1,$$
$$\lambda_{1i} \nabla F_{1i}(x^*)d \geq 0, \ if \ |\lambda_{1i}^*| = 1, i \in I_A(x^*),$$
$$\nabla F_2(x^*)d = 0 \}.$$

Proof of Theorem 1 is given in the appendix. The next theorem is a consequence of Theorem 4 from Fiacco and McCormick [20].

Theorem 2. (Sufficient optimality conditions) *Let (x^*, λ^*) satisfy the first-order necessary optimality conditions (7) and the matrix H be positive definite for all directions $d \in D_1(x^*, \lambda^*)$, $d \neq 0$, that is*

$$d^T H d > 0 \quad \forall d \in D_1(x^*, \lambda^*), d \neq 0.$$

Then x^* is a strict local minimizer of (2), (3).

Corollary 1. *Let (x^*, λ^*) satisfy the first-order necessary conditions (7) and strict complementarity at $x = x^*$ and $\lambda_1 = \lambda_1^*$,*

$$|\lambda_{1i}| < 1 \ if \ i \in I_A(x), \quad (9)$$

and let the Hessian H be positive definite for all $d \in Z(x^*)$,

$$Z(x) := \{ d \neq 0 \mid \nabla F_{1i}(x)d = 0, i \in I_A(x), \nabla F_2(x)d = 0 \}.$$

Then x^* is a strict local minimizer of (2), (3).

2.2 Huber's M-Estimator

Huber's M-estimator was first proposed by Huber [26] in 1964. For getting an idea of the intention behind it, consider an ε-perturbed normal distribution,

$$F = (1 - \varepsilon)\Phi + \varepsilon H,$$

where $\varepsilon \in [0, 1)$ is fixed, $\Phi(t) = (2\pi)^{-\frac{1}{2}} \int_{-\infty}^{t} \exp(-\frac{1}{2}s^2) \, ds$ denotes the standard normal cumulative and H is a contaminating but symmetric distribution. Let further $\mathcal{F}$ denote the set of all such distributions. Huber's M-estimator was designed as a

Gauss–Newton Methods for Robust Parameter Estimation

maximum likelihood estimator corresponding to a least favorable distribution $F_0 \in \mathcal{F}$. This distribution has the density

$$f_0(t) = (1 - \varepsilon)(2\pi)^{-\frac{1}{2}} \exp(-\rho(t)),$$

where $\rho(t) = \rho_\gamma(t)$ is the Huber function as given in (4), and minimizes the Fisher information,

$$I(F) = \int (f'/f)^2 f \, dt,$$

among all $F \in \mathcal{F}$ (see Huber [26]). The error ratio ε and the tuning constant γ are related by

$$(1 - \varepsilon)^{-1} = \int_{-\gamma}^{\gamma} \varphi(t) \, dt + 2\varphi(\gamma)/\gamma,$$

where φ is the standard normal density $\varphi(t) = (2\pi)^{-\frac{1}{2}} \exp(-\frac{1}{2}t^2)$.

From the definition of Huber's function (4) it is obvious that it behaves like the least squares objective function at the center of the distribution, i.e. all observations are given equal weight, while at the extremes it behaves like the l_1 objective function giving less weight to observations that are farther out on the tails. The switch from being quadratic to being linear is located exactly at $|s| = \gamma$ where the least squares part and the l_1 part are linked together smoothly. Thus the Huber objective function is once continuously differentiable if $F_1(x)$ is so. In this way, it compromises between the efficiency of the least squares estimator and the resistance of the least absolute deviations estimator which are also the extreme cases of this estimator for $\gamma = \infty$ and $\gamma = 0$, respectively. The hybrid nature of Huber's M-estimator can be seen clearly due to the following reformulation of the problem (2). (4),

$$\min_{x, v} \tfrac{1}{2} \|v\|_2^2 + \gamma \|F_1(x) - v\|_1, \text{ s.t. } F_2(x) = 0. \tag{10}$$

The problems (2), (4) and (10) are equivalent in the following sense.

Theorem 3. x^* is optimal in (2), (4) if and only if (x^*, v^*) with

$$v_i^* = F_{1i}(x^*) \text{ if } |F_{1i}(x^*)| \leq \gamma,$$
$$v_i^* = \gamma \operatorname{sgn} F_{1i}(x^*) \text{ if } |F_{1i}(x^*)| > \gamma, \ i = 1, \ldots, m_1,$$

is optimal in (10).

Proof follows from the following equality,

$$\frac{1}{2}\|v^*\|_2^2 + \gamma \|F_1(x^*) - v^*\|_1 =$$

$$\frac{1}{2} \sum_{i:|F_{1i}(x^*)| \leq \gamma} F_{1i}(x^*)^2 + \sum_{i:|F_{1i}(x^*)| > \gamma} \left(\frac{1}{2}\gamma^2 + \gamma |F_{1i}(x^*) - \gamma \operatorname{sgn} F_{1i}(x^*)| \right) =$$

$$\frac{1}{2} \sum_{i:|F_{1i}(x^*)| \leq \gamma} F_{1i}(x^*)^2 + \sum_{i:F_{1i}(x^*) > \gamma} \left(\frac{1}{2}\gamma^2 + \gamma F_{1i}(x^*) - \gamma^2 \right) +$$

$$\sum_{i:F_{1i}(x^*)<-\gamma} \left(\frac{1}{2}\gamma^2 - \gamma F_{1i}(x^*) - \gamma^2\right) =$$

$$\frac{1}{2}\sum_{i:|F_{1i}(x^*)|\leq\gamma} F_{1i}(x^*)^2 + \sum_{i:|F_{1i}(x^*)|>\gamma} \left(\gamma|F_{1i}(x^*)| - \frac{1}{2}\gamma^2\right).$$

The hybrid nature of Huber's M-estimator for the unconstrained linear case was studied, e.g., by Mangasarian and Musicant [37]. In particular, it is shown there that there exists a value γ^* such that Huber's M-estimator becomes a least squares estimator for all values $\gamma \geq \gamma^*$. For a sequence $\{\gamma^i\}$ converging to zero, they showed on the other hand that a subsequence $\{x_{i_j}, v_{i_j}\}$ that solves

$$\min_{x,v} \frac{1}{2}\|v\|_2^2 + \gamma\|Ax - b - v\|_1,$$

where A is a real matrix and b a vector of corresponding dimensions, for $\gamma = \gamma^{i_j}$ depends linearly on $\{\gamma^{i_j}\}$, i.e. there exist numbers p, q such that $x_{i_j} = p + q\gamma^{i_j}$ for $\{\gamma^{i_j}\} \to 0$, which implies that for decreasing γ Huber's M-estimator converges linearly to a minimum l_1 norm solution.

Similarly to LAD parameter estimation using the equivalent reformulation of (2), (4), one can derive the optimality conditions for the problem using the optimality conditions of nonlinear programming (e.g. Ben-Tal [8]). First we say that x satisfies a constraint qualification if the Jacobian of constraints at x has full row rank,

$$\text{rank } \nabla F_2(x) = m_2. \tag{11}$$

Theorem 4. (Necessary conditions) *Let x^* be a regular solution of problem (2), (4), that is it satisfies constraint qualification (11). Then there exists a vector of Lagrange multipliers $\lambda^* = [\lambda_k^*]_{k=1,2}, \lambda_k^* \in \mathbb{R}^{m_k}, k = 1, 2$, such that the following conditions (the first-order necessary optimality conditions) are true for the pair (x^*, λ^*):*

$$\nabla F_1(x^*)^T\lambda_1^* + \nabla F_2(x^*)^T\lambda_2^* = 0,$$

$$\lambda_{1i}^* = \begin{cases} F_{1i}(x^*), & |F_{1i}(x^*)| \leq \gamma, \\ \gamma\,\text{sgn}\,(F_{1i}(x^*)), & |F_{1i}(x^*)| > \gamma, \end{cases} \quad i = 1, \dots, m_1. \tag{12}$$

Further, the pair (x^*, λ^*) satisfies the second-order necessary optimality conditions:

$$d^T H d \geq 0, \; H := \sum_{i:|F_{1i}(x^*)|\leq\gamma} \nabla^T F_{1i}(x^*)\nabla F_{1i}(x^*) +$$

$$\left(\sum_{i=1}^{m_1} \lambda_{1i}^* \nabla^2 F_{1i}(x^*) + \sum_{i=1}^{m_2} \lambda_{2i}^* \nabla^2 F_{2i}(x^*)\right),$$

Gauss–Newton Methods for Robust Parameter Estimation 63

for all critical directions d defined by

$$D_H(x^*) = \{d \in \mathbb{R}^{n_x} \mid \nabla F_2(x^*)d = 0\}.$$

Again the proof can be found in the appendix.

Theorem 5. (Sufficient optimality conditions) *Let (x^*, λ^*) satisfy the first-order necessary optimality conditions (12) and the matrix H be positive definite for all directions $d \in D_H(x^*)$, $d \neq 0$, that is*

$$d^T H d > 0 \quad \forall d \in D_H(x^*, \lambda^*), \ d \neq 0.$$

Then x^* is a strict local minimizer of (2), (4).

3 The Gauss–Newton Method

The method of choice for solving parameter estimation problems is a constrained Gauss–Newton method, see e.g. Bock et al. [11]. Starting from a given initial guess $x_0 \in \mathbb{R}^{n_x}$, we proceed iteratively by computing

$$x_{k+1} = x_k + [t_k]\Delta x_k, \tag{13}$$

where Δx_k solves the linearized problem at $x = x_k$,

$$\begin{aligned}
\min_{\Delta x} \ &\rho(F_1(x) + \nabla F_1(x)\Delta x), \\
\text{s.t.} \ &F_2(x) + \nabla F_2(x)\Delta x = 0.
\end{aligned} \tag{14}$$

The scalar t_k is an optional stepsize for globalization of the method. We will briefly mention possible stepsize selection strategies in Sect. 4.

3.1 *Local Convergence of Gauss–Newton for Nonlinear Constrained l_1 Norm Optimization*

Application of the necessary and sufficient optimality conditions according to Theorems 1, 2 to problem (14) linearized at $x = x_k$ with the l_1 norm in the objective function yields that Δx is a solution of this problem if there are vectors v_1, v_2 such that

$$\nabla F_1(x)^T v_1 + \nabla F_2(x)^T v_2 = 0,$$

$$|v_{1i}| \leq 1, \ i = 1, \ldots, m_1, \tag{15}$$

$$v_{1i} = \mathrm{sgn}(F_{1i}(x) + \nabla F_{1i}(x)\Delta x) \text{ if } F_{1i}(x) + \nabla F_{1i}(x)\Delta x \neq 0, i = 1, \ldots, m_1.$$

Hence obviously, (x^*, ν_1, ν_2) is a KKT point of the nonlinear l_1 norm optimization problem if and only if $(0, \nu_1, \nu_2)$ is a KKT point of the corresponding linearized problem (see Kostina [32] and Bock et al. [11]), i.e. x^* is a fixed point of the Gauss–Newton iteration, under assumptions that guarantee the existence and uniqueness of solutions of both the nonlinear and the linearized problems.

With the notations

$$I_{A,lin}(\Delta x) = \{i = 1, \ldots, m_1 : F_{1i}(x) + \nabla F_{1i}(x)\Delta x = 0\},$$

$$I_{N,lin}(\Delta x) = \{i = 1, \ldots, m_1\} \backslash I_{A,lin}(\Delta x),$$

$$F_{1A,lin}(x) = [F_{1i}(x)]_{i \in I_{A,lin}(\Delta x)},$$

$$\nabla F_{1A,lin}(x) = [\nabla F_{1i}(x)]_{i \in I_{A,lin}(\Delta x)},$$

$$\nu_{1A} = [\nu_{1i}(x)]_{i \in I_{A,lin}(\Delta x)},$$

the KKT conditions (15) can be rewritten in the form

$$
\begin{pmatrix}
0 & \nabla F_{1A,lin}(x)^T & \nabla F_2(x)^T \\
\nabla F_{1A,lin}(x) & 0 & 0 \\
\nabla F_2(x) & 0 & 0
\end{pmatrix}
\begin{pmatrix}
\Delta x \\
\nu_{1A} \\
\nu_2
\end{pmatrix} =
$$
$$
- \begin{pmatrix}
\sum_{i \in I_{N,lin}(\Delta x)} \nu_{1i} \nabla F_{1i}(x)^T \\
F_{1A,lin}(x) \\
F_2(x)
\end{pmatrix}.
$$

In order to guarantee existence and uniqueness of the increment Δx we require the condition

$$\text{rank} \nabla F_{A,lin}(x) = n_x, \nabla F_{A,lin}(x) = \begin{pmatrix} \nabla F_{1A,lin}(x) \\ \nabla F_2(x) \end{pmatrix}. \tag{16}$$

On the other hand, the constraint qualification

$$\text{rank} \nabla F_{A,lin}(x) = m_{A,lin} + m_2, \quad m_{A,lin} = |I_{A,lin}|, \tag{17}$$

and strict complementarity,

$$|\nu_{1i}| < 1 \text{ if } i \in I_{A,lin}(\Delta x), \tag{18}$$

in the linearized problem are sufficient for existence and uniqueness of the multipliers ν_{1A} and ν_2. Combining (16) and (17) we get that the increment Δx_k and the multipliers ν_{1A} and ν_2 are unique if the matrix $\nabla F_{A,lin}(x)$ is quadratic and invertible and (18) holds true.

Denote the system of problem functions by $F(x) = \begin{pmatrix} F_1(x) \\ F_2(x) \end{pmatrix}$ and the corresponding full Jacobian by $\nabla F(x) = \begin{pmatrix} \nabla F_1(x) \\ \nabla F_2(x) \end{pmatrix}$. Then under the assumptions (16)–(18) the increment Δx at the point x can be computed by means of a (local) generalized inverse,

$$\Delta x = -\nabla F^+(x) F(x),$$

where a generalized inverse $\nabla F^+(x)$ of the Jacobian $\nabla F(x)$, that is

$$\nabla F^+(x) \nabla F(x) \nabla F^+(x) = \nabla F^+(x),$$

is a permutation of the matrix $(0, \nabla F_{A,lin}(x)^{-1})$ as shown by Bock et al. [11].

A further analysis of the optimality conditions shows that a solution x^* of the nonlinear l_1 norm minimization problem is associated with and identified by an optimal "active set" containing information about zero components of the objective function. It can be shown that, in the neighbourhood of a solution x^* that satisfies (6),

$$\text{rank} \nabla F_A(x) = n_x, \tag{19}$$

and (9) at $x = x^*$, the Gauss–Newton method eventually identifies this optimal "active set" of the nonlinear l_1 problem. Thus, the iterates of the generalized Gauss–Newton method (13), (14), (3) are attracted by a local minimum that satisfies the regularity assumptions and, in approaching such a minimizer x^*, neither the active sets of the problems linearized at x_k will change nor the non-zero components of the objective function will change their signs. Furthermore, these sets are the same as the corresponding optimal sets at x^*. So the full-step method ($t^k \equiv 1$) is essentially the standard Newton method for the system of nonlinear equations

$$F_{A,*}(x) = 0, \quad F_{A,*}(x) := \begin{bmatrix} F_{1i}(x), i \in I_A(x^*) \\ F_2(x) \end{bmatrix},$$

and as a result it has a *quadratic rate of local convergence*, see also Bock et al. [11] and Kostina [32].

3.2 Local Convergence of Gauss–Newton for Nonlinear Constrained Huber Optimization

Application of the necessary and sufficient optimality conditions according to Theorems 4, 5 to problem (14) linearized at x with (4) shows that the solution Δx and the vectors v_1, v_2 satisfy

$$\nabla F_1(x)^T v_1 + \nabla F_2(x)^T v_2 = 0, \tag{20}$$

$$v_{1i} = \begin{cases} F_{1i}(x) + \nabla F_{1i}(x)\Delta x & \text{if } |F_{1i}(x) + \nabla F_{1i}(x)\Delta x| \leq \gamma, \\ \gamma \operatorname{sgn}(F_{1i}(x) + \nabla F_{1i}(x)\Delta x) & \text{if } |F_{1i}(x) + \nabla F_{1i}(x)\Delta x| > \gamma, \end{cases}$$

$$i = 1, \ldots, m_1.$$

Similarly to the l_1 case, (x^*, v_1, v_2) is a KKT point of the nonlinear Huber optimization problem if and only if $(0, v_1, v_2)$ is a KKT point of the corresponding linearized problem, i.e. also for Huber optimization x^* is a fixed point of the Gauss–Newton iteration (under assumptions that guarantee the existence and uniqueness of solutions of both the nonlinear and the linearized problem).

With the partitioning

$$I_{A,lin}(\Delta x) = \{i = 1, \ldots, m_1 : |F_{1i}(x) + \nabla F_{1i}(x)\Delta x| \leq \gamma\},$$

$$I_{N,lin}(\Delta x) = \{i = 1, \ldots, m_1 : |F_{1i}(x) + \nabla F_{1i}(x)\Delta x| > \gamma\}, \tag{21}$$

$$I_{A,lin}(\Delta x) \cup I_{N,lin}(\Delta x) = \{i = 1, \ldots, m_1\}, \ I_{A,lin}(\Delta x) \cap I_{N,lin}(\Delta x) = \emptyset,$$

and the notations

$$D_{lin} = D_{lin}(\Delta x) = \operatorname{diag}(d_{ii}), \quad d_{ii} = \begin{cases} 1, & \text{if } i \in I_{A,lin}(\Delta x), \\ 0, & \text{if } i \in I_{N,lin}(\Delta x), \end{cases}$$

$$F_{1,lin}(x) = \begin{pmatrix} F_{1A,lin}(x) \\ F_{1N,lin}(x) \end{pmatrix},$$

$$F_{1A,lin}(x) = [F_{1i}]_{i \in I_{A,lin}(\Delta x)}, \ F_{1N,lin}(x) = [v_{1i}]_{i \in I_{N,lin}(\Delta x)},$$

the KKT system (20) can be rewritten as

$$\begin{pmatrix} \nabla F_1(x)^T D_{lin} \nabla F_1(x) & \nabla F_2(x)^T \\ \nabla F_2(x) & 0 \end{pmatrix} \begin{pmatrix} \Delta x \\ v_2 \end{pmatrix} = - \begin{pmatrix} \nabla F_1(x)^T F_{1,lin}(x) \\ F_2(x) \end{pmatrix}. \tag{22}$$

Analogously to the l_1 case, we need regularity assumptions that guarantee the existence and uniqueness of the solution of the linearized problem. Obviously, system (22) has a unique solution $\begin{pmatrix} \Delta x \\ v_2 \end{pmatrix}$ if the assumptions (11) and

$$\operatorname{rank} \nabla F_{A,lin}(x) = n_x, \nabla F_{A,lin}(x) := \begin{pmatrix} D_{lin} \nabla F_1(x) \\ \nabla F_2(x) \end{pmatrix}, \tag{23}$$

hold true. In order to guarantee the uniqueness of v_1 we need the uniqueness of the partitioning (21) which implies the strict complementarity condition in the linearized problem:

$$|F_{1i}(x) + \nabla F_{1i}(x)\Delta x| \neq \gamma, i = 1, \ldots, m_1. \tag{24}$$

Gauss–Newton Methods for Robust Parameter Estimation

Under the assumptions (11), (23), and (24) the solution of the linearized problem can again be computed by means of a linear operator $\nabla F^+(x)$,

$$\Delta x = -\nabla F^+(x) F_{A,lin}(x),$$

where $\nabla F^+(x)$ is explicitly given by

$$\nabla F^+(x) = (I,0) \begin{pmatrix} \nabla F_1(x)^T D_{lin} \nabla F_1(x) & \nabla F_2(x)^T \\ \nabla F_2(x) & 0 \end{pmatrix}^{-1} \begin{pmatrix} \nabla F_1(x)^T & 0 \\ 0 & I \end{pmatrix},$$

and is a generalized inverse of $\nabla F_{A,lin}(x)$: $\nabla F^+(x) \nabla F_{A,lin}(x) \nabla F^+(x) = \nabla F^+(x)$.

One can show that the Gauss–Newton method eventually identifies the "optimal partitioning" of the residuals $F_{1i}(x)$ in the neighbourhood of a solution x^* that satisfies certain assumptions:

Theorem 6. *Suppose that x^* with corresponding λ^* is a solution of the problem (2), (4) that satisfies (11),*

$$\mathrm{rank}\nabla F_A(x) = n_x, \nabla F_A(x) := \begin{pmatrix} D(x)\nabla F_1(x) \\ \nabla F_2(x) \end{pmatrix}, \tag{25}$$

$$D(x) = diag(d_{ii}), \quad d_{ii} = \begin{cases} 1, & \text{if } i \in I_A(x), \\ 0, & \text{if } i \in I_N(x), \end{cases}$$

and strict complementarity

$$|F_{1i}(x)| \neq \gamma, i = 1, \ldots, m_1, \tag{26}$$

with $x = x^*$. Then there exists a neighbourhood $\mathcal{D}$ of (x^*, λ^*) such that for all $(x_k, \lambda_k) \in \mathcal{D}$ the linearized Huber problem has a unique solution $(\Delta x_k, v_k)$ whose partitioning $I_{A,lin}(\Delta x_k), I_{N,lin}(\Delta x_k)$ is the same as the partitioning $I_A(x_k), I_N(x_k)$ of the nonlinear problem (2), (4) at $x = x_k$, which is in its turn the same as the partitioning $I_A(x^*), I_N(x^*)$ of the nonlinear problem (2), (4) at $x = x^*$.

Here the partitioning at x is defined as

$$I_A(x) = \{i = 1, \ldots, m_1 : |F_{1i}(x)| \leq \gamma\},$$
$$I_N(x) = \{i = 1, \ldots, m_1 : |F_{1i}(x)| > \gamma\}, \tag{27}$$
$$I_A(x) \cup I_N(x) = \{i = 1, \ldots, m_1\}, \ I_A(x) \cap I_N(x) = \emptyset.$$

In summary, the iterates of the generalized Gauss–Newton method (13), (14), (4) are attracted by a local minimum that satisfies the regularity assumptions mentioned above and, in approaching such a minimizer x^*, the partitioning of the problems linearized at x_k will not change. Furthermore, this partitioning is the same as the corresponding partitioning at x^*. So the full-step method ($t^k \equiv 1$) is essentially the Gauss–Newton method for a modified least squares problem

$$\min \frac{1}{2} \sum_{i \in I_A(x^*)} F_{1i}(x)^2 + \sum_{i \in I_N(x^*)} \gamma \mathrm{sgn}(F_{1i}(x^*)) F_{1i}(x), \quad \text{s.t. } F_2(x) = 0,$$

and as a result it has a *linear rate of local convergence*.

Theorem 7 (Local Contraction Theorem for Huber Optimization). *Let $\mathcal{D}$ be the neighbourhood defined in Theorem 6. Assume further that the following two Lipschitz conditions for ∇F and ∇F^+ hold for all $x, y \in \mathcal{D}$ and all $t \in [0, 1]$,*

$$\frac{\|\nabla F^+(y)(\nabla F_A(x + t(y - x)) - \nabla F_A(x))(y - x)\|}{t\|y - x\|^2} \le \omega(x) \le \omega < \infty,$$

$$\frac{\|(\nabla F^+(y) - \nabla F^+(x)) R_A(x)\|}{\|y - x\|} \le \kappa(x) \le \kappa < 1,$$

where $R_A(x)$ denotes the residuals in the linearized problem:

$$R_A(x) = \begin{pmatrix} R_1(x) \\ R_2(x) \end{pmatrix}, \ R_1(x) = \begin{pmatrix} R_{1A}(x) \\ R_{1N}(x) \end{pmatrix},$$

$$R_{1A}(x) = [F_{1i}(x) + \nabla F_{1i}(x)\Delta x]_{i \in I_A(x)}, R_{1N}(x) = [\gamma \mathrm{sgn}(F_{1i}(x))]_{i \in I_N(x)},$$

$$R_2(x) = F_2(x) + \nabla F_2(x)\Delta x.$$

Assume further that an initial guess $x_0 \in \mathcal{D}$ satisfies

$$\delta_0 = \frac{\omega\|\Delta x_0\|}{2} + \kappa < 1, \quad \mathcal{D}_0 = \bar{B}\left(x_0, \frac{\|\Delta x_0\|}{1 - \delta_0}\right) \subset \mathcal{D}.$$

Then the sequence of iterates $\{x_k\}$ of the Gauss–Newton method (13), (14), (4) is well-defined, remains in $\mathcal{D}$, and converges to a point x^* with $\nabla F^+(x^*) F_A(x^*) = 0$. Furthermore the increments satisfy the following inequality:

$$\|\Delta x_{k+1}\| \le \left(\frac{\omega}{2}\|\Delta x_k\| + \kappa\right) \|\Delta x_k\|,$$

implying the linear convergence of the Gauss–Newton method with rate κ.

The proof of Theorem 7 is given in the appendix.

As in the Gauss–Newton method for l_2 constrained optimization (see Bock [10]), the quantity ω is a measure for the nonlinearity of the problem functions excluding the functions corresponding to "large" measurement errors. Its inverse ω^{-1} characterizes the region of validity of the linearized model.

Gauss–Newton Methods for Robust Parameter Estimation

The quantity κ is an incompatibility constant that describes the compatibility of the problem functions and the measurement data for parameter estimation in case that "large" measurement errors are replaced by $\pm\gamma$. The condition $\kappa < 1$ is necessary for the unknowns to be identifiable from the data. A stationary point x^*, that is a point satisfying the necessary first-order optimality conditions, is statistically stable if $\kappa < 1$ (see Bock [10]). It is interesting to note that

$$\left\| \left[\nabla F^+(x_{k+1}) - \nabla F^+(x_k) \right] R_A(x_k) \right\| \leq \left\| \left[\nabla F^+(x_{k+1}) - \nabla F^+(x_k) \right] \right\| \gamma m_1,$$

and thus in case of Huber optimization the incompatibility constant κ can be reduced by reducing the parameter γ.

4 Globalization of the Gauss–Newton Method

Of course we are not satisfied with an only locally convergent method. It cannot be expected that there is always a good initial guess at hand for the unknowns. In this section we discuss methods for stepsize selection which make the method globally convergent from an arbitrary starting point.

4.1 Line Search Strategies

One possibility to compute the relaxation parameter t_k for the iteration $x_{k+1} = x_k + t_k \Delta x_k$ is a line search procedure. This approach chooses the stepsize t_k such that the new iterate x_{k+1} is in some sense "better" than the current iterate x_k by means of a merit function, i.e. $T(x_{k+1}) < T(x_k)$. A common choice for a merit function is the exact penalty function,

$$T_1(x) = \rho(F_1(x)) + \sum_{i=1}^{m_2} \xi_i |F_{2i}(x)|, \tag{28}$$

where the weights ξ_i have to be chosen sufficiently large.

Under the regularity assumptions for both the original and the linearized problems, (6), (19), and the strict complementarity (9) and (17), (16), and strict complementarity (18) (for l_1) or (11), (23), and (24) and (25), (26) (for Huber), respectively, it can be shown that the increment Δx solving the linearized problem (14) leads to a descent direction of the nonlinear problem, i.e. the compatibility condition

$$\lim_{\varepsilon \to 0} \frac{T_1(x_k + \varepsilon \Delta x) - T_1(x_k)}{\varepsilon} < 0,$$

is satisfied for the exact penalty function (28). Thus the method is globally convergent with stepsize strategies based on $T_1(x)$.

A drawback of this approach is that already for only mildly ill-conditioned problems it can lead to very small stepsizes even in the region of full-step convergence and may thus be very inefficient.

An alternative approach is a stepsize selection based on natural level functions as first introduced by Deuflhard [18]. Bock [10] and Bock et al. [12] introduced the so-called "restrictive monotonicity test." The line search based on the natural level function

$$T_k(x_k + t\Delta x_k) = \|\nabla F^+(x_k) F_A(x_k + t\Delta x_k)\|_2^2$$

is conducted by means of the restrictive monotonicity test: we choose the maximal stepsize $t \le 1$ such that

$$t\|\Delta x_k\| \le \frac{\eta}{\omega(t)}$$

for a given $\eta < 2$, where $\omega(t)$ is an asymptotically correct estimate of the "curvature" of the problem, i.e. a measure for its nonlinearity,

$$\omega(t) = \frac{2\|\nabla F^+(x_k)(F_A(x_k + t\Delta x_k) - (1-t)F_A(x_k))\|}{t^2\|\Delta x_k\|^2}.$$

4.2 Numerical Solution of the Linearized Problem

One of the decisive steps of the method which largely affects its performance is the solution of the linearized problems (14). Since the equality constraints $F_2(x) = 0$ contain the discretized boundary value problem, efficient solution methods must take into account the structure of the constraints induced by discretization methods. Efficient condensing procedures which exploit the block structure of the Jacobian ∇F_2 are described, e.g., by Bock [10] and Bock et al. [11] for multiple shooting, or by Schulz [42] and Cervantes and Biegler [15] for collocation. After this procedure the dimensions of variables and constraints in the linearized problems to be solved at each iteration are significantly reduced.

The "condensed" problem can be formally written in the form

$$\min_s \ \rho_\gamma(A_1 s + c_1), \ \text{s.t.} \ A_2 s + c_2 = 0, \tag{29}$$

$$s \in \mathbb{R}^{r_s}, A_i \in \mathbb{R}^{r_i \times r_s}, i = 1, 2.$$

In case of the l_1 cost functional, problem (29) is a linear programming problem, in case of the Huber cost functional, this problem is a quadratic programming problem, with a very special structure in both cases. This can be clearly seen when considering the dual to problem (29) (case $\gamma = 0$ corresponds to l_1 cost functional):

$$\min_{y_1 \in \mathbb{R}^{r_1}, y_2 \in \mathbb{R}^{r_2}} \ \frac{\gamma}{2} y_1^T y_1 - c_1^T y_1 - c_2^T y_2, \tag{30}$$

$$\text{s.t.} \quad A_1^T y_1 + A_2^T y_2 = 0, \ |y_{1i}| \le 1, i = 1, \dots, r_1.$$

Gauss–Newton Methods for Robust Parameter Estimation

This problem can be efficiently solved using a linear or quadratic programming method tailored to the special structure of problem (30) using so-called "long step" as in Kostina [31] and Kostina and Kostyukova [33] or "flipping bounds" in Koberstein [28] methods. It is interesting to note, that, in the case of the Huber estimator, the optimization problem (29), which can be reformulated as

$$\min_{v,s} \frac{1}{2}||v||_2^2 + \gamma||A_1 s + c_1 - v||_1, \text{ s.t. } A_2 s + c_2 = 0,$$

is related to problems appearing in sparse modeling of signals and images or LASSO problems (see Osborne et al. [38]), thus one can also modify the methods developed for these problems (see an overview article by Bruckstein et al. [13] and references therein) to solve (29).

A violation of the assumed regularity assumption that $\text{rank}(\nabla F_A(x_k))$ should be equal to the number of unknowns means that the measurements judged to be "good" do not provide enough information about the parameters. The parameters are not identifiable from the given data. Hence, a regularization by e.g. a rank reduction is necessary, see Bock et al. [11], or by trust region methods. Alternatively, methods for optimum experimental design can be applied to design optimal experiments that maximize the information gain about the parameters, see Bauer et al. [5,6], Körkel and Kostina [30], and Körkel [29].

5   Numerical Results

5.1   Parameter Estimation for a Chemical Reaction

As a first numerical example we consider the chemical process of the denitrogenization of pyridine, taken from Bock [10].

The reaction coefficients $p_1, \ldots, p_{11}$ are the unknowns of the reaction which are to be estimated from given measurements of concentrations of the species participating in the reaction. The process can be described mathematically by a system of seven ordinary differential equations,

$$\dot{A} = -p_1 A + p_9 B,$$
$$\dot{B} = p_1 A - p_2 B - p_3 BC + p_7 D - p_9 B + p_{10} DF,$$
$$\dot{C} = p_2 B - p_3 BC - 2p_4 CC - p_6 C + p_8 E + p_{10} DF + 2p_{11} EF,$$
$$\dot{D} = p_3 BC - p_5 D - p_7 D - p_{10} DF,$$
$$\dot{E} = p_4 CC + p_5 D - p_8 E - p_{11} EF,$$
$$\dot{F} = p_3 BC + p_4 CC + p_6 C - p_{10} DF - p_{11} EF,$$
$$\dot{G} = p_6 C + p_7 D + p_8 E.$$

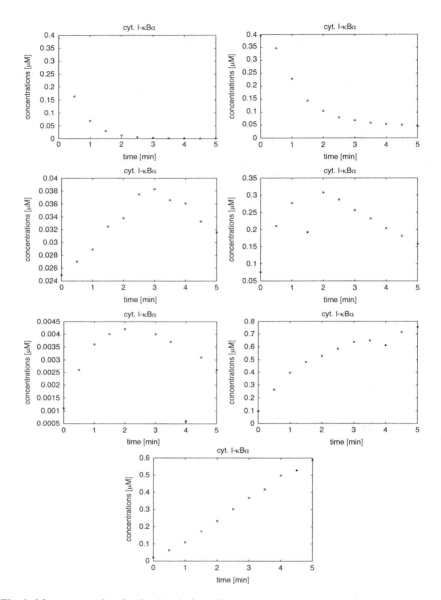

Fig. 1 Measurement data for the chemical reaction

The initial state is $(1.0, 0.0, 0.0, 0.0, 0.0, 0.0, 0.0)^T$. Measurements were taken in the time interval $[0.0, 5.5]$ at points $t_i = 0.5i$, $i = 0, \ldots, 11$.

As measurement data in numerical experiments we chose simulations using "true" parameter values which were corrupted by four outliers. Figure 1 shows the measurements points with the outliers indicated by filled dots.

Gauss–Newton Methods for Robust Parameter Estimation

Table 1 Estimates of the parameters for a chemical reaction and number of iterations

	p_1	p_2	p_3	p_4	p_5	p_6	p_7	p_8	p_9	p_{10}	p_{11}	Iter
"true"	1.810	0.894	29.400	9.210	0.058	2.430	0.0644	5.550	0.0201	0.577	2.150	
l_2	1.812	0.850	29.597	4.467	0.059	2.503	0.112	1.990	0.0203	0.497	8.468	15
l_1	1.810	0.894	29.399	9.209	0.058	2.429	0.0644	5.550	0.0201	0.577	2.149	18
Huber	1.810	0.894	29.393	9.172	0.058	2.430	0.0647	5.551	0.0201	0.576	2.184	13

The results of the parameter estimation using l_2, l_1, and Huber estimators are given in Table 1 and Fig. 2.

Obviously, l_1 and Huber estimation do not differ much while the least squares approximation yields quite different results.

5.2   *Parameter Estimation for the NF-κB Pathway*

In this subsection we consider the NF-κB pathway, which is a benchmark problem from systems biology, cf. Waldherr [43]. We take the equations of the reduced model from Waldherr's thesis, where the involved species are cytosolic I-κBα (x_1), I-κBα mRNA (x_2), total nuclear I-κBα (x_3), and total nuclear NF-κB (x_4),

$$\dot{x}_1 = k_{tl} x_2 - \frac{\alpha (N_{tot} - x_4) x_1}{K_I + x_1} - k_{I,in} x_1 + \frac{k_{I,out} K_N x_3}{K_N + x_4},$$

$$\dot{x}_2 = k_t x_4^2 - \gamma_m x_2,$$

$$\dot{x}_3 = k_{I,in} x_1 - \frac{k_{I,out} K_N x_3}{K_N + x_4} - \frac{k_{NI,out} x_3 x_4}{k_N + x_4},$$

$$\dot{x}_4 = \frac{k_{N,in} K_I (N_{tot} - x_4)}{K_I + x_1} - \frac{K_{NI,out} x_3 x_4}{K_N + x_4},$$

where $K_I = \frac{0.03 + \alpha}{30}$ and $K_N = \frac{0.03 + k_{NI,out}}{30}$. Waldherr [43] claims that the behavior of this model is essentially similar to that of the original model by Krishna et al. [34]. We also take Waldherr's parameter values that lead to damped oscillations in the concentrations of the four species as the "true" values for our simulations. The initial state is $(0.0, 0.0, 0.0, 1.0)$. Measurements were taken in the time interval $[0.0, 300.0]$ at points $t_i = 10.0i$, $i = 0, 30$. As measurement data in numerical experiments we chose simulations using "true" parameter values without artificial measurement errors but only corrupted by few outliers introduced to the measurements of total nuclear NF-κB. Figure 3 shows the measurements points with the outliers indicated by filled dots.

The results of the parameter estimation are presented in Table 2 and Fig. 4.

Again, l_1 and Huber estimators have shown to be more reliable compared to the l_2 estimator in case of data with outliers, although the results of Huber's M-estimator might be still improved by choosing a smaller tuning constant.

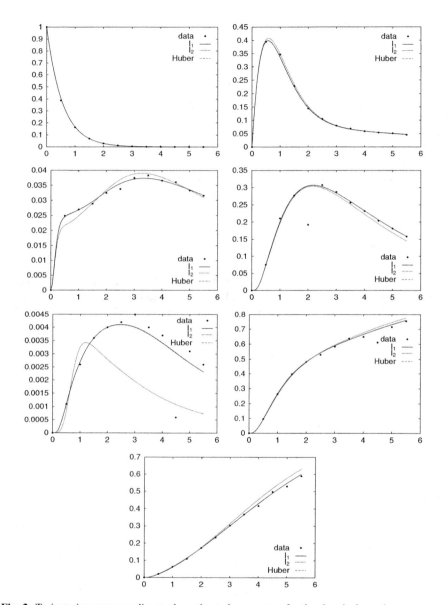

Fig. 2 Trajectories corresponding to the estimated parameters for the chemical reaction

Gauss–Newton Methods for Robust Parameter Estimation

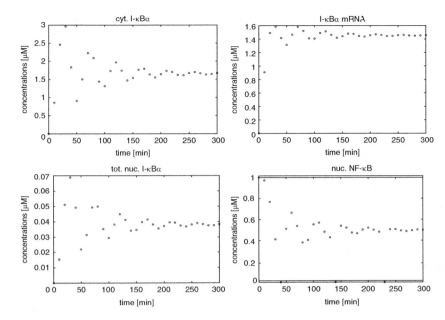

Fig. 3 Measurement data for the NK-κB pathway

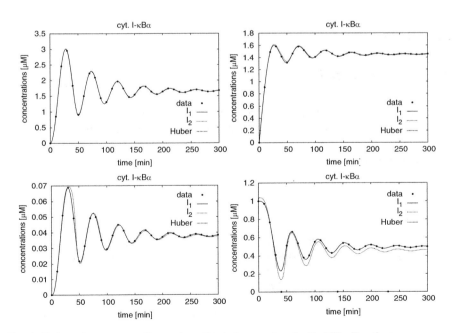

Fig. 4 Trajectories corresponding to the estimated parameters for the NK-κB pathway

Table 2 Estimates of the parameters for KF-κB pathway and number of iterations (three outliers in x_4)

	$k_{N,in}$	$k_{I,in}$	$k_{I,out}$	$k_{NI,out}$	k_t	k_{tl}	α	γ_m	N_{tot}	Iter
"true"	5.400	0.018	0.012	0.830	0.100	0.200	0.525	0.017	1.000	
l_1	5.400	0.018	0.013	0.830	0.100	0.200	0.525	0.017	1.000	6
l_2	6.248	0.020	−0.525	0.981	0.090	0.201	0.446	0.013	1.039	11
Huber	5.406	0.018	0.007	0.831	0.100	0.200	0.524	0.017	1.000	9

6 Conclusion

We showed that the l_1 norm optimization problem and the Huber optimization problem for robust parameter estimation share a common structure. Based on this joint problem formulation, we derived a constrained Gauss–Newton method for the solution of these problems under the assumption that exact derivatives of the problem functions are available. Convergence properties of these methods are analyzed. Further we summarized two possibilities for globalization strategies.

Acknowledgements This research was supported by the German Federal Ministry for Education and Research (BMBF) through the Programme "Mathematics for Innovations in Industry and Public Services", as well as by the LOEWE Center for Synthetic Microbiology (SYNMIKRO).

Appendix

Proof of Theorem 1

We will show that the optimality conditions (7) and (8) result from the optimality conditions derived by Ben-Tal [8] for the nonlinear optimization problem in the form

$$\min_x \ \varphi_0(x), \tag{A.1}$$

$$\varphi_i(x) \leq 0, i \in I = \{1, \ldots, n_I\}, \psi_i(x) = 0, i \in \mathcal{E} = \{1, \ldots, n_{\mathcal{E}}\},$$

which read

- Necessary (second-order) conditions for local minimum: For every $d \in D(x^*)$, there exist nonnegative $y \in \mathbb{R}^{n_I + 1}$, $\mu \in \mathbb{R}^{n_{\mathcal{E}}}$, $(y, \mu) \neq 0$, such that

$$\nabla L(x^*, y, \mu) = 0, \tag{A.2}$$

$$d^T \nabla^2 L(x^*, y, \mu) d \geq 0, \tag{A.3}$$

$$y_i \varphi_i(x^*) = 0, i \in I, \tag{A.4}$$

$$y_i \nabla \varphi_i(x^*) d = 0, i \in I_0^*. \tag{A.5}$$

The following functions and sets are used in the formulations:

(i) The Lagrangian function

$$L(x, y, \mu) = y_0\varphi_0(x) + \sum_{i \in I} y_i\varphi_i(x) + \sum_{i \in \mathcal{E}} \mu_i\psi_i(x)$$

(ii) The set of critical directions at x

$$D(x) = \{d \in \mathbb{R}^{n_x} : \nabla\varphi_i(x)d \le 0, i \in I_0^*; \nabla\psi_i(x)d = 0, i \in \mathcal{E}\} \quad (A.6)$$

where

$$I_0^* = \{0\} \cup I^*, \quad I^* = I^*(x) = \{i \in I : \varphi_i(x) = 0\}$$

Let us note that the optimality conditions of Ben-Tal are formulated without constraint qualification, but the multipliers depend on critical directions.

In what follows we apply the conditions of Ben-Tal to problem (5). The Lagrange function for problem (5) takes the form

$$L(x, u, w, y_0, y_1, y_2, \mu_1, \mu_2) = \quad (A.7)$$

$$y_0 e^T (u + w) - y_1^T u - y_2^T w + \mu_1^T (F_1(x) - u + w) + \mu_2^T F_2(x)$$

with Lagrange multipliers $y_0 \in \mathbb{R}$, $y_1, y_2 \in \mathbb{R}^{m_1}$, $\mu_1 \in \mathbb{R}^{m_1}$, $\mu_2 \in \mathbb{R}^{m_2}$. With the Lagrangian (A.7) the stationarity conditions (A.2) become

$$\nabla_x L(x^*, u^*, w^*, y_0^*, y_1^*, y_2^*, \mu_1^*, \mu_2^*) = (\mu_1^*)^T \nabla F_1(x^*) + (\mu_2^*)^T \nabla F_2(x^*) = 0$$

$$\nabla_u L(x^*, u^*, w^*, y_0^*, y_1^*, y_2^*, \mu_1^*, \mu_2^*) = y_0^* e^T - (y_1^*)^T - (\mu_1^*)^T = 0$$

$$\Rightarrow y_1^* = y_0^* e - \mu_1^* \ge 0$$

$$\nabla_w L(x^*, u^*, w^*, y_0^*, y_1^*, y_2^*, \mu_1^*, \mu_2^*) = y_0^* e^T - (y_2^*)^T + (\mu_1^*)^T = 0$$

$$\Rightarrow y_2^* = y_0^* e + \mu_1^* \ge 0.$$

From the last two relations we get $-y_0^* e \le \mu_1^* \le y_0^* e$. Furthermore, from stationarity it follows that if $y_0^* = 0$ then $\mu_1^* = 0$, $y_1^* = 0$, $y_2^* = 0$ and $(\mu_2^*)^T \nabla F_2(x^*) = 0$. Hence, under regularity assumption (6) $\mu_2^* = 0$, which contradicts the assertion that the Lagrange multipliers are not all identical zero. Consequently, $y_0^* \ne 0$ and we may set $y_0^* = 1$.

Now, let us have a look at the complementarity conditions (A.4), which result in the following relations for the problem (5):

(a) If $u_i^* > 0$ (in this case $F_{1i}(x^*) > 0$, $w_i^* = 0$) then $y_{1i}^* = 0$, and the stationarity conditions imply $\mu_{1i}^* = 1$, $y_{2i}^* = 2$.

(b) If $w_i^* > 0$ (in this case $F_{1i}(x^*) < 0$, $u_i^* = 0$) then $y_{2i}^* = 0$, and the stationarity conditions imply $\mu_{1i}^* = -1$, $y_{1i}^* = 2$.

Taking into account the structure of constraint in (5), it is easy to verify that the stationarity and complementarity conditions for the problem (5) can be re-written as follows:

$$(\mu_1^*)^T \nabla F_1(x^*) + (\mu_2^*)^T \nabla F_2(x^*) = 0,$$

$$-1 \le \mu_{1i}^* \le 1, i = 1, \ldots, m_1, \tag{A.8}$$

$$\mu_{1i}^* = \text{sgn}(F_{1i}(x^*)), \ F_{1i}(x^*) \ne 0, \ i = 1, \ldots, m_1,$$

and conditions (7) follow immediately if we set $\lambda_1^* = \mu_1^*$, $\lambda_2^* = \mu_2^*$. Under (6) the multipliers λ_1^*, λ_2^* that satisfy (A.8) are unique and hence are the same for all critical directions.

Now, let us analyze the conditions (A.5) and the set of critical directions (A.6). The set of critical directions (A.6) for problem (5) can be expressed as

$$D(x, w, u) = \{\Delta x \in \mathbb{R}^n, \Delta u \in \mathbb{R}^{m_1}, \Delta w \in \mathbb{R}^{m_1}, \ :$$

$$e^T(\Delta u + \Delta w) \le 0,$$

$$\Delta w_i \ge 0, \text{ if } w_i = 0; \Delta u_i \ge 0, \text{ if } u_i = 0, i = 1, \ldots, m_1,$$

$$\nabla F_1(x)\Delta x - \Delta u + \Delta w = 0, \nabla F_2(x)\Delta x = 0\}$$

and the conditions (A.5) take the form

$$e^T(\Delta u + \Delta w) = 0,$$

$$\Delta w_i y_{2i} = 0, \text{ if } w_i = 0,$$

$$\Delta u_i y_{1i} = 0, \text{ if } u_i = 0, i = 1, \ldots, m_1.$$

Consider $d \in D_1(x^*, \lambda^*)$ with $\lambda_1^* = \mu_1$, $\lambda_2^* = \mu_2$, set $\Delta x = d$ and make a case-by-case analysis in order to describe the rules for choice of Δu_i, Δw_i, $i = 1, \ldots, m_1$.

(1) $F_{1i}(x^*) > 0$. In this case we have $u_i^* = F_{1i}(x^*)$, $w_i^* = 0$, $\mu_{1i}^* = 1$, $y_{1i}^* = 0$, $y_{2i}^* = 2$. Set

$$\Delta u_i = \nabla F_{1i}(x^*)d, \ \Delta w_i = 0.$$

(2) $F_{1i}(x^*) < 0$. In this case we have $u_i^* = 0$, $w_i^* = -F_{1i}(x^*)$, $\mu_{1i}^* = -1$, $y_{1i}^* = 2$, $y_{2i}^* = 0$. Set

$$\Delta u_i = 0, \ \Delta w_i = -\nabla F_{1i}(x^*)d.$$

Gauss–Newton Methods for Robust Parameter Estimation

(3) $F_{1i}(x^*) = 0$. In this case we have $u_i^* = 0$, $w_i^* = 0$. Consider subcases

(3a) $\mu_{1i}^* = 1$. In this case $y_{1i}^* = 0$, $y_{2i}^* = 2$, $\nabla F_{1i}(x^*)d \geq 0$. Set

$$\Delta u_i = \nabla F_{1i}(x^*)d \geq 0, \ \Delta w_i = 0.$$

(3b) $\mu_{1i}^* = -1$. In this case $y_{1i}^* = 2$, $y_{2i}^* = 0$, $\nabla F_{1i}(x^*)d \leq 0$. Set

$$\Delta u_i = 0, \ \Delta w_i = -\nabla F_{1i}(x^*)d \geq 0.$$

(3c) $|\mu_{1i}^*| < 1$. In this case $y_{1i}^* > 0$, $y_{2i}^* > 0$, $\nabla F_{1i}(x^*)d = 0$. Set

$$\Delta u_i = 0, \ \Delta w_i = 0.$$

Obviously, for this choice of Δx, Δu, Δw we get

$$\Delta w_i y_{2i}^* = \Delta u_i y_{1i}^* = 0, \ i = 1, \dots, m_1,$$
$$\Delta w_i \geq 0, \text{ if } w_i^* = 0; \Delta u_i \geq 0, \text{ if } u_i^* = 0, i = 1, \dots, m_1,$$
$$\nabla F_1(x^*)\Delta x - \Delta u + \Delta w = 0, \ \nabla F_2(x^*)\Delta x = 0.$$

Moreover,

$$e^T(\Delta u + \Delta w) = 0.$$

Indeed, multiplying the stationarity conditions by $(\Delta x^T, \Delta u^T, \Delta w^T)^T$ we get:

$$(\mu_1^*)^T \nabla F_1(x^*)\Delta x + (\mu_2^*)^T \nabla F_2(x^*)\Delta x = 0,$$
$$e^T \Delta u - (y_1^*)^T \Delta u - (\mu_1^*)^T \Delta u = 0,$$
$$e^T \Delta w - (y_2^*)^T \Delta w + (\mu_1^*)^T \Delta w = 0.$$

Since $\nabla F_2(x^*)\Delta x = 0$, we get $(\mu_1^*)^T \nabla F_1(x^*)\Delta x = (\mu_1^*)^T(\Delta u - \Delta w) = 0$. Hence,

$$0 = e^T \Delta u - (y_1^*)^T \Delta u - (\mu_1^*)^T \Delta u + e^T \Delta w - (y_2^*)^T \Delta w + (\mu_1^*)^T \Delta w$$
$$= e^T(\Delta u + \Delta w) - (y_1^*)^T \Delta u - (y_2^*)^T \Delta w = e^T(\Delta u + \Delta w).$$

Thus, the constructed direction $(\Delta x^T, \Delta u^T, \Delta w^T)^T$ is a critical direction in the sense of Ben-Tal, satisfying (A.5), and (8) is a consequence of (A.3). This finishes the proof.

Proof of Theorem 4

Similarly to the proof of Theorem 1, we will show that the optimality conditions (12) result from the optimality conditions of Ben-Tal [8].

The Lagrange function for problem (10) takes the form

$$L(x, v, u, w, y_0, y_1, y_2, \mu_1, \mu_2) = \tag{A.9}$$

$$y_0 \left(\tfrac{1}{2} \|v\|_2^2 + \gamma e^T (u + w) \right) - y_1^T u - y_2^T w + \mu_1^T (F_1(x) - v - u + w) + \mu_2^T F_2(x)$$

with Lagrange multipliers $y_0 \in \mathbb{R}$, $y_1, y_2 \in \mathbb{R}^{m_1}$, $\mu_1 \in \mathbb{R}^{m_1}$, $\mu_2 \in \mathbb{R}^{m_2}$. With the Lagrangian (A.9) the stationarity conditions (A.2) become

$$\nabla_x L(x^*, v^*, u^*, w^*, y_0^*, y_1^*, y_2^*, \mu_1^*, \mu_2^*) = (\mu_1^*)^T \nabla F_1(x^*) + (\mu_2^*)^T \nabla F_2(x^*) = 0$$

$$\nabla_v L(x^*, v^*, u^*, w^*, y_0^*, y_1^*, y_2^*, \mu_1^*, \mu_2^*) = y_0^*(v^*)^T - (\mu_1^*)^T = 0$$

$$\Rightarrow \mu_1^* = y_0^* v^*$$

$$\nabla_u L(x^*, v^*, u^*, w^*, y_0^*, y_1^*, y_2^*, \mu_1^*, \mu_2^*) = y_0^* \gamma e^T - (y_1^*)^T - (\mu_1^*)^T = 0$$

$$\Rightarrow y_1^* = y_0^* \gamma e - \mu_1^* \geq 0$$

$$\nabla_w L(x^*, v^*, u^*, w^*, y_0^*, y_1^*, y_2^*, \mu_1^*, \mu_2^*) = y_0^* \gamma e^T - (y_2^*)^T + (\mu_1^*)^T = 0$$

$$\Rightarrow y_2^* = y_0^* \gamma e + \mu_1^* \geq 0$$

From the last two relations we get $-y_0^* \gamma e \leq \mu_1^* \leq y_0^* \gamma e$. Furthermore, from stationarity it follows that if $y_0^* = 0$ then $\mu_1^* = 0$, $y_1^* = 0$, $y_2^* = 0$ and $(\mu_2^*)^T \nabla F_2(x^*) = 0$. Hence, under the regularity assumption (11) $\mu_2^* = 0$, which contradicts the assertion that the Lagrange multipliers are not all identical zero. Thus, as in the l_1 case $y_0^* \neq 0$ and we may set $y_0^* = 1$.

The complementarity conditions (A.4) for problem (10) take the form:

(a) If $u_i^* > 0$ (in this case $F_{1i}(x^*) - v_i^* > 0$, $w_i^* = 0$) then $y_{1i}^* = 0$, and the stationarity conditions imply $\mu_{1i}^* = \gamma$, $y_{2i}^* = 2\gamma$, $v_i^* = \gamma$.
(b) If $w_i^* > 0$ (in this case $F_{1i}(x^*) - v_i^* < 0$, $u_i^* = 0$) then $y_{2i}^* = 0$, and the stationarity conditions imply $\mu_{1i}^* = -\gamma$, $y_{1i}^* = 2\gamma$, $v_i^* = -\gamma$.

Taking into account the structure of constraints in (10), it is easy to verify that the stationarity and complementarity conditions for the problem (10) can be re-written as follows:

$$(\mu_1^*)^T \nabla F_1(x^*) + (\mu_2^*)^T \nabla F_2(x^*) = 0,$$

$$-\gamma \leq \mu_{1i}^* \leq \gamma, i = 1, \ldots, m_1, \tag{A.10}$$

$$\mu_{1i}^* = \begin{cases} F_{1i}(x^*), & |F_{1i}(x^*)| \leq \gamma, \\ \gamma \operatorname{sgn}(F_{1i}(x^*)), & |F_{1i}(x^*)| > \gamma, \end{cases} \quad i = 1, \ldots, m_1,$$

Gauss–Newton Methods for Robust Parameter Estimation

and conditions (12) follow immediately if we set $\lambda_1^* = \mu_1^*$, $\lambda_2^* = \mu_2^*$. Let us note that $\lambda_1^* = \mu_1^*$ is defined uniquely. Furthermore, the constraint qualification (11) guarantees that the multipliers λ_2^* that satisfy (A.10) are also unique and hence λ_1^*, λ_2^* do not depend on critical directions.

Now, let us analyze the conditions (A.5) and the set of critical directions (A.6). The set of critical directions (A.6) for the problem (10) can be expressed as

$$D(x, v, w, u) = \{\Delta x \in \mathbb{R}^n, \Delta v \in \mathbb{R}^{m_1}, \Delta u \in \mathbb{R}^{m_1}, \Delta w \in \mathbb{R}^{m_1}, :$$

$$(v^*)^T \Delta v + \gamma e^T (\Delta u + \Delta w) \leq 0,$$

$$\Delta w_i \geq 0, \text{ if } w_i = 0; \Delta u_i \geq 0, \text{ if } u_i = 0, i = 1, \ldots, m_1,$$

$$\nabla F_1(x) \Delta x - \Delta v - \Delta u + \Delta w = 0, \nabla F_2(x) \Delta x = 0\}$$

and the conditions (A.5) take the form

$$(v^*)^T \Delta v + \gamma e^T (\Delta u + \Delta w) = 0,$$

$$\Delta w_i y_{2i}^* = 0, \text{ if } w_i = 0,$$

$$\Delta u_i y_{1i}^* = 0, \text{ if } u_i = 0, i = 1, \ldots, m_1.$$

Consider $d \in D_H(x^*, \lambda^*)$ with $\lambda_1^* = \mu_1^*$, $\lambda_2^* = \mu_2^*$, set $\Delta x = d$ and make a case-by-case analysis in order to describe the rules for choice of Δv_i, Δu_i, Δw_i, $i = 1, \ldots, m_1$.

(1) $F_{1i}(x^*) > \gamma$. In this case we have $u_i^* = F_{1i}(x^*) - \gamma$, $w_i^* = 0$, $v_i^* = \gamma$, $\mu_{1i}^* = \gamma$, $y_{1i}^* = 0$, $y_{2i}^* = 2\gamma > 0$. Set

$$\Delta v_i = 0, \quad \Delta u_i = \nabla F_{1i}(x^*)d, \quad \Delta w_i = 0.$$

(2) $F_{1i}(x^*) < -\gamma$. In this case we have $u_i^* = 0$, $w_i^* = \gamma - F_{1i}(x^*)$, $v_i^* = -\gamma$, $\mu_{1i}^* = -\gamma$, $y_{1i}^* = 2\gamma > 0$, $y_{2i}^* = 0$. Set

$$\Delta v_i = 0, \quad \Delta u_i = 0, \quad \Delta w_i = -\nabla F_{1i}(x^*)d.$$

(3) $|F_{1i}(x^*)| \leq \gamma$. In this case we have $u_i^* = 0$, $w_i^* = 0$, $v_i^* = F_{1i}(x^*)$. Consider subcases:

(3a) $|F_{1i}(x^*)| < \gamma$. In this case we have $|\mu_{1i}^*| < \gamma$, $y_{1i}^* > 0$, $y_{2i}^* > 0$. Set

$$\Delta v_i = \nabla F_{1i}(x^*)d, \quad \Delta u_i = 0, \quad \Delta w_i = 0.$$

(3b) $F_{1i}(x^*) = \gamma$. In this case we have $\mu_{1i}^* = \gamma$, $y_{1i}^* = 0$, $y_{2i}^* = 2\gamma > 0$. Set

$$\Delta v_i = \nabla F_{1i}(x^*)d, \quad \Delta u_i = 0, \quad \Delta w_i = 0, \text{ if } \nabla F_{1i}(x^*)d \leq 0;$$

$$\Delta v_i = 0, \quad \Delta u_i = \nabla F_{1i}(x^*)d, \quad \Delta w_i = 0, \text{ if } \nabla F_{1i}(x^*)d > 0;$$

(3c) $F_{1i}(x^*) = -\gamma$. In this case we have $\mu_{1i}^* = -\gamma$, $y_{1i}^* = 2\gamma > 0$, $y_{2i}^* = 0$. Set

$$\Delta v_i = \nabla F_{1i}(x^*)d, \ \Delta u_i = 0, \ \Delta w_i = 0, \ \text{if } \nabla F_{1i}(x^*)d \geq 0;$$
$$\Delta v_i = 0, \ \Delta u_i = 0, \ \Delta w_i = -\nabla F_{1i}(x^*)d, \ \text{if } \nabla F_{1i}(x^*)d < 0;$$

Obviously, for this choice of Δx, Δv, Δu, Δw we get

$$\Delta w_i y_{2i}^* = \Delta u_i y_{1i}^* = 0, \ i = 1, \ldots, m_1,$$
$$\Delta w_i \geq 0, \ \text{if } w_i^* = 0; \Delta u_i \geq 0, \ \text{if } u_i^* = 0, i = 1, \ldots, m_1,$$
$$\nabla F_1(x^*)\Delta x - \Delta v - \Delta u + \Delta w = 0, \ \nabla F_2(x^*)\Delta x = 0.$$

Moreover,

$$(v^*)^T \Delta v + \gamma e^T (\Delta u + \Delta w) = 0.$$

Indeed, multiplying the stationarity conditions by $(\Delta x^T, \Delta v^T, \Delta u^T, \Delta w^T)^T$ we get:

$$(\mu_1^*)^T \nabla F_1(x^*)\Delta x + (\mu_2^*)^T \nabla F_2(x^*)\Delta x = 0,$$
$$(v^*)^T \Delta v - (\mu_1^*)^T \Delta v = 0,$$
$$\gamma e^T \Delta u - (y_1^*)^T \Delta u - (\mu_1^*)^T \Delta u = 0,$$
$$\gamma e^T \Delta w - (y_2^*)^T \Delta w + (\mu_1^*)^T \Delta w = 0.$$

Since $\nabla F_2(x^*)\Delta x = 0$, we get $(\mu_1^*)^T \nabla F_1(x^*)\Delta x = (\mu_1^*)^T (\Delta v + \Delta u - \Delta w) = 0$. Hence,

$$\begin{aligned}
0 &= (v^*)^T \Delta v - (\mu_1^*)^T \Delta v + \gamma e^T \Delta u - (y_1^*)^T \Delta u - (\mu_1^*)^T \Delta u + \\
&\quad \gamma e^T \Delta w - (y_2^*)^T \Delta w + (\mu_1^*)^T \Delta w \\
&= (v^*)^T \Delta v + \gamma e^T (\Delta u + \Delta w) - (y_1^*)^T \Delta u - (y_2^*)^T \Delta w \\
&= (v^*)^T \Delta v + \gamma e^T (\Delta u + \Delta w).
\end{aligned}$$

Thus, the constructed direction $\Delta = (\Delta x^T, \Delta v^T, \Delta u^T, \Delta w^T)^T$ is a critical direction in the sense of Ben-Tal, moreover it satisfies (A.5). Now let us compute

$$\begin{aligned}
&\Delta^T \nabla^2_{(x,v,u,v)^2} L(x^*, v^*, u^*, w^*, y_0^*, y_1^*, y_2^*, \mu_1^*, \mu_2^*)\Delta = \\
&\Delta x^T \nabla^2_{xx} L(x^*, v^*, u^*, w^*, y_0^*, y_1^*, y_2^*, \mu_1^*, \mu_2^*)\Delta x + \\
&\Delta v^T \nabla^2_{vv} L(x^*, v^*, u^*, w^*, y_0^*, y_1^*, y_2^*, \mu_1^*, \mu_2^*)\Delta v \leq
\end{aligned}$$

$$d^T \left(\sum_{i=1}^{m_1} \lambda_{1i}^* \nabla^2 F_{1i}(x^*) + \sum_{i=1}^{m_2} \lambda_{2i}^* \nabla^2 F_{2i}(x^*) + \sum_{i:|F_{1i}(x^*)| \leq \gamma} \nabla F_{1i}^T(x^*) \nabla F_{1i}(x^*) \right) d.$$

Proof of Theorem 7

Proof is based on the following representation of Δx_{k+1} at $x_{k+1} \in \mathcal{D}_0$:

$$\Delta x_{k+1} = -\nabla F^+(x_{k+1}) F_A(x_{k+1}) = -\nabla F^+(x_{k+1}) \begin{pmatrix} F_{1A}(x_{k+1}) \\ F_{1N}(x_{k+1}) \\ F_2(x_{k-1}) \end{pmatrix}$$

$$= -\nabla F^+(x_{k+1}) \left[\begin{pmatrix} F_{1A}(x_{k+1}) \\ F_{1N}(x_{k+1}) \\ F_2(x_{k+1}) \end{pmatrix} - \begin{pmatrix} F_{1A}(x_k) + \nabla F_{1A}(x_k) \Delta x_k \\ F_{1N}(x_k) \\ F_2(x_k) + \nabla F_2(x_k) \Delta x_k \end{pmatrix} \right.$$

$$\left. + \begin{pmatrix} F_{1A}(x_k) + \nabla F_{1A}(x_k) \Delta x_k \\ F_{1N}(x_k) \\ F_2(x_k) + \nabla F_2(x_k) \Delta x_k \end{pmatrix} \right]$$

$$= -\nabla F^+(x_{k+1}) \begin{pmatrix} F_{1A}(x_{k+1}) - F_{1A}(x_k) - \nabla F_{1A}(x_k) \Delta x_k \\ 0 \\ F_2(x_{k+1}) - F_2(x_k) - \nabla F_2(x_{k+1}) \Delta x_k \end{pmatrix}$$

$$+ \nabla F^+(x_{k+1}) \begin{pmatrix} F_{1A}(x_k) + \nabla F_{1A}(x_k) \Delta x_k \\ F_{1N}(x_k) \\ F_2(x_k) + \nabla F_2(x_k) \Delta x_k \end{pmatrix}$$

$$- \nabla F^+(x_k) \begin{pmatrix} F_{1A}(x_k) + \nabla F_{1A}(x_k) \Delta x_k \\ F_{1N}(x_k) \\ F_2(x_k) + \nabla F_2(x_k) \Delta x_k \end{pmatrix}.$$

Hence,

$$\Delta x_{k+1} = -\nabla F^+(x_{k+1}) \begin{pmatrix} F_{1A}(x_{k+1}) - F_{1A}(x_k) - \nabla F_{1A}(x_k) \Delta x_k \\ 0 \\ F_2(x_{k+1}) - F_2(x_k) - \nabla F_2(x_k) \Delta x_k \end{pmatrix} \quad \text{(A.11)}$$

$$- \left[\nabla F^+(x_{k+1}) - \nabla F^+(x_k) \right] R_A(x_k).$$

Here we have used the equality

$$
\nabla F^+(x_k) \begin{pmatrix} F_{1A}(x_k) + \nabla F_{1A}(x_k)\Delta x_k \\ F_{1N}(x_k) \\ F_2(x_k) + \nabla F_2(x_k)\Delta x_k \end{pmatrix}
$$

$$
= -\Delta x_k + \nabla F^+(x_k) \begin{pmatrix} \nabla F_{1A}(x_k) \\ 0 \\ \nabla F_2(x_k) \end{pmatrix} \Delta x_k
$$

$$
= -\Delta x_k + \nabla F^+(x_k) D(x_k) \nabla F_A(x_k)\Delta x_k = -\Delta x_k + \Delta x_k = 0,
$$

and the assumption that the partitioning (27) does not change for $x \in \mathcal{D}$. Using (A.11) and the assumptions of the theorem, we get

$$
\|\Delta x_{k+1}\| \leq \|\nabla F^+(x_{k+1}) \begin{pmatrix} F_{1A}(x_{k+1}) - F_{1A}(x_k) - \nabla F_{1A}(x_k)\Delta x_k \\ 0 \\ F_2(x_{k+1}) - F_2(x_k) - \nabla F_2(x_k)\Delta x_k \end{pmatrix} \|
$$

$$
+ \| \left[\nabla F^+(x_{k+1}) - \nabla F^+(x_k) \right] R_A(x_k) \|
$$

$$
\leq \|\nabla F^+(x_{k+1}) \begin{pmatrix} \int_0^1 (\nabla F_{1A}(x_k + t\Delta x_k) - \nabla F_{1A}(x_k)) \Delta x_k dt \\ 0 \\ \int_0^1 (\nabla F_2(x_k + t\Delta x_k) - \nabla F_2(x_k)) \Delta x_k dt \end{pmatrix} \|
$$

$$
+ \kappa \|\Delta x_k\|
$$

$$
\leq \int_0^1 \|\nabla F^+(x_{k+1}) \begin{pmatrix} \nabla F_{1A}(x_k + t\Delta x_k) - \nabla F_{1A}(x_k)\Delta x_k \\ 0 \\ (\nabla F_2(x_k + t\Delta x_k) - \nabla F_2(x_k)) \Delta x_k \end{pmatrix} \| dt
$$

$$
+ \kappa \|\Delta x_k\|
$$

$$
\leq \int_0^1 t\omega \|\Delta x_k\|^2 dt + \kappa \|\Delta x_k\| = \left(\frac{\omega}{2} \|\Delta x_k\| + \kappa \right) \|\Delta x_k\|.
$$

Thus, we have shown that

$$
\|\Delta x_{k+1}\| \leq \left(\frac{\omega}{2} \|\Delta x_k\| + \kappa \right) \|\Delta x_k\|. \tag{A.12}
$$

With estimate (A.12) all statements of the Theorem can be proven similarly to theorem (3.1.44) in Bock [10].

Gauss–Newton Methods for Robust Parameter Estimation

- The sequences $\{\delta_k\}$, $\delta_k := \frac{\omega}{2}\|\Delta x_k\| + \kappa$, and $\{\|\Delta x_k\|\}$ are monotonously decreasing. Indeed, since $\delta_0 < 1$, then

$$\|\Delta x_1\| \leq \delta_0\|\Delta x_0\| < \|\Delta x_0\|,$$

$$\delta_1 = \frac{\omega}{2}\|\Delta x_1\| + \kappa < \frac{\omega}{2}\|\Delta x_0\| + \kappa = \delta_0.$$

and hence $\delta_{k+1} < \delta_k < \ldots < \delta_1 < \delta_0 < 1$.
- The sequence $\{x_k\}$ remains in D_0. Indeed, since x_0 and $x_1 \in D_0$ we get by induction for $x_{k+1} \in D_0$

$$\|x_{k+2} - x_0\| = \|x_{k+2} - x_{k+1} + x_{k+1} - x_k + \ldots + x_1 - x_0\|$$

$$= \|\Delta x_{k+1} + \Delta x_k + \ldots + \Delta x_0\| \leq \sum_{j=0}^{k+1}\|\Delta x_j\|$$

$$< \left(\delta_0^{k+1} + \delta_0^k + \ldots + 1\right)\|\Delta x_0\| \leq \frac{1}{1 - \delta_0}\|\Delta x_0\|$$

implying $x_{k+2} \in D_0$.
- The sequence $\{x_k\}$ is a Cauchy sequence since

$$\|x_{i+k+1} - x_i\| = \sum_{p=0}^{k}\|\Delta x_{i+p}\| < \sum_{p=0}^{k}\delta_i^p\|\Delta x_i\|$$

$$\leq \frac{1}{1 - \delta_i}\|\Delta x_i\| \leq \frac{\delta_0^i}{1 - \delta_0}\|\Delta x_0\|, \forall k.$$

Hence the sequence $\{x_k\}$ converges: $x_k \to x_\star$, for $k \to \infty$. The point $x^* \in D_0$ is a fixed point of the Gauss–Newton iteration: $\Delta x_k \to 0 = \Delta x_\star := \nabla F^+(x^*)F_A(x^*)$ $(k \to \infty)$.

References

1. D.A. Andrews, F. Bickel, F. Hampel, P. Huber, W. Rogers, and J. Tukey, *Robust Estimates of Location: Survey and Advanves*, Princeton University Press, 1972.
2. U. Ascher, Collocation for two-point boundary value problems revisited, *SIAM Journal on Numerical Analysis*, 23(3), pp. 596–609, 1986.
3. I. Barrodale and F.D.K. Roberts, An improved algorithm for discrete l_1 linear approximation, *SIAM Journal on Numerical Analysis* 10, pp. 839–848, 1973.
4. I. Barrodale and F.D.K. Roberts, Algorithm 478: Solution of an overdetermined system of linear equations in the l_1 norm, *Communications of the ACM* 17, pp. 319–320, 1974.
5. I. Bauer, H. G. Bock, S. Körkel, and J. P. Schlöder, Numerical methods for initial value problems and derivative generation for DAE models with application to optimum experimental design of chemical processes, in F. Keil, W. Mackens, H. Voss, and J. Werther, eds., *Scientific Computing in Chemical Engineering II*, 2, pp. 282–289, Springer, Berlin-Heidelberg, 1999.

6. I. Bauer, H. G. Bock, S. Körkel, and J. P. Schlöder, Numerical methods for optimum experimental design in DAE systems, *Journal of Computational and Applied Mathematics* 120, pp. 1–25, 2000.
7. A.E. Beaton and J.W. Tukey, The Fitting of Power Series, Meaning Polynomials, Illustrated on Band-Spectroscopic Data, *Technometrics* 16, pp. 147–185, 1974.
8. A. Ben-Tal, Second-Order and Related Extremality Conditions in Nonlinear Programming, *Journal of Optimization Theory and Applications* 31(2), pp. 143–165, 1980.
9. H.G. Bock, Numerical treatment of inverse problems in chemical reaction kinetics, in K.-H. Ebert, P. Deuflhard, and W. Jäger, eds., *Modelling of Chemical Reaction Systems*, volume 18 of *Springer Series in Chemical Physics*, Springer Verlag, pp. 102–125, 1981.
10. H.G. Bock, *Randwertproblemmethoden zur Parameteridentifizierung in Systemen nichtlinearer Differentialgleichungen*, Bonner Mathematische Schriften, Nr. 183, Bonn, 1987.
11. H.G. Bock, E.A. Kostina, and J.P. Schlöder, Numerical Methods for Parameter Estimation in Nonlinear Differential Algebraic Equations, *GAMM Mitteilungen* 30(2), pp. 376–408, 2007.
12. H.G. Bock, E. Kostina, and J.P. Schlöder, On the Role of Natural Level Functions to Achieve Global Convergence for Damped Newton Methods, in M.J.D. Powell and S. Scholtes, eds., *System Modelling and Optimization. Methods, Theory and Applications*, Kluwer, Boston, pp. 51–74, 2000.
13. A.M. Bruckstein, D.L. Donoho, and M. Elad, From Sparse Solutions of Systems of Equations to Sparse Modeling of Signals and Images, *SIAM Review* 51(1), pp. 34–81, 2009.
14. R. Bulirsch, *Die Mehrzielmethode zur numerischen Lösung von nichtlinearen Randwertproblemen und Aufgaben der optimalen Steuerung*, Technical report, Carl-Cranz-Gesellschaft, 1971.
15. A. Cervantes and L.T. Biegler, Large-scale DAE optimization using a simultaneous NLP formulation, *AIChE Journal* 44(5), pp. 1038–1050, 2004.
16. W. Chauvenet, *A Manual of Spherical and Practical Astronomy Vol. II - Theory and Use of Astronomical Instruments*, J.B. Lippincott & Co., 1868.
17. J.E. Dennis and R.E. Welsch, Techniques for nonlinear least squares and robust regression, *Communications in Statistics - Simulation and Computation* 7, pp. 345–359, 1978.
18. P. Deuflhard, A Modified Newton Method for the Solution of Ill-Conditioned Systems of Nonlinear Equations with Application to Multiple Shooting, *Numerical Mathematics* 22, pp. 289–315, 1974.
19. R.C. Fair, On the robust estimation of econometric models, *Annals of Economic and Social Measurement* 3, pp. 667–677, 1974.
20. A.V. Fiacco and G.P. McCormick, Nonlinear Programming, Sequential Unconstrained Minimization Techniques, SIAM, 1990.
21. C.F. Gauss, *Theoria Motus Corporum Coelestium in Sectionibus Conicis Solem Ambientium*, F. Perthes & J.H. Besser, 1809.
22. J.E. Gentle, Least absolute value estimation: an introduction, *Communications in Statistics - Simulation and Computation* 6, pp. 313–328, 1977.
23. F.R. Hampel, Robust Estimation: A Condensed Partial Survey, *Zeitschrift für Wahrscheinlichkeitstheorie und verwandte Gebiete* 27, pp. 87–104, 1973.
24. M.J. Hinich and P.P. Talwar, A Simple Method for Robust Regression, *Journal of the American Statistical Association* 70, pp. 113–119, 1975.
25. P.E. Holland and R.E. Welsch, Robust regression using iteratively reweighted least-squares, *Communications in Statistics - Theory and Methods* 6, pp. 813–827, 1977.
26. P.J. Huber, Robust Estimation of a Location Parameter, *The Annals of Mathematical Statistics* 35, pp. 73–101, 1964.
27. P.J. Huber and E.M. Ronchetti, *Robust Statistics - Second Edition*, John Wiley & Sons, 2009.
28. A. Koberstein, *The Dual Simplex method, Techniques for a Fast and Stable Implementation*, PhD Thesis, U Paderborn, 2005.
29. S. Körkel and E. A. Kostina, Numerical Methods for Nonlinear Experimental Design, in H. G. Bock, E. A. Kostina, H. X. Phu, and R. Rannacher, eds., *Modeling, Simulation and Optimization of Complex Processes, Proceedings of the International Conference on High Performance Scientific Computing, 2003, Hanoi, Vietnam*. Springer, 2004.

30. S. Körkel, *Numerische Methoden für Optimale Versuchsplanungsprobleme bei nichtlinearen DAE-Modellen*, PhD thesis, U Heidelberg, 2002.
31. E. Kostina, The Long Step Rule in the Bounded-Variable Dual Simplex Method: Numerical Experiments, *Mathematical Methods of Operations Research* 3(**55**), pp. 413–429, 2002.
32. E. Kostina, Robust Parameter Estimation in Dynamic Systems, *Optimization and Engineering* 5(**4**), pp. 461–484, 2004.
33. E. Kostina and O. I. Kostyukova, *A Primal-Dual Active-Set Method for Convex Quadratic Programming*, Technical Report, IWR, U Heidelberg, 2003.
34. S. Krishna, M.H. Jensen, and K. Sneppen, Minimal model of spiky oscillations in NF-κB signaling, *Proceedings of the National Academy of Sciences of the United States of America* 103(**29**), pp. 10840–10845, 2006.
35. P.S. Laplace, *Théorie Analytique des Probabilités*, V. Courcier, 1814.
36. A .M. Legendre, *Nouvelles méthodes pour la détermination des orbites des comètes*, Firmin Didot, 1805.
37. O.L. Mangasarian and D.R. Musicant, Robust Linear and Vector Support Regression, *IEEE Transactions on Pattern Analysis and Machine Intelligence* 22(**9**), pp. 1–6, 2000.
38. M.R. Osborne, B. Presnell, and B.A. Turlach, On the LASSO and Its Dual, *Journal of Computational and Graphical Statistics* 9(**2**), pp. 319–337, 2000.
39. B. Peirce, Criterion for the rejection of doubtful observations, *The Astronomical Journal* 45, pp. 161–163, 1852.
40. J.R. Rice and J.S. White, Norms for Smoothing and Estimation, *SIAM Review* 6, pp. 243–256, 1964.
41. J.A. Scales and A. Gersztenkorn, Robust methods in inverse theory, *Inverse Problems* 4, pp. 1071–1091, 1988.
42. V.H. Schulz, *Ein effizientes Kollokationsverfahren zur numerischen Behandlung von Mehrpunktrandwertaufgaben in der Parameteridentifizierung und Optimalen Steuerung*, Diploma thesis, U Augsburg, 1990.
43. S. Waldherr, *Uncertainty and robustness analysis of biochemical reaction networks via convex optimisation and robust control theory*, PhD thesis, University of Stuttgart, 2009.

An Optimal Scanning Sensor Activation Policy for Parameter Estimation of Distributed Systems

Dariusz Uciński

Abstract A technique is proposed to solve an optimal node activation problem in sensor networks whose measurements are supposed to be used to estimate unknown parameters of the underlying process model in the form of a partial differential equation. Given a partition of the observation horizon into a finite number of consecutive intervals, the problem is set up to select nodes which will be active over each interval while the others will remain dormant such that the log-determinant of the resulting Fisher information matrix associated with the estimated parameters is maximized. The search for the optimal solution is performed using the branch-and-bound method in which an extremely simple and efficient technique is employed to produce an upper bound to the maximum objective function. Its idea consists in solving a relaxed problem through the application of a simplicial decomposition algorithm in which the restricted master problem is solved using a multiplicative algorithm for D-optimal design. The performance evaluation of the technique is additionally presented by means of simulations.

1 Introduction

1.1 Distributed Sensor Networks and Sensor Management

Distributed sensor networks have recently become an important research area regarding spatio-temporal phenomena [10, 11, 15, 35, 72, 99]. This is because dramatic progress in hardware, sensor and wireless networking technologies enables large-scale deployment of superior data acquisition systems with adjusting

---

D. Uciński (✉)
Institute of Control and Computation Engineering, University of Zielona Góra,
ul. Podgórna 50, 65–246 Zielona Góra, Poland
e-mail: d.ucinski@issi.uz.zgora.pl

H.G. Bock et al. (eds.), *Model Based Parameter Estimation*, Contributions
in Mathematical and Computational Sciences 4, DOI 10.1007/978-3-642-30367-8_4,
© Springer-Verlag Berlin Heidelberg 2013

resolutions. A sensor network may comprise thousands of inexpensive, miniature and low-power sensor nodes that can be deployed throughout a physical space and connect through a multi-hop wireless network, providing dense sensing close to physical phenomena. The nodes process and locally communicate the collected information, as well as coordinate actions with each other.

Sensor networks have recently come into prominence because they hold the potential to revolutionize a wide spectrum of both civilian and military applications, including monitoring microclimates and wildlife habitats, tracking chemical plumes, traffic surveillance, industrial and manufacturing automation, building and structures monitoring, and many others. What makes sensor networks so attractive is their miniaturization, low cost, low power radio and autonomous ad hoc connectivity, which basically eliminates the need for any human intervention.

The design, implementation and operation for a sensor network requires the confluence of many disciplines, including signal processing, networking and protocols, embedded systems, information management and distributed algorithms. There are, however, a number of other problems that make the design of applications for sensor networks a difficult task. In this context, power limitation is a central issue when nodes are powered by batteries. To reduce power consumption, nodes are provided with a "sleep" mode in which they consume only a fraction of the power when awake. What is more, an additional reason for not using all the available sensors could be the reduction of the observation system complexity and the cost of operation and maintenance [97]. Management of sleeping and awake modes is thus of utmost importance as certainly the selection of active and dormant sensor may have a dramatic effect on the performance possibilities. This becomes even more critical when the observed process has a high dynamics in space and time and then it is desirable to have a sensor management policy so as to activate at a given time moment only sensors which provide the most valuable information about the observed process. Such a scanning strategy of taking measurements can be also interpreted in terms of a group of sensors which are mobile in the sense that after performing measurements at fixed spatial positions (i.e., exactly in the same way as stationary sensors) on a time interval of nonzero length, they can change their locations to continue the measurements at more informative points, and the time necessary for this change may be neglected. In this way, we deal with a kind of mobility in sensor networks.

Scanning drastically expands the spectrum of the network's capabilities. Moreover, assuming that each node possesses a certain amount of decision making autonomy gives rise to a dynamic system with a considerable amount of flexibility, depending on the extent to which the nodes can cooperate in order to perform a mission. This flexibility, for example, allows us to handle a large number of dynamic events with a much smaller number of nodes [10, 11, 15, 55, 72]. Naturally, mobility implies an additional layer of complexity. For example, if communication connectivity is to be maintained, we must ensure that each node remains within the range of at least some other nodes. We must also take into account that mobility consumes a considerable amount of energy, which amplifies the need for various forms of power control. However, the complexity of the resulting sensor management problem is

An Optimal Scanning Sensor Activation Policy

compensated by a number of benefits. Specifically, observations are not assigned to fixed spatial positions, but the network is capable of tracking points which provide at a given time moment best information about the spatio-temporal process.

1.2 PDE Models in Configuring Sensor Networks

In a typical sensor network application, sensors are supposed to be deployed so as to monitor a region and collect the most valuable information from the observed system. The quality of sensor deployment can be quantified by the appropriate performance indices and optimum sensor node configurations can thus be sought. The resulting observation strategies concern optimally scheduling active and dormant nodes. Up to now, approaches aiming at guaranteeing a dense region coverage or satisfactory network connectivity have dominated this line of research and abstracted away from the mathematical description of the physical processes underlying the observed phenomena. In this way, much information is lost which could potentially be used to make the operation of the sensor network more efficient and yield substantial gains in the functionality of the whole observation system.

The observed processes in question are often termed *distributed parameter systems* (DPSs) as their states depend not only on time, but also on spatial coordinates. Appropriate mathematical modeling of DPSs most often yields partial differential equations (PDEs). It goes without saying that such models involve using rather sophisticated mathematical methods. This explains why so few attempts have been made at exploiting them in the context of sensor networks. But in recompense for this effort, we would be in a position to describe the process more accurately and to implement more effective control strategies.

For the last 40 years, DPSs have occupied an important place in control and system theories. They are now an established area of research with a long list of journal articles, conference proceedings and numerous textbooks [6, 16, 44, 56]. It is intriguing that for a long time, due to the inherent impossibility of observing the system states over the entire spatial domain, one of the topics of importance for specialists in control theory has been the problem of selecting the number and location of sensors and actuators for the control and state/parameter estimation in such systems. A number of sensor location methods were invented and supported by a sound theory, but they are not directly fit to the emerging technology of sensor networks. The present paper is intended as a step towards bridging this gap and meeting the needs created in the context of PDE model calibration based on observations from a sensor network with scanning nodes.

1.3 Sensor Location for Parameter Estimation in DPSs

The importance of sensor planning has been recognized in many application domains prior to the invention of sensor networks, e.g., regarding air quality monitoring systems, groundwater-resources management, recovery of valuable minerals and hydrocarbon, model calibration in meteorology and oceanography, chemical engineering, hazardous environments and smart materials [5, 12, 17, 36, 37, 51, 52, 62, 74, 81]. The operation and control of such systems usually requires precise information on the parameters which condition the accuracy of the underlying mathematical model, but that information is only available through a limited number of possibly expensive sensors. Over the past years, this limitation has stimulated laborious research on the development of strategies for efficient sensor placement (for reviews, see papers [41, 97] and comprehensive monographs [79, 81]). Nevertheless, although the need for systematic methods was widely recognized, most techniques communicated by various authors usually rely on exhaustive search over a predefined set of candidates and the combinatorial nature of the design problem is taken into account very occasionally [97]. Needless to say that this approach, which is feasible for a relatively small number of possible locations, soon becomes useless as the number of possible location candidates increases. It goes without saying that these problems are compounded by the passage to scanning observations. The number of competing solutions grows exponentially with the numbers of sensors and switching instants. Complicated general search algorithms of discrete optimization can readily consume appreciable computer time and space, too.

1.4 Related Approaches

Instead of the wasteful and tedious exhaustive search of the solution space, which constitutes quite a naive approach, approaches originating in statistical optimum experimental design [3, 24, 60, 64, 83, 84, 98] and its extensions to models for dynamic systems, especially in the context of the optimal choice of sampling instants and input signals [27, 28, 31, 47, 75] turn out extremely useful. In this vein, various computational schemes have been developed to attack directly the original problem or its convenient approximation. The adopted optimization criteria are essentially the same, i.e., various scalar measures of performance based on the Fisher information matrix (FIM) associated with the parameters to be identified are maximized. The underlying idea is to express the goodness of parameter estimates in terms of the covariance matrix of the estimates. For sensor-location purposes, one assumes that an unbiased and efficient (or minimum-variance) estimator is employed. This leads to a great simplification since the Cramér–Rao lower bound for the aforementioned covariance matrix is merely the inverse of the FIM, which can be computed with relative ease, even though the exact covariance matrix of a particular estimator is very difficult to obtain.

As regards dynamic DPSs, the first treatment of this type for the sensor-location problem was proposed by Uspenskii and Fedorov [93] who maximized the D-optimality criterion, being the determinant of the FIM associated with the estimated parameters characterizing the source term in a simple one-dimensional linear diffusion equation. The authors observed that the linear dependence of the observed outputs on these parameters makes it possible to directly apply the machinery of optimum experimental design theory. The delineated approach was extended by Rafajłowicz [66] to cover a class of DPSs described by linear hyperbolic equations with known eigenfunctions and unknown eigenvalues. The aim was to find conditions for the optimality of the measurement design and the spectral density of the stochastic input. It was indicated that common numerical procedures from classical experimental design for linear regression models could be adopted to find optimal sensor location. Moreover, the demonstrated optimality conditions imply that the optimal input comprises a finite number of sinusoidal signals and that optimal sensor positions are not difficult to find in some cases. A similar problem was studied in [67] in a more general framework of DPSs which can be described in terms of Green's functions.

Over the past two decades, this methodology has been substantially refined to extend its applicability. A comprehensive treatment of both theoretical and algorithmic aspects of the resulting sensor location strategies is contained in the monograph [81]. The potential of the approach for generalizations was exploited, e.g., by Patan and Patan [58] who developed a fault detection scheme for DPSs based on the maximization of the power of a parametric hypothesis test regarding the nominal state of a given DPS. The approach based on maximization of the determinant of the appropriate FIM is by no means restricted to theoretical considerations and there are examples which do confirm its effectiveness in practical applications. Thus, in [50] a given number of stationary sensors were optimally located using nonlinear programming techniques for a biotechnological system consisting of a bubble column loop fermenter. On the other hand, Sun [74] advocates using optimum experimental design techniques to solve inverse problems in groundwater modelling. How to monitor the water quality around a landfill place is an example of such a network design. Nonlinear programming techniques are also used there to find numerical approximations to the respective exact solutions.

A similar approach was used in [38, 39] for on-orbit modal identification of large space structures. Although the respective models are not PDEs, but their discretized versions obtained through the finite-element method, the proposed solutions can still be of interest owing to the striking similitude of both the formulations. A fast and efficient approach was delineated for reducing a relatively large initial candidate sensor-location set to a much smaller optimum set which retains the linear independence of the target modes and does not lead to a substantial deterioration in the accuracy of modal-response estimates, which is quantified by the determinant of the FIM. Some improvements on this approach by incorporating basic elements of tabu search were proposed by Kincaid and Padula [40].

A related optimality criterion was given by Point et al. [61] who investigated maximization of the Gram determinant being a measure of the independence of the sensitivity functions evaluated at sensor locations. The authors argue that such a procedure guarantees that the parameters are identifiable and the correlation between the sensor outputs is minimized. The form of the criterion itself resembles the D-optimality criterion, but the counterpart of the FIM takes on much larger dimensions, which suggests that the approach may involve more cumbersome calculations. Nevertheless, the delineated technique was successfully applied to a laboratory-scale, catalytic fixed-bed reactor [94].

At this juncture, it should be noted that spatial design methods related to the design of monitoring networks are also of great interest to statisticians and a vast amount of literature on the subject already exists [49, 53, 54] contributing to the research field of spatial statistics [14] motivated by practical problems in agriculture, geology, meteorology, environmental sciences and economics. However, the models considered in the statistical literature are quite different from the dynamic models described by PDEs discussed here. Spatiotemporal data are not considered in this context and the main purpose is to model the spatial process by a spatial random field, incorporate prior knowledge and select the best subset of points of a desired cardinality to best represent the field in question. The motivation is a need to interpolate the observed behaviour of a process at unobserved spatial locations, as well as to design a network of optimal observation locations which allows an accurate representation of the process. The field itself is modelled by some multivariate distribution, usually Gaussian [2]. Designs for spatial trend and variogram estimation can be considered. The basic theory of optimal design for spatial random fields is outlined in the excellent monograph by Müller [49] which bridges the gap between spatial statistics and classical optimum experimental design theory. The optimal design problem can also be formulated in terms of information-based criteria whose application amounts to maximizing the amount of information (of the Kullback–Leibler type) to be gained from an experiment [45]. However, the applicability of all those fine statistical results in the engineering context discussed here is not clear for now and more detailed research into this direction should be pursued in the near future (specifically, generalizations regarding time dynamics are not obvious, although in [45] some promising attempts have been made).

Sensors can be mounted on various platforms and, as an appealing alternative to stationary sensors, these platforms can be highly dynamic in motion. Recent technological advances in communication systems and the growing ease in making small, low power and inexpensive mobile systems now make it feasible to deploy a group of networked vehicles in a number of environments [10, 11, 15, 55, 72]. The complexity of the resulting design problem is compensated by a number of benefits. Specifically, sensors are not assigned to fixed positions which are optimal only on the average, but are capable of tracking points which provide at a given time instant best information about the parameters to be identified. Consequently, by actively reconfiguring a sensor system we can expect the minimal value of an adopted design criterion to be lower than the one for the stationary case. In the seminal article [68], the D-optimality criterion is considered and an optimal time-dependent measure is sought, rather than the trajectories themselves. On the other hand, Uciński [80, 81, 90], apart from generalizations of Rafajłowicz's results, develops some computational algorithms based on the FIM. He reduces the problem to a state-constrained optimal-control one for which solutions are obtained via the methods of successive linearizations which is capable of handling various constraints imposed on sensor motions. In turn, the work [86] was intended as an attempt to properly

An Optimal Scanning Sensor Activation Policy

formulate and solve the time-optimal problem for moving sensors which observe the state of a DPS so as to estimate some of its parameters. In the paper by [87], a similar technique was presented so as to make the Hessian of the parameter estimation cost well conditioned subject an additional constraint imposed on the achievable D-efficiency of the solutions. This line of research was extended in the monograph by [73] towards applications involving sensor networks. We should also pass a reference to works by [18, 19, 21, 22, 22, 34, 89], and [20] focused on on-line optimal guidance of actuator/sensor network nodes for control and/or state estimation, which demonstrate that inclusion of a DPS model into the optimization setting can substantially improve the quality of the network performance.

Communications on selecting scanning sensor positions, which is the closest to the main topic of this paper, are rather limited. The problem was first exposed in [88, 91] and then more thoroughly examined in Chap. 4 of the monograph [81]. The key idea of that approach which originates in the seminal work [23] (see also [13], Chap. 4 of [24] and [85]) is to operate on the density of sensors per unit area instead of the positions of individual sensors. Such conditions permits to relax the discrete optimization problem in context and to replace it by its continuous approximation. This transformation is warranted in the case of a relatively large number of measurement sensors and a predefined set of switching times. Mathematically, the relevant procedure involves looking for a family of "optimal" probability measures defined on subsets of the set of feasible measurement points. In spite of its somewhat abstract assumptions, the resulting algorithms of the exchange type to solve the scanning sensor guidance problem are very easy to implement and extremely efficient. In turn, the scanning measurement problem with free switching times was attacked in [57] by treating it as an optimal discrete-valued control problem which is then transformed into an equivalent continuous-valued optimal-control formulation. In principle, the latter method is only applicable to situations when the number of scanning sensors to be scheduled is rather moderate, as it does not prevent the drawback of the "curse of dimensionality" inherent to the original combinatorial problem (enormous amounts of memory storage and time are still involved when the number of sensors is high). It offers, however, a possibility of selecting switching moments in continuous time based on the application of commonplace nonlinear programming algorithms.

1.5 Results in This Contribution

The purpose of the investigations undertaken here, for which preliminary results (without proofs) were reported in [82], was to establish a practical approach to selection of activated sensors over consecutive time stages which, while being independent of a particular model of the dynamic DPS in question, would be versatile enough to cope with practical sensor networks consisting of a large number of nodes. Specifically, given I sensor network nodes, we partition the entire observation horizon into fixed consecutive subintervals and allow only n of them (typically, n is much smaller than I) to be activated over each time subinterval. Consequently, the problem at each stage is to divide the I available nodes between n activated sites and the remaining $I - n$ dormant sites so as to maximize the determinant of the FIM associated with the parameters to be estimated. As was already mentioned, selecting the best subset of nodes to be activated constitutes an inherently discrete large-scale resource allocation problem whose solution may be prohibitively time-consuming. That is why an efficient guided search algorithm based on the branch-and-bound method is developed, which implicitly enumerates all the feasible sensor configurations, using relaxed optimization problems that involve no integer constraints.

It should be emphasized here that this idea is not novel, since branch-and-bound constitutes one of the most frequent approaches to solve discrete optimization problems and it has indeed been used in the context of network design, cf., e.g., [8]. What is more, it was already studied in [92] in quite a similar vein, but for stationary sensors (i.e., the positions of the active sensors were fixed over the whole observation horizon). Extensions to the scanning sensor setting are not that straightforward, however, especially due to the requirement that the number of active sensors over each time stage be always the same. Moreover, the size of the search space grows exponentially with the number of time partitions. The main contribution of this paper consists in the development of a simple, yet powerful, computational scheme to obtain upper bounds to the optimal values of the D-optimality criterion for the restricted problems. These bounds are obtained by relaxing the 0–1 constraints on the design variables, thereby allowing them to take any value in the interval $[0, 1]$ and resulting in a concave problem of determinant maximization over the set of all linear combinations of a finite number of nonnegative definite matrices, subject to additional linear constraints on the coefficients of those combinations. In order to solve it numerically, optimality conditions are first derived and discussed, because they take a surprisingly simple form involving an interesting separability principle for the set of locations at which the weights achieve their upper bounds and the ones at which the weights are zero. Then an algorithm is proposed which can be interpreted as a simplicial decomposition one with the restricted master problem solved by an uncomplicated multiplicative weight optimization algorithm. The resulting procedure is guaranteed to produce iterates converging to the solution of the relaxed restricted problem. To illustrate the use of our algorithm, we report some numerical experience on a sensor network design problem regarding a two-dimensional diffusion process.

1.6 Notation

Given a set H, $|H|$ and $\bar{H}$ signify its cardinality and closure, respectively. We use $\mathbb{R}$ to denote the set of real numbers and $\mathbb{R}_+$ to denote the set of nonnegative real numbers. The n-dimensional Euclidean vector space is denoted by $\mathbb{R}^n$, and the Euclidean matrix space of real matrices with n rows and k columns is denoted by $\mathbb{R}^{m \times k}$. We will write $\mathbb{S}^n$ for the subspace of $\mathbb{R}^{n \times n}$ consisting of all symmetric matrices. In $\mathbb{S}^n$ two sets are of special importance: the cone of nonnegative definite matrices and the cone of positive definite matrices, denoted by $\mathbb{S}_+^n$ and $\mathbb{S}_{++}^n$, respectively. The curly inequality symbol $\succeq$ and its strict form $\succ$ are used to denote the Löwner partial ordering of symmetric matrices: For $A, B \in \mathbb{S}^n$, we have

$$A \succeq B \iff A - B \in \mathbb{S}_+^n,$$
$$A \succ B \iff A - B \in \mathbb{S}_{++}^n.$$

We call a point of the form $\alpha_1 u_1 + \cdots + \alpha_\ell u_\ell$, where $\alpha_1 + \cdots + \alpha_\ell = 1$ and $\alpha_i \geq 0$, $i = 1, \ldots, \ell$, a convex combination of the points $u_1, \ldots, u_\ell$ (it can be thought of as a mixture or a weighted average of the points, with α_i the fraction of u_i in the mixture). Given a set of points U, $\text{co}(U)$ stands for its convex hull, i.e., the set of all convex combinations of elements of U,

$$\text{co}(U) = \left\{ \sum_{i=1}^{\ell} \alpha_i u_i \ \middle| \ u_i \in U, \ \alpha_i \geq 0, \ i = 1, \ldots, \ell; \ \sum_{i=1}^{\ell} \alpha_i = 1, \ \ell = 1, 2, 3, \ldots \right\}.$$

The probability (or canonical) simplex in $\mathbb{R}^n$ is defined as

$$P_n = \text{co}(\{e_1, \ldots, e_n\}) = \left\{ p \in \mathbb{R}_+^n \ \middle| \ \sum_{i=1}^{n} p_i = 1 \right\},$$

where e_j is the usual unit basis vector along the j-th coordinate of $\mathbb{R}^n$.

Finally, recall that for any $A \in \mathbb{R}^{n \times n}$ which may depend on a parameter β, there holds

$$\frac{\partial}{\partial \beta} \ln \det(A) = \text{trace}\left(A^{-1} \frac{\partial A}{\partial \beta} \right)$$

whenever A is nonsingular.

2   D-Optimal Sensor Activation Problem

Let $\Omega \subset \mathbb{R}^d$ be a bounded spatial domain with sufficiently smooth boundary Γ, and let $T = (0, t_f]$ be a bounded time interval. Consider a distributed parameter system (DPS) whose scalar state at a spatial point $x \in \bar{\Omega} \subset \mathbb{R}^a$ and time instant $t \in \bar{T}$ is denoted by $y(x, t)$. Mathematically, the system state is governed by the partial differential equation (PDE)

$$\frac{\partial y}{\partial t} = \mathcal{F}(x, t, y, \theta) \quad \text{in } \Omega \times T, \tag{1}$$

where $\mathcal{F}$ is a well-posed, possibly nonlinear, differential operator which involves first- and second-order spatial derivatives and may include terms accounting for forcing inputs specified a priori. The PDE (1) is accompanied by the appropriate boundary and initial conditions

$$\mathcal{B}(x,t,y,\theta) = 0 \qquad \text{on } \Gamma \times T, \tag{2}$$

$$y = y_0 \qquad \text{in } \Omega \times \{t = 0\}, \tag{3}$$

respectively, $\mathcal{B}$ being an operator acting on the boundary Γ and $y_0 = y_0(x)$ a given function. Conditions (2) and (3) complement (1) such that the existence of a sufficiently smooth and unique solution is guaranteed. We assume that the forms of $\mathcal{F}$ and $\mathcal{B}$ are given explicitly up to an m-dimensional vector of unknown constant parameters θ which must be estimated using observations of the system. The implicit dependence of the state y on the parameter vector θ will be reflected by the notation $y(x,t;\theta)$.

Let us form an arbitrary partition of the time interval T by choosing points $t_0 < t_1 < \cdots < t_K = t_f$ defining subintervals $T_k = (t_{k-1}, t_k]$, $k = 1, \ldots, K$ called *scanning stages*. We then consider that the state y is constantly observed by n sensors which will possibly be changing their locations at the beginning of every time subinterval, but will be remaining stationary for the duration of each of the subintervals. Thus the observations proceed in K stages and the instantaneous sensor configuration ζ can be viewed as follows:

$$\zeta(t) = (x_k^1, \ldots, x_k^n) \quad \text{for } t \in T_k, \quad k = 1, \ldots, K, \tag{4}$$

where $x_k^j \in X \subset \mathbb{R}^d$ stands for the location of the j-th sensor on the subinterval T_k, X being the part of the spatial domain $\bar{\Omega}$ where the measurements can be taken. Accordingly, the discrete-continuous observations provided by n scanning pointwise sensors are defined by the scalar output

$$z_{kj}(t) = y(x_k^j, t; \theta) + \varepsilon(x_k^j, t), \quad j = 1, \ldots, n \tag{5}$$

for $t \in T_k$, $k = 1, \ldots, K$, where $\varepsilon(x_k^j, t)$ denotes the measurement noise. This relatively simple conceptual framework involves no loss of generality since it can be easily generalized to incorporate, e.g., multiresponse systems or inaccessibility of state measurements, cf. [81, p. 95].

It is customary to assume that the measurement noise is zero-mean, Gaussian, spatial uncorrelated and white [1, 56, 65], i.e.,

$$e\{\varepsilon(x_k^j, t)\varepsilon(x_{k'}^{j'}, t')\} = \sigma^2 \delta_{kk'} \delta_{jj'} \delta(t - t'), \tag{6}$$

where σ^2 defines the intensity of the noise, δ_{ij} and $\delta(\,\cdot\,)$ standing for the Kronecker and Dirac delta functions, respectively. Although white noise is a physically impossible process, it constitutes a reasonable approximation to a disturbance whose

An Optimal Scanning Sensor Activation Policy

adjacent samples are uncorrelated at all time instants for which the time increment exceeds some value which is small compared with the time constants of the DPS.

The most widely used formulation of the parameter estimation problem is as follows: Given the model (1)–(3) and the outcomes of the measurements $z_{kj}(\cdot)$, $k = 1, \ldots, K$ and $j = 1, \ldots, n$, estimate θ by $\widehat{\theta}$, a global minimizer of the output least-squares error criterion

$$\mathcal{J}(\vartheta) = \sum_{k=1}^{K} \sum_{j=1}^{n} \int_{T_k} \left\{ z_{kj}(t) - y(x_k^j, t; \vartheta) \right\}^2 dt \qquad (7)$$

where $y(\cdot, \cdot; \vartheta)$ denotes the solution to (1)–(3) for a given value of the parameter vector ϑ. In practice, a regularized version of the above problem is often considered by adding to $\mathcal{J}(\vartheta)$ a term imposing stability or a-priori information or both [4, 96].

Inevitably, the covariance matrix $\mathrm{cov}(\widehat{\theta})$ of the above least-squares estimator depends on the sensor locations x_k^j. This fact suggests that we may attempt to select them so as to yield best estimates of the system parameters. To form a basis for the comparison of different locations, a quantitative measure Ψ of the "goodness" of particular sensor configurations is required. Such a measure is customarily based on the concept of the *Fisher Information Matrix* (FIM) which is widely used in optimum experimental design theory for lumped systems [3, 24, 60, 64, 98]. In our setting, the FIM is given by [65]

$$M(\zeta) = \sum_{k=1}^{K} \sum_{j=1}^{n} \int_{T_k} g(x_k^j, t) g(x_k^j, t)^{\mathsf{T}} dt, \qquad (8)$$

where

$$g(x, t) = \left[\frac{\partial y(x, t; \vartheta)}{\partial \vartheta_1}, \ldots, \frac{\partial y(x, t; \vartheta)}{\partial \vartheta_m} \right]_{\vartheta = \theta^0}^{\mathsf{T}} \qquad (9)$$

stands for the so-called *sensitivity vector*, θ^0 being a prior estimate to the unknown parameter vector θ [66, 67, 74, 81]. The rationale behind this choice is the fact that, up to a constant scalar multiplier, the inverse of the FIM constitutes a good approximation of $\mathrm{cov}(\widehat{\theta})$ provided that the time horizon is large, the nonlinearity of the model with respect to its parameters is mild, and the measurement errors are independently distributed and have small magnitudes [24, 98].

As for a specific form of Ψ, various options exist [3, 24, 98], but the most popular criterion, called the D-optimality criterion, is the log-determinant of the FIM:

$$\Psi(M) = \log \det(M). \qquad (10)$$

The resulting D-optimum sensor configuration leads to the minimum volume of the uncertainty ellipsoid for the estimates.

The introduction of an optimality criterion renders it possible to formulate the sensor location problem as maximization of the performance measure $\mathcal{R}(\zeta) := \Psi[M(\zeta)]$ with respect to all feasible ζ. This apparently simple formulation may lead to the conclusion that the only question remaining is that of selecting an appropriate solver from a library of numerical optimization routines. Unfortunately, an in-depth analysis reveals complications which accompany this way of thinking.

A key difficulty in developing successful numerical techniques for sensor location is that the number of sensors to be placed in a given region may be quite large (this is a common situation as far as sensor networks are considered). When trying to treat the task as a constrained nonlinear programming problem, the actual number of variables is even doubled or tripled, since the position of each sensor is determined by its two or three spatial coordinates, so that the resulting problem is rather of large scale. What is more, a desired global extremum is usually hidden among many poorer local extrema. Consequently, to directly find a numerical solution may be extremely difficult. Additionally, a technical complication might also be the sensor clusterization which constitutes a price to pay for the simplifying assumption that the measurement noise is spatially uncorrelated. This means that in an optimal solution different sensors often tend to take measurements at one point, and this is acceptable in applications rather occasionally.

In the literature, a common remedy for the last predicament is to guess a priori a set of I possible candidate locations, where $I > n$, and then to seek for each time interval T_k the best subset of n locations from among the I possible, so that the problem thus reduces to a combinatorial one. In other words, the problem is to divide at each scanning stage the I available sites between n gauged sites and the remaining $I - n$ ungauged sites so as to maximize the determinant of the FIM associated with the parameters to be estimated. Equivalently, this setting can be perfectly seen as the selection of n activated sensors within the framework of an I-node sensor network mentioned in the Introduction. This formulation will be also adopted in what follows.

Specifically, let x^i, $i = 1, \ldots, I$ denote given spatial positions of sensor network nodes. Now that our design criterion has been established, the problem is to find optimal allocations of n active sensors to x^i, $i = 1, \ldots, I$ during K stages so as to maximize the value of the design criterion incurred by the allocation. In order to formulate this mathematically, introduce for each location x^i variables v_{ki}, $k = 1, \ldots, K$ which take the value 1 or 0 depending on whether a sensor located at x^i is active over the time interval T_k, respectively. The FIM in (8) can then be rewritten as

$$M(v) = \sum_{k=1}^{K} \sum_{i=1}^{I} v_{ki} M_{ki}, \tag{11}$$

where

$$v = (v_1, \ldots, v_K), \tag{12}$$

$$v_k = (v_{k1}, \ldots, v_{kI}), \quad k = 1, \ldots, K \tag{13}$$

and

$$M_{ki} = \int_{T_k} g(x^i, t) g(x^i, t)^\mathsf{T} \, dt. \tag{14}$$

It is straightforward to verify that the $m \times m$ matrices M_{ki} are nonnegative definite and, therefore, so is $M(v)$.

Then our design problem is rephrased as follows:

> *Problem P:* Find a sequence $v = (v_{11}, \ldots, v_{1I}, \ldots, v_{K1}, \ldots, v_{KI}) \in \mathbb{R}^{KI}$ to maximize
>
> $$\mathcal{P}(v) = \log \det(M(v)) \tag{15}$$
>
> subject to the constraints
>
> $$\sum_{i=1}^{I} v_{ki} = n, \qquad k = 1, \ldots, K, \tag{16}$$
>
> $$v_{ki} \in \{0, 1\}, \quad i = 1, \ldots, I, \quad k = 1, \ldots, K. \tag{17}$$

This constitutes a 0–1 integer programming problem which necessitates an ingenious solution. In what follows, we propose to solve it using the branch-and-bound method which is a standard technique to solve integer-programming problems.

3 Branch-and-Bound Search

3.1 Key Idea

The branch-and-bound (BB) constitutes a general algorithmic technique for finding optimal solutions of various optimization problems, especially discrete or combinatorial [7, 25]. If applied carefully, it can lead to algorithms that run reasonably fast on average.

Principally, the BB method is a tree-search algorithm combined with a rule for pruning subtrees. Suppose we wish to maximize an objective function $\mathcal{P}(v)$ over a finite set V of admissible values of the argument v called the feasible region. The BB then progresses by iteratively applying two procedures: branching and bounding. *Branching* starts with smartly covering the feasible region by two or more smaller feasible subregions (ideally, partitioning into disjoint subregions). It is then repeated recursively to each of the subregions until no more division is possible, which leads to a progressively finer partition of V. The consecutively produced subregions naturally generate a tree structure called the BB tree. Its nodes correspond to the constructed subregions, with the feasible set V as the root node and the singleton solutions $\{v\}$, $v \in V$ as terminal nodes. In turn, the core of *bounding* is a fast method of finding upper and lower bounds to the maximum value of the objective function over a feasible subdomain. The idea is to use these bounds to economize computation by eliminating nodes of the BB tree that have no chance of containing an optimal solution. If the upper bound for a subregion V_A from the search tree is lower than the lower bound for any other (previously examined) subregion V_B, then V_A and all its descendant nodes may be safely discarded from the search. This step, termed *pruning*, is usually implemented by maintaining a global variable that records the maximum lower bound encountered among all subregions examined so far. Any node whose upper bound is lower than this value need not be considered further and thereby can be eliminated. It may happen that the lower bound for a node matches its upper bound. That value is then the maximum of the function within the corresponding subregion and the node is said to be solved. The search proceeds until all nodes have been solved or pruned, or until some specified threshold is met between the best solution found and the upper bounds on all unsolved problems.

In what follows, we will use the symbol E to denote the set $\{1, \ldots, K\} \times \{1, \ldots, I\}$ of all possible pairs (k, i) of the indices identifying a scanning stage and a sensor location. Our implementation of BB for Problem P involves the partition of the feasible set

$$
V = \left\{ v = (v_{11}, \ldots, v_{1I}, \ldots, v_{K1}, \ldots, v_{KI}) \mid \right.
$$

$$
\left. \sum_{i=1}^{I} v_{ki} = n, \ k = 1, \ldots, K, \quad v_{ki} = 0 \text{ or } 1, \ \forall (k, i) \in E \right\}, \tag{18}
$$

into subsets. It is customary to select subsets of the form [7]:

$$
V(E_0, E_1) = \{ v \in V \mid v_{ki} = 0, \ \forall (k, i) \in E_0, \ v_{ki} = 1, \ \forall (k, i) \in E_1 \}, \tag{19}
$$

where E_0 and E_1 are disjoint subsets of E. Consequently, $V(E_0, E_1)$ is the subset of V such that sensors are active at the stages and locations with indices in E_1, sensors are dormant at the stages and locations with indices in E_0, and sensors may be active or dormant at the remaining stages and locations.

Each subset $V(E_0, E_1)$ is identified with a node in the BB tree. The key assumption in the BB method is that for every nonterminal node $V(E_0, E_1)$, i.e., the node for which $E_0 \cup E_1 \neq E$, there is an algorithm that determines an upper bound $\bar{\mathcal{P}}(E_0, E_1)$ to the maximum design criterion over $V(E_0, E_1)$, i.e.,

$$
\bar{\mathcal{P}}(E_0, E_1) \geq \max_{v \in V(E_0, E_1)} \mathcal{P}(v), \tag{20}
$$

An Optimal Scanning Sensor Activation Policy 103

and a feasible solution $\underline{v} \in V$ for which $\mathcal{P}(\underline{v})$ can serve as a lower bound to the maximum design criterion over V. We may compute $\bar{\mathcal{P}}(E_0, E_1)$ by solving the following relaxed problem:

Problem $\mathbf{R}(E_0, E_1)$: Find a sequence $\bar{v}$ to maximize (15) subject to the constraints

$$\sum_{i=1}^{I} v_{ki} = n, \qquad k = 1, \dots, K, \tag{21}$$

$$v_{ki} = 0, \quad (k,i) \in E_0, \tag{22}$$

$$v_{ki} = 1, \quad (k,i) \in E_1, \tag{23}$$

$$0 \le v_{ki} \le 1, \quad (k,i) \in E \setminus (E_0 \cup E_1). \tag{24}$$

In Problem $R(E_0, E_1)$ all 0–1 constraints on the variables v_{ki} are relaxed by allowing them to take any value in the interval $[0, 1]$, except that the variables v_{ki}, $(k,i) \in E_0 \cup E_1$ are fixed at either 0 or 1. A simple and efficient method for its solution is given in Sect. 4. As a result of its application, we set $\bar{\mathcal{P}}(E_0, E_1) = \mathcal{P}(\bar{v})$.

As for $\underline{v}$, we can specify it as the best feasible solution (i.e., an element of V) found so far. If no solution has been found yet, we can either set the lower bound to $-\infty$, or use an initial guess about the optimal solution (experience provides evidence that the latter choice leads to much more rapid convergence).

3.2 Branching Rule and BB Algorithm

The result of solving Problem $R(E_0, E_1)$ can serve as a basis to construct a branching rule for the binary BB tree. We adopt here the approach in which the tree node/subset $V(E_0, E_1)$ is expanded (i.e., partitioned) by first picking out all fractional values from among the values of the relaxed variables, and then rounding to 0 and 1 a value which is the most distant from both 0 and 1. Specifically, we apply the following steps:

1. Determine

$$(k_\star, i_\star) = \arg \min_{(k,i) \in E \setminus (E_0 \cup E_1)} |v_{ki} - 0.5|. \tag{25}$$

(In case there are several minimizers, randomly pick one of them.)
2. Partition $V(E_0, E_1)$ into $V(E_0 \cup \{(k_\star, i_\star)\}, E_1)$ and $V(E_0, E_1 \cup \{(k_\star, i_\star)\})$ whereby two descendants of the node in question are defined.

A recursive application of the branching rule starts from the root of the BB tree, which corresponds to the trivial subset $V(\emptyset, \emptyset) = V$ and the fully relaxed problem.

Each node of the BB tree corresponds to a continuous relaxed problem, $R(E_0, E_1)$, while each edge corresponds to fixing one relaxed variable at 0 or 1.

The above scheme has to be complemented with a search strategy to incrementally explore all the nodes of the BB tree. Here we use a common depth-first technique [69,70] which always expands the deepest node in the current fringe of the search tree. The reason behind this decision is that the search proceeds immediately to the deepest level of the search tree, where the nodes have no successors [26]. In this way, lower bounds on the optimal solution can be found or improved as fast as possible.

A recursive version of the resulting depth-first branch-and-bound is implemented in Algorithm 1. The operators involved in this implementation are as follows:

- SINGULARITY-TEST(E_0, E_1) returns true only if expansion of the current node will result in a singular FIM, see Sect. 4.2 for details.
- RELAXED-SOLUTION(E_0, E_1) returns a solution to Problem $R(E_0, E_1)$.
- DET-FIM(v) returns the log-determinant of the FIM corresponding to v.
- INTEGRAL-TEST(v) returns true only if the current solution v is integral.
- INDEX-BRANCH(v) returns the pair of indices defined by (25).

Algorithm 1 A recursive version of the depth-first branch-and-bound method. It uses two global variables, *LOWER* and *v_best*, which are respectively the maximal value of the FIM determinant over feasible solutions found so far and the solution at which it is attained

1: **procedure** RECURSIVE-DFBB(E_0, E_1)
2:     **if** $E_0 \cup E_1 = E$ **then**   ▷ Deepest level of the BB tree has been attained
3:          $det\_v \leftarrow$ DET-FIM($v(E_0, E_1)$)
4:          **if** $det\_v > LOWER$ **then**
5:              $v\_best \leftarrow v(E_0, E_1)$
6:              $LOWER \leftarrow det\_v$
7:          **end if**
8:          **return**
9:     **end if**
10:     **if** SINGULARITY-TEST(E_0, E_1) **then**
11:          **return**   ▷ Only zero determinants can be expected
12:     **end if**
13:     $v\_relaxed \leftarrow$ RELAXED-SOLUTION(E_0, E_1)
14:     $det\_relaxed \leftarrow$ DET-FIM($v\_relaxed$)
15:     **if** $det\_relaxed \leq LOWER$ **then**
16:          **return**   ▷ Pruning
17:     **else if** INTEGRAL-TEST($v\_relaxed$) **then**
18:          $v\_best \leftarrow v\_relaxed$
19:          $LOWER \leftarrow det\_relaxed$
20:          **return**   ▷ Relaxed solution is integral
21:     **else**
22:          $(k_\star, i_\star) \leftarrow$ INDEX-BRANCH($v\_relaxed$)   ▷ Partition into two descendants
23:          RECURSIVE-DFBB($E_0 \cup \{(k_\star, i_\star)\}$, E_1)
24:          RECURSIVE-DFBB(E_0, $E_1 \cup \{(k_\star, i_\star)\}$)
25:     **end if**
26: **end procedure**

4 Simplicial Decomposition for Solving the Relaxed Problem

4.1 Optimality Conditions

The non-leaf nodes of the BB tree are processed by relaxing the original combinatorial problem, which directly leads to Problem $R(E_0, E_1)$. This section provides a detailed exposition of a simplicial decomposition method which is particularly suited for its solution.

For notational convenience, for each $k = 1, \ldots, K$ we define the sets

$$I_1(k) = \left\{ i \mid 1 \leq i \leq I \text{ and } (k, i) \in E_1 \right\}, \tag{26}$$

$$I_0(k) = \left\{ i \mid 1 \leq i \leq I \text{ and } (k, i) \in E_0 \right\}, \tag{27}$$

$$\mathcal{K} = \left\{ k \mid 1 \leq k \leq K \text{ and } | I_1(k) \cup I_0(k)| < I \right\}. \tag{28}$$

Thus, at Stage k, sets $I_1(k)$ and $I_0(k)$ consist of the indices of sites for which the associated variables v_{ki} are fixed at 1 and 0, respectively, and hence cannot be qualified as relaxed, while $\mathcal{K}$ contains the indices of the stages at which there are relaxed variables.

Consider any bijection ς from $\{1, \ldots, L\}$ to $\mathcal{K}$, where $L = |\mathcal{K}|$. For each $l = 1, \ldots, L$, setting

$$r_l = n - | I_1(\varsigma(l))|, \tag{29}$$

$$q_l = I - | I_1(\varsigma(l)) \cup I_0(\varsigma(l))|, \tag{30}$$

we can then construct a bijection ϱ_l from $\{1, \ldots, q_l\}$ to $I \setminus (I_1(\varsigma(l)) \cup I_0(\varsigma(l)))$. Thus, q_l stands for the number of relaxed variables at Stage $\varsigma(l)$ and r_l signifies the number of sensors which still remain to be activated at the same stage.

Accordingly, we can replace the relaxed variables v_{ki}, $(k, i) \in E \setminus (E_0 \cup E_1)$ by w_{lj} such that $w_{lj} = v_{\varsigma(l), \varrho_l(j)}$, $j = 1, \ldots, q_l$ and $l = 1, \ldots, L$. This leads to the following formulation:

Problem $\mathbf{R'}(E_0, E_1)$: Find a sequence

$$w = (w_{1,1}, \ldots, w_{1,q_1}, \ldots, w_{L,1}, \ldots, w_{L,q_L}) \in \mathbb{R}^f \tag{31}$$

where $f = \sum_{l=1}^{L} q_l$, to maximize

$$Q(w) = \log \det\big(G(w)\big) \tag{32}$$

---

subject to the constraints

$$\sum_{j=1}^{q_l} w_{lj} = r_l, \quad l = 1, \dots, L, \tag{33}$$

$$0 \le w_{lj} \le 1, \quad j = 1, \dots, q_l, \quad l = 1, \dots, L, \tag{34}$$

where

$$G(w) = A + \sum_{l=1}^{L} \sum_{j=1}^{q_l} w_{lj} S_{lj}, \qquad A = \sum_{k=1}^{K} \sum_{i \in I_1(k)} M_{ki}, \tag{35}$$

$$S_{lj} = M_{\varsigma(l), \varrho_l(j)}, \quad j = 1, \dots, q_l, \quad l = 1, \dots, \tag{36}$$

(Note that at Stage $\varsigma(l)$ we have that $|I_1(\varsigma(l))|$ sensors have already been activated at locations x^i, $i \in I_1(\varsigma(l))$, and thus a decision about the activation of r_l remaining sensors has to be made.)

In the sequel, W will stand for the set of all vectors w of the form (31) satisfying (33) and (34). Note that it forms a polygon in $\mathbb{R}^f$. Recall that the log-determinant is concave and strictly concave over the cones $\mathbb{S}_+^m$ and $\mathbb{S}_{++}^m$, respectively, cf. [9, 64]. Thus the objective function (32) is concave as the composition of the log-determinant with an affine mapping, see [9, p. 79]. We wish to maximize it over the polyhedral set W. If the FIM corresponding to an optimal solution $w^\star$ is nonsingular, then an intriguing form of the optimality conditions can be derived.

Proposition 1. *Suppose that the matrix $G(w^\star)$ is nonsingular for some $w^\star \in W$. The vector $w^\star$ constitutes a global solution to Problem $R'(E_0, E_1)$ if, and only if, there exist numbers $\lambda_l^\star$, $l = 1, \dots, L$ such that*

$$\varphi(l, j, w^\star) \begin{cases} \ge \lambda_l^\star & \text{if } w_{lj}^\star = 1, \\ = \lambda_l^\star & \text{if } 0 < w_{lj}^\star < 1, \\ \le \lambda_l^\star & \text{if } w_{lj}^\star = 0, \end{cases} \tag{37}$$

where

$$\varphi(l, j, w) = \text{trace}\big[G^{-1}(w) S_{lj}\big], \quad j = 1, \dots, q_l, \quad l = 1, \dots, L. \tag{38}$$

Proposition 1 reveals one characteristic feature of the optimal solutions, namely that, when identifying them, the function φ turns out to be crucial and optimality means separability of the components of $w^\star$ in terms of the values of this

An Optimal Scanning Sensor Activation Policy

function. Specifically, for each $l = 1, \ldots, L$ the values of $\varphi(l, \cdot, w^\star)$ for the indices corresponding to the fractional components of $w_{lj}^\star$, $j = 1, \ldots, q_l$ must be equal to some constant $\lambda_l^\star$, whereas for the components taking the value 0 or the value 1 the values of $\varphi(l, \cdot, w^\star)$ must be no larger and no smaller than $\lambda_l^\star$, respectively.

4.2 Handling Singular Information Matrices

Note that the assumption that $G(w)$ is nonsingular can be dropped, since there is a very simple method to check whether or not the current relaxed problem will lead to a FIM which is nonsingular.

Proposition 2. *The FIM corresponding to the solution to Problem $R'(E_0, E_1)$ is singular if and only if so is $G(\bar{w})$, where*

$$\bar{w} = (\underbrace{r_1/q_1, \ldots, r_1/q_1}_{q_1 \; times}, \ldots, \underbrace{r_L/q_L, \ldots, r_L/q_L}_{q_L \; times}). \tag{39}$$

Consequently, a test of the singularity of $G(\bar{w}) = A + \sum_{l=1}^{L}(r_l/q_l)\sum_{j=1}^{q_l} S_{lj}$ can be built into the BB procedure in order to drop the corresponding node from further considerations and forego the examination of its descendants. Otherwise, the vector (39) may serve as a good starting point for the simplicial decomposition algorithm outlined in what follows.

Remark 1. A solution to Problem $R'(E_0, E_1)$ is not necessarily unique. Note, however, that for nonsingular cases (after all, pruning discards such cases from further consideration), the resulting FIM is unique. Indeed, Problem $R'(E_0, E_1)$ can equivalently be viewed as maximization of the log-determinant over the convex and compact set of matrices $\mathfrak{M} = \{G(w) \mid \sum_{j=1}^{q_l} w_{lj} = r_l, \; 0 \le w_{lj} \le 1, \; j = 1, \ldots, q_l, \; \text{and} \; l = 1, \ldots, L\}$. But the log-determinant is strictly concave over the cone of positive-definite matrices, $\mathbb{S}_{++}^m$, which constitutes the interior of $\mathbb{S}_+^m$ relative to $\mathbb{S}^m$, and this fact implies the unicity of the optimal FIM.

4.3 Simplicial Decomposition

Simplicial Decomposition (SD) constitutes an important class of methods for solving large-scale continuous problems in mathematical programming with convex feasible sets [7, 32, 59]. In the original framework, where a concave objective function is to be maximized over a bounded polyhedron, it iterates by alternately solving a linear programming subproblem (the so-called *column generation problem*) which generates an extreme point of the polyhedron, and a nonlinear *restricted master problem* (RMP) which finds the maximum of the objective function over the

108   D. Uciński

convex hull (a simplex) of previously defined extreme points. This basic strategy of simplicial decomposition has appeared in numerous references [29, 30, 95], where possible improvements and extensions have also been discussed. A principal characteristic of an SD method is that the sequence of successive solutions to the master problem tends to a solution to the original problem in such a way that the objective function strictly monotonically approaches its optimal value.

Problem $R'(E_0, E_1)$ is perfectly suited for the application of the SD scheme. In this case, it boils down to Algorithm 2. Here $\nabla Q(\boldsymbol{w})$ signifies the gradient of Q at $\boldsymbol{w}$, and it is easy to check that

$$\frac{\partial Q(\boldsymbol{w})}{\partial w_{lj}} = \text{trace}(\boldsymbol{G}^{-1}(\boldsymbol{w})\boldsymbol{S}_{lj}) \tag{40}$$

Since we deal with maximization of a concave function Q over a bounded polyhedral set W, the convergence of Algorithm 2 in a finite number of RMP steps is automatically guaranteed [7,32, p. 221]. Observe that Step 3 implements the *column dropping rule* [59], according to which any extreme point with zero weight in the expression of $w^{(\tau)}$ as a convex combination of elements in $Z^{(\tau)}$ is removed. This rule makes the number of elements in successive sets $Z^{(\tau)}$ reasonably low.

Algorithm 2 Algorithm model for simplicial decomposition

Step 0.   (Initialization)
Set
$$\boldsymbol{w}^{(0)} = (\underbrace{r_1/q_1, \ldots, r_1/q_1}_{q_1 \text{ times}}, \ldots, \underbrace{r_L/q_L, \ldots, r_L/q_L}_{q_L \text{ times}}).$$

and $Z^{(0)} = \{\boldsymbol{w}^{(0)}\}$. Select $0 < \epsilon \ll 1$, a parameter used in the stopping rule, and set $\tau = 0$.

Step 1.   (Solution of the column generation subproblem)
Determine
$$z = \arg\max_{\boldsymbol{w} \in W} \nabla Q(\boldsymbol{w}^{(\tau)})^{\mathsf{T}}(\boldsymbol{w} - \boldsymbol{w}^{(\tau)}). \tag{41}$$

Step 2.   (Termination check)
If $\nabla Q(\boldsymbol{w}^{(\tau)})^{\mathsf{T}}(z - \boldsymbol{w}^{(\tau)}) \leq \epsilon$, then STOP and $\boldsymbol{w}^{(\tau)}$ is optimal. Otherwise, set $Z^{(\tau+1)} = Z^{(\tau)} \cup \{z\}$.

Step 3.   (Solution of the restricted master problem)
Find
$$\boldsymbol{w}^{(\tau+1)} = \arg\max_{\boldsymbol{w} \in \text{co}(Z^{(\tau+1)})} Q(\boldsymbol{w}) \tag{42}$$

and purge $Z^{(\tau+1)}$ of all extreme points with zero weight in the expression of $\boldsymbol{w}^{(\tau+1)}$ as a convex combination of elements in $Z^{(\tau+1)}$. Increment τ by one and go back to Step 1.

The SD algorithm may be viewed as a form of modular nonlinear programming, provided that one has an effective computer code for solving the restricted master problem, as well as access to a code which can take advantage of the linearity of the column generation subproblem [30]. The former issue will be addressed in the next subsection, where an extremely simple and efficient multiplicative algorithm for

An Optimal Scanning Sensor Activation Policy   109

weight optimization will be discussed. In turn, the latter issue can be easily settled, as in the linear programming problem of Step 1 the feasible region W is defined by L equality constraints (33) and $\sum_{l=1}^{L} q_l$ bound constraints (34). Note, however, that the positive definiteness of the matrix $G^{-1}(w^{(\tau)})$ and the nonnegative definiteness of S_{lj} taken in conjunction with (40) imply that $\partial Q(w^{(\tau)})/\partial w_{lj} \geq 0$. Hence it is easily seen that the column generation subproblem has an explicit solution where for each $l = 1, \ldots, L$ the relaxed weights w_{lj} corresponding to r_l largest values selected from among $\partial Q(w^{(\tau)})/\partial w_{lj}, j = 1, \ldots, q_l$ are set to unity, the others being zero. Consequently, no specialized linear programming code is needed. This greatly simplifies the implementation and constitutes a clear advantage of the presented approach.

4.4   Solution of the Restricted Master Problem

Suppose that in the $(\tau + 1)$-th iteration of Algorithm 2, we have

$$Z^{(\tau+1)} = \{z^1, \ldots, z^s\}, \tag{43}$$

possibly with $s < \tau + 1$ owing to the built-in deletion mechanism of points in $Z^{(i)}$, $1 \leq i \leq \tau$, which did not contribute to the convex combinations yielding the corresponding iterates $w^{(\ell)}$. Step 3 of Algorithm 2 involves maximization of the design criterion (32) over

$$\mathrm{co}(Z^{(\tau+1)}) = \left\{ \sum_{\ell=1}^{s} \alpha_\ell z^\ell \,\Big|\, \alpha_\ell \geq 0, \ \ell = 1, \ldots, s, \ \sum_{\ell=1}^{s} \alpha_\ell = 1 \right\}. \tag{44}$$

From the representation of any $w \in \mathrm{co}(Z^{(\tau+1)})$ as

$$w = \sum_{\ell=1}^{s} \alpha_\ell z^\ell, \tag{45}$$

or, in component-wise form,

$$w_{lj} = \sum_{\ell=1}^{s} \alpha_\ell z_{lj}^\ell, \quad j = 1, \ldots, q_l, \quad l = 1, \ldots, L, \tag{46}$$

it follows that

$$G(w) = A + \sum_{l=1}^{L} \sum_{j=1}^{q_l} w_{lj} S_{lj} = \sum_{\ell=1}^{s} \alpha_\ell \left(A + \sum_{l=1}^{L} \sum_{j=1}^{q_l} z_{lj}^\ell S_{lj} \right) = \sum_{\ell=1}^{s} \alpha_\ell G(z_\ell). \tag{47}$$

From this, we see that the RMP can equivalently be formulated as the following problem:

Problem $\mathbf{P_{RMP}}$: Find the sequence of weights $\boldsymbol{\alpha} = (\alpha_1, \dots, \alpha_s)$ to maximize

$$\mathcal{T}(\boldsymbol{\alpha}) = \log \det(\boldsymbol{H}(\boldsymbol{\alpha})) \tag{48}$$

subject to the constraints

$$\sum_{\ell=1}^{s} \alpha_\ell = 1, \tag{49}$$

$$\alpha_\ell \geq 0, \quad \ell = 1, \dots, s, \tag{50}$$

where

$$\boldsymbol{H}(\boldsymbol{\alpha}) = \sum_{\ell=1}^{s} \alpha_\ell \boldsymbol{Q}_\ell, \qquad \boldsymbol{Q}_\ell = \boldsymbol{G}(z_\ell). \tag{51}$$

Basically, since the constraints (49) and (50) define the probability simplex P_s in $\mathbb{R}^s$, i.e., a very nice convex feasible domain, it is intuitively appealing to determine optimal weights using a numerical algorithm specialized for solving convex optimization problems. But another, much simpler technique can be employed to suitably guide weight calculation. It fully exploits the specific form of the objective function (48) by giving Problem $\mathrm{P_{RMP}}$ an equivalent probabilistic formulation. Specifically, the nonnegativeness of the weights z_{lj}^ℓ and the nonnegative definiteness of the matrices $\boldsymbol{A}$ and $\boldsymbol{S}_{lj}$ for $j = 1, \dots, q_l$ and $l = 1, \dots, L$, imply that $\boldsymbol{Q}_\ell \succeq 0$, $\ell = 1, \dots, s$. Defining X as a discrete random variable which may take values in the set $\{1, \dots, s\}$ and treating the weights α_ℓ, $\ell = 1, \dots, s$ as the probabilities attached to its possible numerical values, i.e.,

$$p_X(\ell) = \mathbb{P}(X = \ell) = \alpha_\ell, \quad \ell = 1, \dots, s, \tag{52}$$

we can interpret p_X as the probability mass function (pmf) of X and $\boldsymbol{H}(\boldsymbol{\alpha}) = \sum_{\ell=1}^{s} \alpha_\ell \boldsymbol{Q}_\ell$ in (48) as the $\mathbb{P}$-weighted mean of the function $Q : \ell \mapsto \boldsymbol{Q}_\ell$. Therefore, Problem $\mathcal{P}_{\mathrm{RMP}}$ can be thought of as that of finding a probability mass function maximizing the log-determinant of the mean of Q. This formulation has captured close attention in optimum experimental design theory, where various characterizations of optimal solutions and efficient computational schemes have been proposed [3, 24, 98]. In particular, in the case of the D-optimality criterion studied here, we can prove the following conditions for global optimality [81, 92]:

An Optimal Scanning Sensor Activation Policy 111

Proposition 3. *Suppose that the matrix* $H(\alpha^\star)$ *is nonsingular for some* $\alpha^\star \in P_s$. *The vector* $\alpha^\star$ *constitutes a global solution to Problem* P_{RMP} *if and only if*

$$\psi(\ell, \alpha^\star) \begin{cases} = m & \text{if } \alpha_\ell^\star > 0, \\ \leq m & \text{if } \alpha_\ell^\star = 0 \end{cases} \tag{53}$$

for each $\ell = 1, \dots, s$, *where*

$$\psi(\ell, \alpha) = \text{trace}\big[H^{-1}(\alpha)Q_\ell\big], \quad \ell = 1, \dots, s. \tag{54}$$

A very simple and numerically effective sequential procedure was devised and analysed in [60, 71, 76–78] for the case of rank-one matrices Q_ℓ, which was then extended to the general case in [81, p. 62]. Its version adapted to the RMP proceeds as summarized in Algorithm 3. Clear advantages here are ease of implementation and negligible additional memory requirements.

Algorithm 3 Algorithm model for the restricted master problem

---

Step 0: (Initialization)

Select weights $\alpha_\ell^{(0)} > 0, \ell = 1, \dots, s$ which determine the initial pmf $p_{\chi^{(0)}}$ for which we must have $\mathcal{T}(\alpha^{(0)}) > -\infty$, e.g., set $\alpha_\ell^{(0)} = 1/s, \ell = 1, \dots, s$. Choose $0 < \eta \ll 1$, a parameter used in the stopping rule. Set $\kappa = 0$.

Step 1: (Termination check)

If

$$\frac{\psi(\ell, \alpha^{(\kappa)})}{m} < 1 + \eta, \quad \ell = 1, \dots, s, \tag{55}$$

then STOP.

Step 2: (Multiplicative update)

Evaluate

$$\alpha_\ell^{(\kappa+1)} = \alpha_\ell^{(\kappa)} \frac{\psi(\ell, \alpha^{(\kappa)})}{m}, \quad \ell = 1, \dots, s. \tag{56}$$

Increment κ by one and go to Step 1.

The idea is reminiscent of the EM algorithm used for maximum likelihood estimation [43]. The properties of this computational scheme are considered in some detail in [81]. Suffice it to say here that Algorithm 3 is globally convergent regardless of the choice of initial weights (they must only be all nonzero and correspond to a nonsingular FIM). Indeed, we have the following result [81, p. 65]:

Proposition 4. *Assume that* $\{\alpha^{(\kappa)}\}$ *is a sequence of iterates constructed by Algorithm 3. Then the sequence* $\{\mathcal{T}(\alpha^{(\kappa)})\}$ *is monotone increasing and*

$$\lim_{\kappa \to \infty} \mathcal{T}(\alpha^{(\kappa)}) = \max_{\alpha \in P_s} \mathcal{T}(\alpha). \tag{57}$$

112   D. Uciński

The basic scheme of Algorithm 3 can be refined to incorporate various improvements which make convergence much faster. For example, produced solutions often happen to contain many insignificant weights α_ℓ, which results from a limited accuracy of computations and the interruption of Algorithm 3 after a finite number of steps. In practice, these weights may well be disregarded since setting them as zeros and distributing the sum of their values among the remaining weights (so as not to violate (49)) involves a negligible change in the value of the performance measure $\mathcal{T}(\boldsymbol{\alpha}^{(\kappa)})$. The sum of the weights removed can be distributed among the other weights for which $\psi(\ell, \boldsymbol{\alpha}^{(\kappa)}) > m$, and additionally, in a manner proportional to $\psi(\ell, \boldsymbol{\alpha}^{(\kappa)}) - m$.

Another improvement is due to [63] where a simple method was proposed to identify elements of $Z^{(\tau+1)}$ which do not contribute to the sought optimal convex combination in $\mathrm{co}(Z^{(\tau+1)})$. It can be generalized to the general case considered here and used during the search to discard such useless points "on the fly," thereby substantially reducing the problem dimensionality.

5   Numerical Example

Consider the following numerical example serving as a test for the solution techniques proposed in this paper. The point of departure is the two-dimensional diffusion equation

$$\frac{\partial y(\boldsymbol{x},t)}{\partial t} = \frac{\partial}{\partial x_1}\left(\kappa(\boldsymbol{x};\boldsymbol{\theta})\frac{\partial y(\boldsymbol{x},t)}{\partial x_1}\right) + \frac{\partial}{\partial x_2}\left(\kappa(\boldsymbol{x};\boldsymbol{\theta})\frac{\partial y(\boldsymbol{x},t)}{\partial x_2}\right) + u(\boldsymbol{x},t),$$

$$\boldsymbol{x} \in \Omega = (0,1) \times (0,1), \quad t \in (0,1) \tag{58}$$

subject to the conditions

$$y(\boldsymbol{x},0) = 0, \quad \boldsymbol{x} \in \Omega,$$
$$y(\boldsymbol{x},t) = 0, \quad (\boldsymbol{x},t) \in \partial\Omega \times T.$$

The diffusion coefficient to be identified has the form

$$\kappa(\boldsymbol{x};\boldsymbol{\theta}) = \theta_1 + \theta_2 x_1 + \theta_3 x_2. \tag{59}$$

As regards the forcing term in our model, it has the form

$$u(\boldsymbol{x},t) = 20 \exp\left(-50(x_1 - t)^2\right) \tag{60}$$

which mimics the action of a line source whose support is constantly oriented along the x_2-axis and moves with constant speed from the left to the right boundary of Ω. Our purpose is to estimate κ (i.e., the parameters θ_1, θ_2 and θ_3) as accurately

An Optimal Scanning Sensor Activation Policy

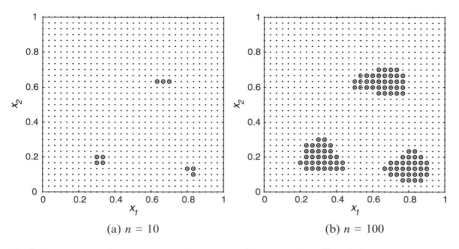

Fig. 1 D-optimal configurations of active sensors without scanning ($K = 1$, $\bar{i} = 900$)

as possible based on the measurements made by n scanning sensors. In the case considered, the D-optimum design criterion was chosen as the measure of the estimation accuracy. A uniform mesh of $I = 30 \times 30 = 900$ points was thus assumed as the set of sites where sensor network nodes are placed (they are marked with dots in Figs. 1–3). Our task consists in selecting the best from them which will be activated.

The determination of the information matrix requires the knowledge of the sensitivity coefficients (9) with components

$$g_1(x,t) = \left.\frac{\partial y(x,t;\vartheta)}{\partial \vartheta_1}\right|_{\vartheta=\theta^0}, \quad g_2(x,t) = \left.\frac{\partial y(x,t;\vartheta)}{\partial \vartheta_2}\right|_{\vartheta=\theta^0},$$

$$g_3(x,t) = \left.\frac{\partial y(x,t;\vartheta)}{\partial \vartheta_3}\right|_{\vartheta=\theta^0},$$

computed at the assumed nominal values $\theta_1^0 = 0.1$, $\theta_2^0 = -0.05$ and $\theta_3^0 = 0.2$. They can be obtained by solving the following system of PDEs:

$$\begin{cases} \dfrac{\partial y}{\partial t} = \nabla \cdot (\kappa \nabla y) + u, \\ \dfrac{\partial g_1}{\partial t} = \nabla \cdot \nabla y + \nabla \cdot (\kappa \nabla g_1), \\ \dfrac{\partial g_2}{\partial t} = \nabla \cdot (x_1 \nabla y) + \nabla \cdot (\kappa \nabla g_2), \\ \dfrac{\partial g_3}{\partial t} = \nabla \cdot (x_2 \nabla y) + \nabla \cdot (\kappa \nabla g_3). \end{cases} \quad (61)$$

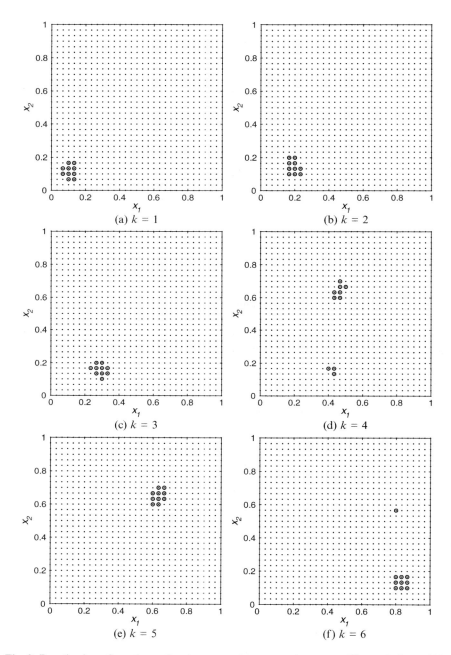

Fig. 2 D-optimal configurations of active sensors at consecutive stages ($K = 6$, $I = 900$, $n = 10$)

An Optimal Scanning Sensor Activation Policy

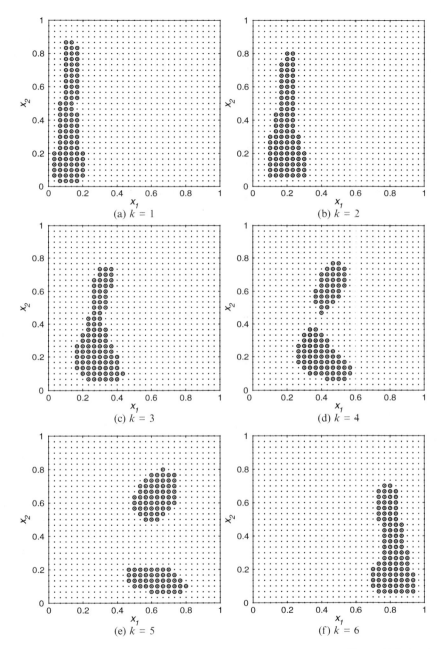

Fig. 3 D-optimal configurations of active sensors at consecutive stages ($K = 6$, $I = 900$, $n = 100$)

The first equation constitutes the original state equation (58) and the second, third and fourth equations result from its differentiation with respect to θ_1, θ_2 and θ_3, respectively. The initial and Dirichlet boundary conditions for all the four equations are homogeneous (i.e., they are zero).

The solution to (61) is found numerically using the MATLAB PDE toolbox. Although it is impossible to employ its handy graphical user interface (it is tailored to single PDEs, and not to systems of PDEs), we can still solve (61) using command-line functions.

First, based on the technique outlined in [92], D-optimal positions of active sensors were determined for the case of $n = 10$ and $n = 100$ sensors to be activated, cf. Fig. 1 (open circles denote the best points for locating measurement sensors). The numbers of feasible solutions are $\binom{900}{10}$ and $\binom{900}{100}$, which exceed 9×10^{22} and 9×10^{134}, respectively. Then, using the approach described in this paper, the observation horizon was decomposed into $K = 6$ stages of equal length. This expanded the sizes of the solutions spaces to $\binom{900}{10}^6 > 5 \times 10^{137}$ and $\binom{900}{100}^6 > 6 \times 10^{809}$, respectively. In spite of that, it was still possible to obtain the optimal solutions displayed in Figs. 2 and 3 in no more than twenty seconds on a low-cost laptop (Pentium IV, 2.40 GHz, 496 MB RAM) running Windows XP Professional and MATLAB 7.1. The parameters required by Algorithms 2 and 3 were set as $\epsilon = 0.01$ and $\eta = 0.01$, respectively. This surprisingly fine performance was possible owing to a good initial lower bound on the optimal solution which permitted to economize computation by eliminating nodes of the BB tree that have no chance of containing an optimal solution. It was obtained by solving a fully relaxed problem and activating sensors which corresponded to the largest weights. Otherwise, the solutions obtained by literally following Algorithm 1 consumed about 15 min of the CPU time, which is not bad, either.

Observe that the values of the diffusion coefficient $\kappa(x)$ in the upper left of Ω are greater than those in the lower right. This means that the state changes during the system evolution are quicker when we move up and to the left (on the other hand, the system would have reached the steady state there earlier). This fact explains the form of the configurations obtained—the sensors switch to new positions so as to follow the moving source while measuring the state in the regions where the DPS is the most sensitive with respect to the unknown parameter κ, finally terminating the movement in the lower right part of the domain Ω.

Finally, remark that, as was expected, the scanning observations substantially influence the attainable values of the D-optimality criterion since the ratios of these values for scanning and stationary cases are 10.075 and 5.438 for $n = 10$ and $n = 100$, respectively.

6 Conclusions

The problem of finding optimal activation policies for sensor network nodes deployed throughout a spatial domain in view of accurate parameter estimation for parameter distributed systems was examined on the assumption that the modes of the node operation can be changed at given time moments, which corresponds to the so-called scanning observations. Ordinarily, the task is reduced to seeking best subsets of locations from among all the possible ones. Numerical algorithms for the construction of optimum sensor configurations by searching over a list of candidate locations customarily involve an iterative improvement of the initial sensor configuration. The combinatorial nature of the problem so formulated implies that with a long list of candidate points, complicated search algorithms can readily consume appreciable computer time and space. In contrast to this approach, a guided search using a branch-and-bound technique was proposed here. This constitutes quite a natural option, but the main problem when trying to implement it has been the lack of a low-cost procedure to obtain upper bounds to the optimal values of the D-optimality criterion. The main contribution of this paper consists in the development of an efficient computational scheme to produce such bounds. This was possible by adapting a specialized multiplicative algorithm for determinant maximization, which is in common use by statisticians concerned with optimum experimental design. The link to plug this algorithm into the proposed scheme was a simplicial decomposition being perfectly suited for large-scale problems which can be encountered here. The idea of its application in the context of D-optimally activating sensor nodes is an extension of the approach set forth in [92]. Consequently, the proposed method can be implemented with great ease and experience provides evidence that, with this tool, even large-scale design problems can be solved using an off-the-shelf PC.

In addition to the algorithmic part of the paper, a characterization of the optimal solutions to specific subproblems discussed in Proposition 1 was obtained. The corresponding proof proceeds here based on the Karush–Kuhn–Tucker conditions.

Let us remark that an alternative approach to select a best n-element subset active sensors from among a given I-element set of all candidate sensors could be to employ an exchange algorithm. Typically, algorithms of this type begin with an n-point starting sensor configuration which then sequentially evolves through addition of new elements selected from among vacant sites and deletion of sites at which sensors have provisionally been planned to reside, in an effort to improve the value of the adopted design criterion [48]. Accordingly, a one-point exchange procedure was used in [91] and further developed in [81, p. 105] in a scanning sensor setting, based on the concept of replication-free designs set forth in [23]. Note, however, that this approach is warranted only when the number of activated sensors is comparatively high. Perhaps an efficient extension of this idea could be to adapt the fast algorithm based on multiple simultaneous exchanges, which was developed in [42]. A step in this direction was made in [46] who refined it and applied the resulting "sort-and-cut" technique to solve an E-optimum sensor selection problem. It is beyond doubt that all these approaches outperform the BB technique proposed here as far as the running time is concerned. On should note, however, that exchange algorithms are only capable of finding globally competitive solutions (i.e., nearly optimal ones), with an explicit trade of global optimality for speed. The approach presented here is superior in the sense that it always produces global maxima and, what is more, does it within tolerable time.

118   D. Uciński

Acknowledgements This work was supported by the Polish Ministry of Science and Higher Education under grant No. N N519 2971 33.

Appendix 1: Proof of Proposition 1

Problem R$'(E_0, E_1)$ can be rewritten as follows: Find $w^\star \in \mathbb{R}^f$ to minimize

$$\widetilde{Q}(w) = -\log \det\big(G(w)\big) \tag{62}$$

subject to the constraints

$$\sum_{j=1}^{q_l} w_{lj} - r_l = 0, \quad l = 1, \dots, L, \tag{63}$$

$$-w_{lj} \leq 0, \quad j = 1, \dots, q_l, \quad l = 1, \dots, L \tag{64}$$

$$w_{lj} - 1 \leq 0, \quad j = 1, \dots, q_l, \quad l = 1, \dots, L. \tag{65}$$

Associating the dual variables

$$\lambda_l \in \mathbb{R}, \quad \mu_{li} \in \mathbb{R}_+, \quad \nu_{lj} \in \mathbb{R}_+, \quad j = 1, \dots, q_l, \quad l = 1, \dots, L \tag{66}$$

with constraints (63)–(65), respectively, we define the Lagrangian of (62)–(65) as

$$\mathcal{L}(w, \lambda, \mu, \nu) = -\log \det(G(w))$$

$$+ \sum_{l=1}^{L} \lambda_l \left(\sum_{j=1}^{q_l} w_{lj} - r_l \right) - \sum_{l=1}^{L} \sum_{j=1}^{q_l} \mu_{lj} w_{lj} + \sum_{l=1}^{L} \sum_{j=1}^{q_l} \nu_{lj} \left(w_{lj} - 1 \right). \tag{67}$$

An easy computation shows that

$$\frac{\partial \mathcal{L}}{\partial w_{lj}} = -\varphi(l, j, w) + \lambda_l - \mu_{lj} + \nu_{lj}. \tag{68}$$

Let us examine the first-order Karush–Kuhn–Tucker (KKT) conditions for our problem [7]:

$$-\varphi(l, j, w) + \lambda_l - \mu_{lj} + \nu_{lj} = 0, \qquad j = 1, \dots, q_l, \quad l = 1, \dots, L, \tag{69}$$

$$\mu_j w_{lj} = 0, \qquad j = 1, \dots, q_l, \quad l = 1, \dots, L, \tag{70}$$

$$\nu_{lj} \left(w_{lj} - 1 \right) = 0, \qquad j = 1, \dots, q_l, \quad l = 1, \dots, L, \tag{71}$$

$$\mu_{lj} \geq 0, \qquad j = 1, \ldots, q_l, \quad l = 1, \ldots, L, \qquad (72)$$

$$v_{lj} \geq 0, \qquad j = 1, \ldots, q_l, \quad l = 1, \ldots, L, \qquad (73)$$

$$0 \leq w_{lj} \leq 1, \qquad j = 1, \ldots, q_l, \quad l = 1, \ldots, L, \qquad (74)$$

$$\sum_{j=1}^{q_l} w_{lj} = r_l, \qquad l = 1, \ldots, L. \qquad (75)$$

For problems with constraints which are both linear and linearly independent, as is the case here, the KKT conditions are necessary for optimality. Additionally, the convexity of the objective function (62) implies that they also become sufficient. Consequently, the optimality of $w^\star$ amounts to the existence of some values of λ_l, μ_{lj} and v_{lj}, $j = 1, \ldots, q_l$ and $l = 1, \ldots, L$, denoted by $\lambda_l^\star$, $\mu_{lj}^\star$ and $v_{lj}^\star$, $j = 1, \ldots, q_l$ and $l = 1, \ldots, L$, respectively, such that (69)–(75) are satisfied.

Suppose that $w_{lj}^\star = 1$ for some pair of indices (l, j). Then from (70) it follows that $\mu_{lj}^\star = 0$ and, therefore, (69) reduces to

$$\varphi(l, j, w^\star) = \lambda_l^\star + v_{lj}^\star \geq \lambda_l^\star, \qquad (76)$$

the last inequality owing to (73). In turn, on account of (71), the assumption $w_{lj}^\star = 0$ yields $v_{lj}^\star = 0$, and then (69) simplifies to

$$\varphi(l, j, w^\star) = \lambda_l^\star - \mu_{lj}^\star \leq \lambda_l^\star, \qquad (77)$$

which is owing to (72). Finally, by (70) and (71), the assumption $0 < w_{lj}^\star < 1$ clearly forces $\mu_{lj}^\star = v_{lj}^\star = 0$, for which (69) gives

$$\varphi(l, j, w^\star) = \lambda_l^\star, \qquad (78)$$

Conversely, having found $w^\star \in \mathbb{R}^f$ and $\lambda_l^\star \in \mathbb{R}, l = 1, \ldots, L$ for which (37) is fulfilled, we can define

$$\mu_{lj}^\star = \max(\lambda_l^\star - \varphi(l, j, w^\star), 0), \quad v_{lj}^\star = \max(\varphi(l, j, w^\star) - \lambda_l^\star, 0),$$

$$j = 1, \ldots, q_l, \quad l = 1, \ldots, L, \qquad (79)$$

which guarantees the satisfaction of (69)–(75). This means that $w^\star$ is a KKT point and this is equivalent to its global optimality.

Appendix 2: Proof of Proposition 2

Setting $c = \max\{q_1/r_1, \ldots, q_L/r_L\}$, observe that the following Löwner ordering holds true for any $w \in \mathbb{R}_+^f$:

$$0 \preceq \boldsymbol{G}(\boldsymbol{w}) = \boldsymbol{A} + \sum_{l=1}^{L}\sum_{j=1}^{q_l} w_{lj}\boldsymbol{S}_{lj} \preceq c\left(\boldsymbol{A} + \sum_{l=1}^{L}\sum_{j=1}^{q_l} \bar{w}_{lj}\boldsymbol{S}_{lj}\right) = c\,\boldsymbol{G}(\bar{\boldsymbol{w}}). \quad (80)$$

A fundamental property of the determinant is that it preserves this monotonicity [33, Corr. 7.7.4], which gives

$$0 \le \det\big(\boldsymbol{G}(\boldsymbol{w})\big) \le c^m \det\big(\boldsymbol{G}(\bar{\boldsymbol{w}})\big). \quad (81)$$

This makes our claim obvious.

References

1. Amouroux, M., Babary, J.P.: Sensor and control location problems. In: M.G. Singh (ed.) Systems & Control Encyclopedia. Theory, Technology, Applications, vol. 6, pp. 4238–4245. Pergamon Press, Oxford (1988)
2. Armstrong, M.: Basic Linear Geostatistics. Springer-Verlag, Berlin (1998)
3. Atkinson, A.C., Donev, A.N., Tobias, R.D.: Optimum Experimental Designs, with SAS. Oxford University Press, Oxford (2007)
4. Banks, H.T., Kunisch, K.: Estimation Techniques for Distributed Parameter Systems. Systems & Control: Foundations & Applications. Birkhäuser, Boston (1989)
5. Banks, H.T., Smith, R.C., Wang, Y.: Smart Material Structures: Modeling, Estimation and Control. Research in Applied Mathematics. Masson, Paris (1996)
6. Bensoussan, A., Da Prato, G., Delfour, M.C., Mitter, S.K.: Representation and Control of Infinite Dimensional Systems, 2nd edn. Birkhäuser, Boston (2007)
7. Bertsekas, D.P.: Nonlinear Programming, 2nd edn. Optimization and Computation Series. Athena Scientific, Belmont, MA (1999)
8. Boer, E.P.J., Hendrix, E.M.T., Rasch, D.A.M.K.: Optimization of monitoring networks for estimation of the semivariance function. In: A.C. Atkinson, P. Hackl, W. Müller (eds.) *mODa 6*, Proc. 6th Int. Workshop on Model-Oriented Data Analysis, Puchberg/Schneeberg, Austria, 2001, pp. 21–28. Physica-Verlag, Heidelberg (2001)
9. Boyd, S., Vandenberghe, L.: Convex Optimization. Cambridge University Press, Cambridge (2004)
10. Cassandras, C.G., Li, W.: Sensor networks and cooperative control. European Journal of Control **11**(4–5), 436–463 (2005)
11. Chong, C.Y., Kumar, S.P.: Sensor networks: Evolution, opportunities, and challenges. Proceedings of the IEEE **91**(8), 1247–1256 (2003)
12. Christofides, P.D.: Nonlinear and Robust Control of PDE Systems: Methods and Applications to Transport-Reaction Processes. Systems & Control: Foundations & Applications. Birkhäuser, Boston (2001)
13. Cook, D., Fedorov, V.: Constrained optimization of experimental design. Statistics **26**, 129–178 (1995)
14. Cressie, N.A.C.: Statistics for Spatial Data, revised edn. John Wiley & Sons, New York (1993)
15. Culler, D., Estrin, D., Srivastava, M.: Overview of sensor networks. IEEE Computer **37**(8), 41–49 (2004)
16. Curtain, R.F., Zwart, H.: An Introduction to Infinite-Dimensional Linear Systems Theory. Texts in Applied Mathematics. Springer-Verlag, New York (1995)
17. Daescu, D.N., Navon, I.M.: Adaptive observations in the context of 4D-Var data assimilation. Meteorology and Atmospheric Physics **85**, 205–226 (2004)

18. Demetriou, M.A.: Detection and containment policy of moving source in 2D diffusion processes using sensor/actuator network. In: Proceedings of the European Control Conference 2007, Kos, Greece, July 2–5 (2006). Published on CD-ROM
19. Demetriou, M.A.: Power management of sensor networks for detection of a moving source in 2-D spatial domains. In: Proceedings of the 2006 American Control Conference, Minneapolis, MN, June 14–16 (2006). Published on CD-ROM
20. Demetriou, M.A.: Process estimation and moving source detection in 2-D diffusion processes by scheduling of sensor networks. In: Proceedings of the 2007 American Control Conference, New York City, USA, July 11–13 (2007). Published on CD-ROM
21. Demetriou, M.A.: Natural consensus filters for second order infinite dimensional systems. Systems & Control Letters **58**(12), 826–833 (2009)
22. Demetriou, M.A., Hussein, I.I.: Estimation of spatially distributed processes using mobile spatially distributed sensor network. SIAM Journal on Control and Optimization **48**(1), 266–291 (2009). DOI 10.1137/060677884
23. Fedorov, V.V.: Optimal design with bounded density: Optimization algorithms of the exchange type. Journal of Statistical Planning and Inference **22**, 1–13 (1989)
24. Fedorov, V.V., Hackl, P.: Model-Oriented Design of Experiments. Lecture Notes in Statistics. Springer-Verlag, New York (1997)
25. Floudas, C.A.: Mixed integer nonlinear programming, MINLP. In: C.A. Floudas, P.M. Pardalos (eds.) Encyclopedia of Optimization, vol. 3, pp. 401–414. Kluwer Academic Publishers, Dordrecht, The Netherlands (2001)
26. Gerdts, M.: Solving mixed-integer optimal control problems by branch&bound: A case study from automobile test-driving with gear shift. Journal of Optimization Theory and Applications **26**, 1–18 (2005)
27. Gevers, M.: Identification for control: From the early achievements to the revival of experiment design. European Journal of Control **11**(4–5), 335–352 (2005)
28. Goodwin, G.C., Payne, R.L.: Dynamic System Identification. Experiment Design and Data Analysis. Mathematics in Science and Engineering. Academic Press, New York (1977)
29. Hearn, D.W., Lawphongpanich, S., Ventura, J.A.: Finiteness in restricted simplicial decomposition. Operations Research Letters **4**(3), 125–130 (1985)
30. Hearn, D.W., Lawphongpanich, S., Ventura, J.A.: Restricted simplicial decomposition: Computation and extensions. Mathematical Programming Study **31**, 99–118 (1987)
31. Hjalmarsson, H.: From experiment design to closed-loop control. Automatica **41**, 393–438 (2005)
32. von Hohenbalken, B.: Simplicial decomposition in nonlinear programming algorithms. Mathematical Programming **13**, 49–68 (1977)
33. Horn, R.A., Johnson, C.R.: Matrix Analysis. Cambridge University Press, Cambridge, UK (1986)
34. Hussein, I.I., Demetriou, M.A.: Estimation of distributed processes using mobile spatially distributed sensors. In: Proceedings of the 2007 American Control Conference, New York, USA, July 11–13 (2007). Published on CD-ROM
35. Jain, N., Agrawal, D.P.: Current trends in wireless sensor network design. International Journal of Distributed Sensor Networks **1**, 101–122 (2005)
36. Jeremić, A., Nehorai, A.: Design of chemical sensor arrays for monitoring disposal sites on the ocean floor. IEEE Transactions on Oceanic Engineering **23**(4), 334–343 (1998)
37. Jeremić, A., Nehorai, A.: Landmine detection and localization using chemical sensor array processing. IEEE Transactions on Signal Processing **48**(5), 1295–1305 (2000)
38. Kammer, D.C.: Sensor placement for on-orbit modal identification and correlation of large space structures. In: Proc. American Control Conf., San Diego, California, 23–25 May 1990, vol. 3, pp. 2984–2990 (1990)
39. Kammer, D.C.: Effects of noise on sensor placement for on-orbit modal identification of large space structures. Transactions of the ASME **114**, 436–443 (1992)
40. Kincaid, R.K., Padula, S.L.: D-optimal designs for sensor and actuator locations. Computers & Operations Research **29**, 701–713 (2002)

41. Kubrusly, C.S., Malebranche, H.: Sensors and controllers location in distributed systems — A survey. Automatica **21**(2), 117–128 (1985)
42. Lam, R.L.H., Welch, W.J., Young, S.S.: Uniform coverage designs for molecule selection. Technometrics **44**(2), 99–109 (2002)
43. Lange, K.: Numerical Analysis for Statisticians. Springer-Verlag, New York (1999)
44. Lasiecka, I., Triggiani, R.: Control Theory for Partial Differential Equations: Continuous and Approximation Theories, *Encyclopedia of Mathematics and Its Applications*, vol. I and II. Cambridge University Press, Cambridge (2000)
45. Le, N.D., Zidek, J.V.: Statistical Analysis of Environmental Space-Time Processes. Springer-Verlag, New York (2006)
46. Liu, C.Q., Ding, Y., Chen, Y.: Optimal coordinate sensor placements for estimating mean and variance components of variation sources. IEE Transactions **37**, 877–889 (2005)
47. Ljung, L.: System Identification: Theory for the User, 2nd edn. Prentice Hall, Upper Saddle River, NJ (1999)
48. Meyer, R.K., Nachtsheim, C.J.: The coordinate-exchange algorithm for constructing exact optimal experimental designs. Technometrics **37**(1), 60–69 (1995)
49. Müller, W.G.: Collecting Spatial Data. Optimum Design of Experiments for Random Fields, 2nd revised edn. Contributions to Statistics. Physica-Verlag, Heidelberg (2001)
50. Munack, A.: Optimal sensor allocation for identification of unknown parameters in a bubble-column loop bioreactor. In: A.V. Balakrishnan, M. Thoma (eds.) Analysis and Optimization of Systems, Part 2, Lecture Notes in Control and Information Sciences, volume 63, pp. 415–433. Springer-Verlag, Berlin (1984)
51. Navon, I.M.: Practical and theoretical aspects of adjoint parameter estimation and identifiability in meteorology and oceanography. Dynamics of Atmospheres and Oceans **27**, 55–79 (1997)
52. Nehorai, A., Porat, B., Paldi, E.: Detection and localization of vapor-emitting sources. IEEE Transactions on Signal Processing **43**(1), 243–253 (1995)
53. Nychka, D., Piegorsch, W.W., Cox, L.H. (eds.): Case Studies in Environmental Statistics. Lecture Notes in Statistics, volume 132. Springer-Verlag, New York (1998)
54. Nychka, D., Saltzman, N.: Design of air-quality monitoring networks. In: D. Nychka, W.W. Piegorsch, L.H. Cox (eds.) Case Studies in Environmental Statistics, Lecture Notes in Statistics, volume 132, pp. 51–76. Springer-Verlag, New York (1998)
55. Ögren, P., Fiorelli, E., Leonard, N.E.: Cooperative control of mobile sensor networks: Adaptive gradient climbing in a distributed environment. IEEE Transactions on Automatic Control **49**(8), 1292–1302 (2004)
56. Omatu, S., Seinfeld, J.H.: Distributed Parameter Systems: Theory and Applications. Oxford Mathematical Monographs. Oxford University Press, New York (1989)
57. Patan, M.: Optimal activation policies for continous scanning observations in parameter estimation of distributed systems. International Journal of Systems Science **37**(11), 763–775 (2006)
58. Patan, M., Patan, K.: Optimal observation strategies for model-based fault detection in distributed systems. International Journal of Control **78**(18), 1497–1510 (2005)
59. Patriksson, M.: Simplicial decomposition algorithms. In: C.A. Floudas, P.M. Pardalos (eds.) Encyclopedia of Optimization, vol. 5, pp. 205–212. Kluwer Academic Publishers, Dordrecht, The Netherlands (2001)
60. Pázman, A.: Foundations of Optimum Experimental Design. Mathematics and Its Applications. D. Reidel Publishing Company, Dordrecht, The Netherlands (1986)
61. Point, N., Vande Wouwer, A., Remy, M.: Practical issues in distributed parameter estimation: Gradient computation and optimal experiment design. Control Engineering Practice **4**(11), 1553–1562 (1996)
62. Porat, B., Nehorai, A.: Localizing vapor-emitting sources by moving sensors. IEEE Transactions on Signal Processing **44**(4), 1018–1021 (1996)
63. Pronzato, L.: Removing non-optimal support points in D-optimum design algorithms. Statistics & Probability Letters **63**, 223–228 (2003)

64. Pukelsheim, F.: Optimal Design of Experiments. Probability and Mathematical Statistics. John Wiley & Sons, New York (1993)
65. Quereshi, Z.H., Ng, T.S., Goodwin, G.C.: Optimum experimental design for identification of distributed parameter systems. International Journal of Control **31**(1), 21–29 (1980)
66. Rafajłowicz, E.: Design of experiments for eigenvalue identification in distributed-parameter systems. International Journal of Control **34**(6), 1079–1094 (1981)
67. Rafajłowicz, E.: Optimal experiment design for identification of linear distributed-parameter systems: Frequency domain approach. IEEE Transactions on Automatic Control **28**(7), 806–808 (1983)
68. Rafajłowicz, E.: Optimum choice of moving sensor trajectories for distributed parameter system identification. International Journal of Control **43**(5), 1441–1451 (1986)
69. Reinefeld, A.: Heuristic search. In: C.A. Floudas, P.M. Pardalos (eds.) Encyclopedia of Optimization, vol. 2, pp. 409–411. Kluwer Academic Publishers, Dordrecht, The Netherlands (2001)
70. Russell, S.J., Norvig, P.: Artificial Intelligence: A Modern Approach, 2nd edn. Pearson Education International, Upper Saddle River, NJ (2003)
71. Silvey, S.D., Titterington, D.M., Torsney, B.: An algorithm for optimal designs on a finite design space. Communications in Statistics — Theory and Methods **14**, 1379–1389 (1978)
72. Sinopoli, B., Sharp, C., Schenato, L., Schaffert, S., Sastry, S.S.: Distributed control applications within sensor networks. Proceedings of the IEEE **91**(8), 1235–1246 (2003)
73. Song, Z., Chen, Y., Sastry, C.R., Tas, N.C.: Optimal Observation for Cyber-physical Systems: A Fisher-information-matrix-based Approach. Springer-Verlag, London (2009)
74. Sun, N.Z.: Inverse Problems in Groundwater Modeling. Theory and Applications of Transport in Porous Media. Kluwer Academic Publishers, Dordrecht, The Netherlands (1994)
75. Titterington, D.M.: Aspects of optimal design in dynamic systems. Technometrics **22**(3), 287–299 (1980)
76. Torsney, B.: Computing optimising distributions with applications in design, estimation and image processing. In: Y. Dodge, V.V. Fedorov, H.P. Wynn (eds.) Optimal Design and Analysis of Experiments, pp. 316–370. Elsevier, Amsterdam (1988)
77. Torsney, B., Mandal, S.: Construction of constrained optimal designs. In: A. Atkinson, B. Bogacka, A. Zhigljavsky (eds.) Optimum Design 2000, chap. 14, pp. 141–152. Kluwer Academic Publishers, Dordrecht, The Netherlands (2001)
78. Torsney, B., Mandal, S.: Multiplicative algorithms for constructing optimizing distributions: Further developments. In: A. Di Bucchianico, H. Läuter, H.P. Wynn (eds.) *mODa 7*, Proc. 7th Int. Workshop on Model-Oriented Data Analysis, Heeze, The Netherlands, 2004, pp. 163–171. Physica-Verlag, Heidelberg (2004)
79. Uciński, D.: Measurement Optimization for Parameter Estimation in Distributed Systems. Technical University Press, Zielona Góra (1999)
80. Uciński, D.: Optimal sensor location for parameter estimation of distributed processes. International Journal of Control **73**(13), 1235–1248 (2000)
81. Uciński, D.: Optimal Measurement Methods for Distributed-Parameter System Identification. CRC Press, Boca Raton, FL (2005)
82. Uciński, D.: D-optimum sensor activity scheduling for distributed parameter systems. In: Preprints of the 15th IFAC Symposium on System Identification, Saint-Malo, France, July 6–8, (2009). Published on CD-ROM
83. Uciński, D., Atkinson, A.C.: Experimental design for time-dependent models with correlated observations. Studies in Nonlinear Dynamics & Econometrics **8**(2) (2004). Article No. 13
84. Uciński, D., Bogacka, B.: T-optimum designs for discrimination between two multivariate dynamic models. Journal of the Royal Statistical Society: Series B (Statistical Methodology) **67**, 3–18 (2005)
85. Uciński, D., Bogacka, B.: A constrained optimum experimental design problem for model discrimination with a continuously varying factor. Journal of Statistical Planning and Inference **137**(12), 4048–4065 (2007)

86. Uciński, D., Chen, Y.: Time-optimal path planning of moving sensors for parameter estimation of distributed systems. In: Proc. 44th IEEE Conference on Decision and Control, and the European Control Conference 2005, Seville, Spain (2005). Published on CD-ROM

87. Uciński, D., Chen, Y.: Sensor motion planning in distributed parameter systems using Turing's measure of conditioning. In: Proc. 45th IEEE Conference on Decision and Control, San Diego, CA (2006). Published on CD-ROM

88. Uciński, D., Demetriou, M.A.: An approach to the optimal scanning measurement problem using optimum experimental design. In: Proc. American Control Conference, Boston, MA (2004). Published on CD-ROM

89. Uciński, D., Demetriou, M.A.: Resource-constrained sensor routing for optimal observation of distributed parameter systems. In: Proc. 18th International Symposium on Mathematical Theory of Networks and Systems, Blacksburg, VA, July 28–August 1 (2008). Published on CD-ROM

90. Uciński, D., Korbicz, J.: Optimal sensor allocation for parameter estimation in distributed systems. Journal of Inverse and Ill-Posed Problems 9(3), 301–317 (2001)

91. Uciński, D., Patan, M.: Optimal location of discrete scanning sensors for parameter estimation of distributed systems. In: Proc. 15th IFAC World Congress, Barcelona, Spain, 22–26 July 2002 (2002). Published on CD-ROM

92. Uciński, D., Patan, M.: D-optimal design of a monitoring network for parameter estimation of distributed systems. Journal of Global Optimization 39, 291–322 (2007)

93. Uspenskii, A.B., Fedorov, V.V.: Computational Aspects of the Least-Squares Method in the Analysis and Design of Regression Experiments. Moscow University Press, Moscow (1975). (In Russian)

94. Vande Wouwer, A., Point, N., Porteman, S., Remy, M.: On a practical criterion for optimal sensor configuration — Application to a fixed-bed reactor. In: Proc. 14th IFAC World Congress, Beijing, China, 5–9 July, 1999, vol. I: Modeling, Identification, Signal Processing II, Adaptive Control, pp. 37–42 (1999)

95. Ventura, J.A., Hearn, D.W.: Restricted simplicial decomposition for convex constrained problems. Mathematical Programming 59, 71–85 (1993)

96. Vogel, C.R.: Computational Methods for Inverse Problems. Frontiers in Applied Mathematics. Society for Industrial and Applied Mathematics, Philadelphia (2002)

97. van de Wal, M., de Jager, B.: A review of methods for input/output selection. Automatica 37, 487–510 (2001)

98. Walter, É., Pronzato, L.: Identification of Parametric Models from Experimental Data. Communications and Control Engineering. Springer-Verlag, Berlin (1997)

99. Zhao, F., Guibas, L.J.: Wireless Sensor Networks: An Information Processing Approach. Morgan Kaufmann Publishers, Amsterdam (2004)

Interaction Between Experiment, Modeling and Simulation of Spatial Aspects in the JAK2/STAT5 Signaling Pathway

Elfriede Friedmann, Andrea C. Pfeifer, Rebecca Neumann, Ursula Klingmüller, and Rolf Rannacher

Abstract In this article, we describe how two different approaches, mathematical modeling and quantitative measurements, can be combined to gain new insights on one of the most extensively studied signal transduction pathways, the Janus kinase (JAK)/signal transducer and activator of transcription (STAT) pathway. We present two data-based models, which describe the intracellular signaling in a single cell where the signal is transduced from the extracellular domain to the nucleus where STATs regulate gene expression. The two models, one without and one with spatial aspects, have been developed to perform modeling based simulations to find out the way of transport of the STATs in the cytoplasm, and whether cell shape plays a role in this pathway. In this paper we will focus on the parameters of the models, which can be measured with specific techniques in different cell lines, how the unknown parameters are estimated, what the limits of these techniques and how accurate the determinations are.

Keywords JAK2/STAT5 signaling pathway • Parameter estimation • Reaction diffusion equations

---

E. Friedmann (✉) · R. Neumann · R. Rannacher
Department of Applied Mathematics, University Heidelberg, Im Neuenheimer Feld 293, 69120 Heidelberg, Germany
e-mail: friedmann@iwr.uni-heidelberg.de; rebecca.neumann@iwr.uni-heidelberg.de; rannacher@iwr.uni-heidelberg.de

A.C. Pfeifer · U. Klingmüller
Division Systems Biology of Signal Transduction, DKFZ-ZMBH Alliance, German Cancer Research Center, Im Neuenheimer Feld 280, 69120 Heidelberg, Germany
e-mail: U.Klingmueller@dkfz-heidelberg.de

H.G. Bock et al. (eds.), *Model Based Parameter Estimation*, Contributions in Mathematical and Computational Sciences 4, DOI 10.1007/978-3-642-30367-8_5, © Springer-Verlag Berlin Heidelberg 2013

1 Biological Question

Fundamental progress in systems biology can only be achieved if experimentalists and theoreticians closely collaborate. Mathematical models cannot be formulated precisely without detailed knowledge of the experiments while complex biological systems can often not be understood fully without mathematical interpretation of the dynamic processes involved.

In multicellular organisms communication between cells is frequently mediated by signal molecules secreted to the extracellular space, which bind to cell surface receptors. The signal has to be transmitted from the extracellular domain of the cell surface receptors to the nucleus and thereby regulates gene expression.

One of the most extensively studied signal transduction pathways is the JAK/STAT pathway [3]. Several members of the signal transducer and activator of transcription (STAT) protein family have been implicated in various cancers. Briefly, after binding of ligand to the receptor two receptor associated Janus kinases (JAK) transphosphorylate each other and subsequently tyrosine phosphorylate the cytoplasmic domain of the receptor. STAT proteins can then bind to the phospho-tyrosine residues via their SH2 domains and are phosphorylated by JAK. Phosphorylated STATs dissociate from the receptor, dimerize, move to the nucleus and regulate transcription of target genes.

How transport from the site of STAT phosphorylation at the plasma membrane to the site of action in the nucleus is mediated is still unclear. Whether STATs freely diffuse through the cytoplasm to reach the nuclear envelope or are actively transported along the cytoskeleton remains a subject of debate. Moreover, it is not known if STATs can in addition be phosphorylated by membrane-bound kinases on endosomes present in the cytosol, which would reduce the distance between the site of phosphorylation and nuclear envelope.

To address these questions, we investigate the erythropoietin (Epo)-regulated JAK2/STAT5 pathway dynamics in two geometrically different cell types. The fibroblast cell line NIH3T3 grows attached to surfaces, forming several tens of microns long and often branched stretches of cytoplasm surrounding an elliptical nucleus of approximately $10 \mu m$ in diameter, resulting in potentially large distances between plasma membrane and nuclear envelope. On the other hand, CFU-E (colony-forming unit erythroid stage) cells are primary cells isolated from mouse embryonic livers. They are precursors of red blood cells, growing in suspension not attached to surfaces, and showing a spherical shape with a diameter of roughly $10 \mu m$ for the whole cell and $8 \mu m$ for the nucleus, resulting in a fairly small cytoplasmic volume and only a short distance from plasma membrane to the nuclear envelope (Fig. 1).

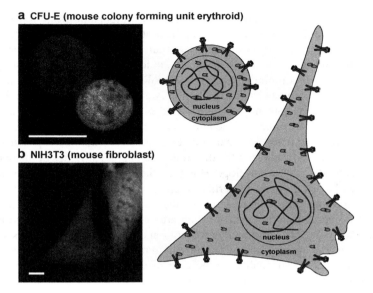

Fig. 1 *Two different cell types used for this study.* (**a**) CFU-E cells and (**b**) NIH3T3-EpoR cell expressing STAT5-GFP imaged on a confocal microscope. Scale bar, 10 μm. Schematic representations of the cells also indicate localization of EpoR (brace like structure) on the plasma membrane with Epo (*black*) bound as well as STAT5 (*small rings*) in the cytoplasm in monomeric and dimeric form as well as bound to the receptor. Nuclear STAT5 is also shown bound to chromatin (*dark grey line*). JAK2 is omitted for simplification. Cells are drawn approximately to scale, proteins are overrepresented

2 Biological Data

A major limitation in systems biology remains the lack of sufficient high-quality, quantitative data for different variables of systems under investigation. To overcome this constraint, we have based our mathematical modeling on experimental data acquired by different experimental techniques, all generating quantitative data of high quality. Nevertheless, the restrictions of each method have to be assessed carefully to avoid misinterpretations of data. In addition, it is advisable to establish standard procedures for cell culture, sample preparation and experimental setup to guarantee comparable results [7].

2.1 Quantitative Immunoblotting Data

To measure dynamic changes in total protein concentrations as well as transient protein modifications such as tyrosine phosphorylation immunoblotting is commonly used. Proteins in cell extracts are separated by gel electrophoresis. Subsequently, they are electrophoretically transferred to a membrane on which they can be detected with antibodies specific for the protein or protein modification. Under certain circumstances it is necessary to purify the protein of interest from the remaining cell extract prior to immunoblotting by immunoprecipitation. This is especially useful, if proteins are only present in the cell at low concentrations or less specific antibodies are used for immunoblotting. For EpoR, JAK2 and STAT5 a combination of immunoprecipitation and immunoblotting is used (Fig. 2a). To reduce the error of the procedure recombinant proteins are used as calibrators [5], [6].This allows us to determine a rough estimate of the number of protein molecules per cell.

When necessary, a crude separation of cytoplasm and nucleoplasm can be achieved by taking advantage of the different detergent sensitivities of the outer cell membrane and the nuclear envelope. However, this biochemical separation is never complete and leads to losses of certain fractions of the compartments that are difficult to determine. Furthermore, large numbers of cells are required if quantitative immunoblotting and immunoprecipitation are combined (on average one million cells per data point). If additionally a separation of cellular compartments, e.g. the cytoplasm and the nucleus, is required the cell number has to be further increased to obtain data of sufficient quality. In the case of established cell lines this usually only poses a handling problem, but if primary cells like CFU-E cells are used, experimentalists are strongly limited by the number of animals that can and should be sacrificed for an experiment. Furthermore, these biochemical methods only yield average data from large cell population. Asynchronous dynamics of single cells can be lost through averaging.

2.2 Live Cell Imaging

For the presented study it was necessary to determine the cell size and shape of CFU-E and NIH3T3 cells in addition to the biochemically measured dynamic changes of protein concentrations and protein modifications. Cellular features such as cell size and shape or more detailed information on protein localization can only be assessed by microscopy, which offers a high spatial resolution at the single cell level.

We have determined the size of CFU-E cells and the CFU-E cell nucleus by transmitted light microscopy, measuring the diameter of the cell and the nucleus and assuming the cell shape to be a perfect sphere. For NIH3T3 cells, cells have been enzymatically detached from their growth surfaces and assumed to adopt spherical shape in suspension and also measured by transmitted light microscopy. Alternatively, NIH3T3 cells expressing fluorescently labeled protein have been imaged by confocal microscopy. For volume measurements by confocal microscopy a series of images have been acquired, moving the sample in z-direction. This image stack covering the whole cell in x, y and z has been analyzed and the volume calculated from the number of voxels. Thus, by expressing fluorescently labeled proteins with specific localizations the sizes of different cellular compartments can be determined.

Interaction Between Experiment, Modeling and Simulation of Spatial Aspects 129

To measure the changes of STAT5 localization over time after addition of Epo, STAT5 labeled with green fluorescent protein (STAT5-GFP) has been followed in single cells by timelapse microscopy (Fig. 2b). This allows one to quantify the ratio of nuclear to cytosolic protein, detect potential concentration inhomogeneities as well as absolute protein concentrations if properly calibrated. The sampling rate that can be achieved is more than one image per minute if necessary. Sampling rate as well as duration of the experiment have to be adjusted depending on the goal of the experiment. In general, again depending on the strength of the signal, 200–500 images can be acquired of a sample, with a sampling rate between less than a second up to hours.

However, a few restrictions have to be taken into account. Live cell imaging requires the expression of fluorescently labeled proteins. This is usually achieved by genetically tagging the protein of interest with a fluorescent protein and expressing it in addition to the unlabeled protein already present in the cells. This can lead to severe alterations in the pathway dynamics and has to be tested carefully. For microscopy, cells have to be immobilized on thin optical glass surfaces. This is easily done with adherent cells like the NIH3T3 cells by growing them in culturing dishes with a glass bottom and directly placing the culturing dish on the microscope, usually equipped with an incubation chamber to maintain the appropriate environment for the cells. In the case of cells growing in suspension such as the CFU-Es imaging of living cells over time is rather challenging and has not been done for this study. In the future, this might be possible as commercial solutions for imaging suspension cells are being developed.

2.3 Measurement of Diffusion and Transport Kinetics

2.3.1 Fluorescence Correlation Spectroscopy

The protein localization in cells in steady state conveys a very static picture. However, many proteins are constantly in rapid movement. Fluorescence Correlation Spectroscopy (FCS) [8] is a widely applied method to measure movement of proteins caused by diffusion or reaction–diffusion systems (Fig. 2d). Fluctuations of fluorescently-labeled proteins within the focal volume of a confocal or two-photon microscope are recorded and analyzed by temporal autocorrelation.

To achieve a high level of fluctuations, the concentration of the observed protein should be kept low. This is often not the case for standard mammalian expression systems but can be accomplished by using an inducible expression system. Alternatively, a large fraction of the fluorescence can be photobleached before the FCS measurement. However, this can cause artefacts due to photodamage caused by the bleaching process. Furthermore, spatial inhomogeneities of the cell as well as photophysical characteristics of the fluorophore can cause difficulties during data acquisition and analysis. This can lead to a fairly high variability of parameters determined from FCS data.

The diffusion coefficient of STAT5-GFP in the cytoplasm of fibroblasts was measured by FCS and determined to be approximately $15\,\mu m^2/s$. No significant difference for the diffusion coefficient of STAT5-GFP in starved cells and cells stimulated by Epo could be detected. For CFU-E cells a similar diffusion coefficient was assumed.

2.3.2 Fluorescence Recovery After Photobleaching

In the case of FCS usually only faster processes such as diffusion or binding kinetics can be studied. In addition, proteins can also be transported from one cellular compartment to another. These protein dynamics in addition to more rapid molecule movements—can be assessed by fluorescence recovery after photobleaching (FRAP). In a typical FRAP experiment, the steady state fluorescence distribution is perturbed by photobleaching part of the fluorescence with a strong laser pulse and the exchange of bleached and unbleached proteins is monitored over time.

Here, we have measured nuclear import and export rates of unphosphorylated STAT5 in NIH3T3 cells ([3] and Fig. 2c). As standard FRAP analysis is only applicable to steady state situations at least on the time scale of the experiment and recovery exchange between nucleus and cytoplasm is on a similar time scale as protein phosphorylation after Epo stimulation only nucleocytoplasmic shuttling of unphosphorylated STAT5 in serum starved cells was measured. FRAP has not been performed in CFU-E cells as this would require stable maintenance of non-adherent cells on the microscope to allow for analysis of the measurement as discussed above.

3 Modeling

In this section we will consider two models describing the same biological processes, one without spatial resolution described by a system of ordinary differential equations (ODE), and the other with partially considered spatial resolution, which is described by a mixed system containing ordinary and partial differential equations (ODE and PDE). Parameter estimation was performed for the ODE model and the same parameter values are used in the combined ODE/PDE model.

3.1 Ordinary Differential Equations Model

In our models, we focus only on a part of the JAK2/STAT5 signaling pathway: after binding of the hormone erythropoietin (Epo) to its receptor (EpoR), STAT5 is phosphorylated at the EpoR by JAK2 with rate r_{act}, dimerizes and diffuses through the cytoplasm to the nucleus. STAT5 is imported into the nucleus with rate r_{imp} and

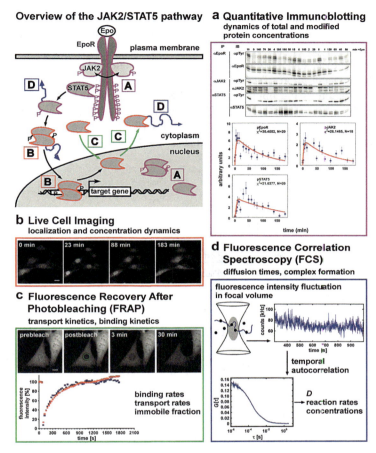

Fig. 2 *Overview of experimental methods.* Several quantitative techniques were combined to generate data for mathematical modeling. An overview of the JAK-STAT signaling pathway is shown in the *top left panel*. *Colored frames* and *arrows* represent the protein species and reactions monitored by the individual techniques. *Bold letter in boxes* refer to the respective techniques. (**a**) Cytoplasmic extracts of NIH3T3-EpoR cells were subjected to immunoprecipitation with antibodies against EpoR and JAK2, or STAT5, and analyzed by quantitative immunoblotting. Samples are randomized to reduce correlated errors. Processed data points (*blue points*) and fits of an ODE model to the data (*red curve*) are shown below. (**b**) Nuclear accumulation of STAT5-GFP in NIH3T3-EpoR cells after addition of Epo is investigated by timelapse microscopy on a confocal microscope. (**c**) Fluorescence recovery after photobleaching (FRAP) is used to analyze nucleocytoplasmic shuttling of STAT5-GFP. Nuclear STAT5-GFP in NIH3T3-EpoR cells was photobleached (*green circle*) and recovery of nuclear fluorescence was observed. Scale bar, 10 μm. Two representative processed data sets are shown for unphosphorylated STAT5-GFP. (**d**) Mobility of STAT5-GFP in the cytoplasm was measured by fluorescence correlation spectroscopy (FCS). Intensity fluctuations in the focal volume (represented in *green*) were analyzed by temporal autocorrelation. Diffusion coefficients were estimated by fitting a one-component model to the data

pSTAT5 with rate r_{imp2}. The processes in the nucleus are described by four linear delay equations including the dephosphorylation of phosphorylated STAT5, which then is exported back to the cytoplasm with rate r_{exp}.

We denote by u_0 the concentration of unphosphorylated STAT5 in the cytoplasm, by u_1 the concentration of the phosphorylated STAT5 in the cytoplasm, by u_2 the concentration of the unphosphorylated STAT5 in the nucleus, and by u_3 the concentration of the phosphorylated STAT5 in the nucleus. The variables $u_4, \ldots, u_7$ are introduced as fictitious concentrations to describe the processes in the nucleus by linear delay equations. Our model consists of following equations:

$$u_0'(t) = -\frac{r_{act}}{v_{cyt}} \cdot pJAK(t) \cdot u_0(t) - \frac{r_{imp}}{v_{cyt}} \cdot u_0(t) + \frac{r_{exp}}{v_{cyt}} \cdot u_2(t) \tag{1}$$

$$u_1'(t) = \frac{r_{act}}{v_{cyt}} pJAK(t) \cdot u_0(t) - \frac{r_{imp2}}{v_{cyt}} \cdot u_1(t) \tag{2}$$

$$u_2'(t) = \frac{r_{delay}}{v_{nuc}} \cdot u_7(t) - \frac{r_{exp}}{v_{nuc}} \cdot u_2(t) + \frac{r_{imp}}{v_{nuc}} \cdot u_0(t) \tag{3}$$

$$u_3'(t) = \frac{r_{imp2}}{v_{nuc}} \cdot u_1(t) - \frac{r_{delay}}{v_{nuc}} \cdot u_3(t) \tag{4}$$

$$u_4'(t) = \frac{r_{delay}}{v_{nuc}} (u_3(t) - u_4(t)) \tag{5}$$

$$u_5'(t) = \frac{r_{delay}}{v_{nuc}} (u_4(t) - u_5(t)) \tag{6}$$

$$u_6'(t) = \frac{r_{delay}}{v_{nuc}} (u_5(t) - u_6(t)) \tag{7}$$

$$u_7'(t) = \frac{r_{delay}}{v_{nuc}} (u_6(t) - u_7(t)). \tag{8}$$

This model is considered for two different cell types: a spherical shaped CFU-E and a NIH3T3 fibroblast cell. Therefore we obtain two sets of parameters, one for each cell type where the initial values are chosen as:

$$u_1(0) = u_3(0) = u_4(0) = u_5(0) = u_6(0) = u_7(0) = 0 \tag{9}$$

$$\text{CFU-E:} \quad u_0(0) = 50 \, \hat{a} \cdot mol/\mu m^3 \tag{10}$$

$$u_2(0) = 18 \, \hat{a} \cdot mol/\mu m^3 \tag{11}$$

$$\text{NIH3T3:} \quad u_0(0) = 16 \, \hat{a} \cdot mol/\mu m^3 \tag{12}$$

$$u_2(0) = 20 \, \hat{a} \cdot mol/\mu m^3 \tag{13}$$

The number of molecules per compartment is determined using a combination of immunoprecipitation and immunoblotting as described in Sect. 2.1 Here, $\hat{a}$ is the Avogadro constant, so that the unit $\hat{a} \cdot mol$ is the number of molecules.

Interaction Between Experiment, Modeling and Simulation of Spatial Aspects 133

The parameters v_{cyt} and v_{nuc}, which represent the average volume of the cytoplasm and nucleus, are measured by transmitted light microscopy (Sect. 2.2). We use the values $v_{cyt} = 429\,\mu m^3$, $v_{nuc} = 268\,\mu m^3$ for the CFU-E cell and $v_{cyt} = 1,758\,\mu m^3$, $v_{nuc} = 366\,\mu m^3$ for the NIH3T3 cell. The nuclear import and export rates can be measured only for the unphosphorylated STAT5 in NIH3T3 cells by FRAP experiments (Sect. 2.3.2): $r_{imp} \approx 0,1\,s^{-1}$ and $r_{exp} \approx 0,09598\,s^{-1}$. The import rate of the unphosphorylated STAT5 in the CFU-E cells is assumed to be approximately the same as in the NIH3T3 cells, so that the same value for r_{imp} ($\approx 0,1\,s^{-1}$ is used for both cell types in our model.

Another input function in this model is the phosphorylation function $pJAK(t)$, the evolution in time of the activated cytoplasmic domain of the receptor. In the experiments the STAT5 molecules are phosphorylated through a controlled input of Epo that activates the receptor-associated kinase JAK2, which then phosphorylates STAT5. The biological processes at the receptor can be modeled as an additional receptor module, which makes our systems of equations and the parameter set much larger. Since the concentration of phosphorylated JAK2 molecules can be measured, we simplify our model and use the data points for pJAK2 interpolated with a smooth spline as input (Fig. 3). For our model, we assume that the number of JAK2 molecules at the plasma membrane is equal to the number of EpoR molecules on the plasma membrane.

Our system of equations (1)–(13) is a homogenous linear system for the unknown $u = (u_0, .., u_7)$, which can be formulated in a general form as follows:

$$u'(t) = f(t, u(t))$$
$$u(0) = b. \tag{14}$$

Theorem 1. *For the given set of data the system (14) is well-posed.*

The proof of this theorem is based on standard techniques. The right hand side $f(t, u(t))$ is Lipschitz, so that we can apply Picard–Lindelöf's theorem and get local existence, uniqueness and Lipschitz continuity with respect to data. This local result can then be extended to a global one due to the linearity of $f(t, u(t))$. The function $f(t, u(t))$ can be written as $f(t, u(t)) = A(t)u(t)$, where $A(t)$ is negative definite for the considered parameters, which gives global (even exponential) stability. In a fothcoming paper we will show the non-negativity, boundedness and stability of the solution.

3.2 Parameter Estimation

Parameters that can not be experimentally measured for our model for the CFU-E cell are the phosphorylation rate of STAT5 r_{act}, its export rate r_{exp}, the import rate of phosphorylated STAT5 r_{imp2}, and the time delay for the processes in the nucleus r_{delay}. For the model in a NIH3T3 cell we have only three unknown parameters:

	CFU-E	CFU-E min	CFU-E max
r_{act} (min^{-1})	11	9	15
r_{imp2} (min^{-1})	58	47	71
r_{exp} (min^{-1})	225	194	260
r_{delay} (min^{-1})	265	263	266

Table 1 Deviation rates of the parameters used in the model for the CFU-E cell

	NIH3T3	NIH3T3 min	NIH3T3 max
r_{act} (min^{-1})	187	171	204
r_{imp2} (min^{-1})	1010	886	1151
r_{delay} (min^{-1})	194	162	232

Table 2 Deviation rates of the parameters used in the model for the NIH3T3 cell

r_{act}, r_{imp2} and r_{delay}. Due to the FRAP experiments described in Sect. 2.3.2 the export rate of Stat5 r_{exp} has been measured. To determine the unknown parameters, parameter estimation has been performed using the software PottersWheel [1], developed to perform data-based modeling of partially observed and noisy systems like signal transduction pathways.

For each cell type two series of experiments are performed. For the CFU-E cells the observables in both experiments are: the phosphorylation function pJAK(t), the phosphorylated STAT5 $u_1(t)$, the total STAT5 $u_0(t) + u_1(t)$ in the cytoplasm, and in the nucleus $u_2(t) + u_3(t)$. In the experiments with the NIH3T3 cells either the phosphorylation function pJAK(t), the phosphorylated STAT5 in the cytoplasm $u_1(t)$, the phosphorylated STAT5 in the nucleus $u_3(t)$ and the total STAT5 in the nucleus $u_2(t) + u_3(t)$ are observed, or the phosphorylation function pJAK(t), the un- and phosphorylated STAT5 separately in the cytoplasm $u_0(t)$, $u_1(t)$ and in the nucleus $u_2(t)$, $u_3(t)$. The data for the parameter estimation have been generated by quantitative immunoblotting where the specific components have been identified via antibodies. For an interpretation of the errors of biochemical data see [5], [6].

The parameter estimation process belongs to the class of non-linear least square problems. The merit function χ^2, which is optimized in PottersWheel to fit the model $y = y(t; p)$, is given by

$$\chi^2(p) = \Sigma_{i=1}^{N} \left(\frac{y_i - y(t_i; p)}{\sigma_i} \right)^2. \tag{15}$$

Here y_i are data point i with standard deviation σ_i and $y(t_i; p)$ being the model value at time point i for parameter values p. As the measurement errors are normally distributed, the minimization of the weighted least-square error corresponds to applying a Maximum Likelihood estimator for the unknown parameters [1]. For the fitting one can choose between powerful deterministic and stochastic optimization algorithms. Our data have been fitted 500 times in a fit sequence using the same dataset where the initial value in the consequent fit is chosen as the best fit in the previous sequence with randomly disturbed parameters. Only the best 30% have been analyzed further. The results are listed in Tables 1 and 2 (first column).

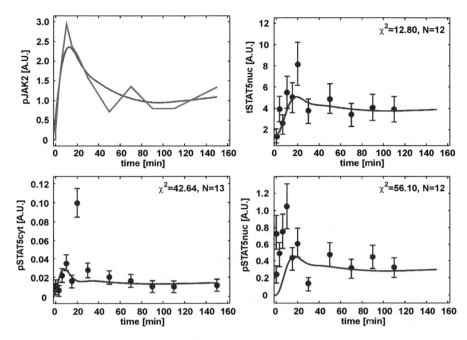

Fig. 3 *Data-based ODE modeling of STAT5 phosphorylation in NIH3T3-EpoR cells.* NIH3T3-EpoR cells were serum-starved for 5 h and stimulated with Epo. Cells were fractionated in nuclear and cytoplasmic extracts. Extracts were subjected to immunoprecipitation with antibodies against EpoR and JAK2 (cytoplasm only), or STAT5 (cytoplasm and nucleus), and analyzed by quantitative immunoblotting. Processed data points (*blue points*) and fits of the mathematical model to the data (*red curves*) are shown above. A spline through the data points for phosphorylated JAK2 (*upper left panel*) was used as input

For the interpretation of the simulation results it is important to determine the relative deviation of the parameters: For the NIH3T3 fibroblast cell the relative deviation for the parameter describing the phosphorylation of STAT5 r_{act} is less than 10% and up to 20% for the delay time r_{delay} in the delay reactions considered. For the parameter set of the CFU-E cell the smallest relative deviation (0.61%) is obtained for the export rate of unphosphorylated STAT5 r_{exp} and the largest relative deviation (27%) is obtained for the phosphorylation rate r_{act}. The relative deviation of the time delay remains approximately the same for both models. How these fits could be further improved is discussed in Sect. 5.

The graph in Fig. 3 represents an exemplary dataset from NIH3T3 cells. Multi-experiment fitting was performed with PottersWheel.

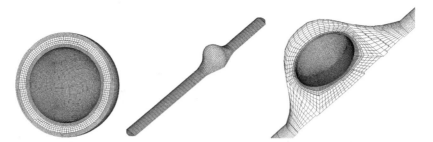

Fig. 4 Constructed grid for the CFU-E cell, NIH3T3 cell and a section through the nucleus of the NIH3T3 cell

3.3 Reaction Diffusion Model

To address the biological question described above, about the way of transport of the STAT5 molecules in the cytoplasm, we add diffusion for unphosphorylated and phosphorylated STAT5 in the cytoplasm. Different transport processes can be modeled by diffusion equations involving different diffusion coefficients. At first, we model free diffusion using a constant diffusion coefficient based on measurements by fluorescence correlation spectroscopy (FCS, Sect. 2.3.1). The diffusion coefficient $D = 15\,\mu m^2/s = 900\,\mu m^2/min$ is used in the simulations. The additional transport of the molecules along the microtubules is modeled for the NIH3T3 cell through an anisotropic diffusion coefficient whereas the mainstream direction of STAT5 movement is set in the y-direction of the cell.

To model the reaction diffusion equations, we consider two different geometries, spherical for a CFU-E cell and a long tube with an ellipsoidal body for a NIH3T3 cell (Fig. 4). To answer the biological question only the cytoplasm has to be dissolved spatially. The processes in the nucleus such as DNA binding and dephosphorylation of STAT5 do not have to be known in detail. For their description it is sufficient to use time delays as black box elements. As already described in the previous section, phosphorylation as well as nuclear import and export of STAT5 only occurs on the boundary of our considered computational domain, the cytoplasm. For this specific question, we therefore obtain a mixed system of differential equations: two diffusion equations with Robin boundary conditions and six ODE, two of them are coupled to the PDEs through the import terms and the other four describe the processes in the nucleus by linear delay equations:

Cytoplasm: Ω_{cyt}

$$\partial_t u_0(t,x) = D\Delta u_0(t,x) \tag{16}$$

$$\partial_t u_1(t,x) = D\Delta u_1(t,x) \tag{17}$$

Nucleus: Ω_{nuc}

$$u_2'(t) = \frac{r_{\text{delay}}}{v_{\text{nuc}}} \cdot u_7(t) - \frac{r_{\text{exp}}}{v_{\text{nuc}}} \cdot u_2(t) + \frac{r_{\text{imp}}}{v_{\text{nuc}}} \cdot \frac{1}{|\partial\Omega_{\text{nuc}}|} \cdot \int_{\partial\Omega_{nuc}} u_0(t,s)\,\mathrm{d}s \quad (18)$$

$$u_3'(t) = \frac{r_{\text{imp2}}}{v_{\text{nuc}}} \cdot \frac{1}{|\partial\Omega_{\text{nuc}}|} \cdot \int_{\partial\Omega_{nuc}} u_1(t,s)\,\mathrm{d}s - \frac{r_{\text{delay}}}{v_{\text{nuc}}} \cdot u_3(t) \quad (19)$$

$$u_4'(t) = \frac{r_{\text{delay}}}{v_{\text{nuc}}}(u_3(t) - u_4(t)) \quad (20)$$

$$u_5'(t) = \frac{r_{\text{delay}}}{v_{\text{nuc}}}(u_4(t) - u_5(t)) \quad (21)$$

$$u_6'(t) = \frac{r_{\text{delay}}}{v_{\text{nuc}}}(u_5(t) - u_6(t)) \quad (22)$$

$$u_7'(t) = \frac{r_{\text{delay}}}{v_{\text{nuc}}}(u_6(t) - u_7(t)). \quad (23)$$

The initial conditions are the same as used in the previous section. We only have to pay attention that the concentrations of unphosphorylated and phosphorylated STAT molecules in the cytoplasm are now space dependent ($u_0(t,x)$ and $u_1(t,x)$). The phosphorylation, import and export of molecules enter through the Robin boundary conditions:

$$D\partial_n u_0(t, \partial\Omega_{\text{cyt}}) = -\frac{r_{\text{act}}}{|\partial\Omega_{\text{cyt}}|} \cdot \text{pJAK}(t,x) \cdot u_0(t,x) \quad (24)$$

$$D\partial_n u_0(t, \partial\Omega_{\text{nuc}}) = -\frac{r_{\text{imp}}}{|\partial\Omega_{\text{nuc}}|} \cdot u_0(t,x) + \frac{r_{\text{exp}}}{|\partial\Omega_{\text{nuc}}|} \cdot u_2(t) \quad (25)$$

$$D\partial_n u_1(t, \partial\Omega_{\text{cyt}}) = \frac{r_{\text{act}}}{|\partial\Omega_{\text{cyt}}|} \cdot \text{pJAK}(t,x) \cdot u_0(t,x) \quad (26)$$

$$D\partial_n u_1(t, \partial\Omega_{\text{nuc}}) = -\frac{r_{\text{imp2}}}{|\partial\Omega_{\text{nuc}}|} \cdot u_1(t,x), \quad (27)$$

where $\partial\Omega_{\text{cyt}}$ represents the outer boundary of the cell, i.e. the membrane and $\partial\Omega_{\text{nuc}}$ the boundary of the nucleus.

At this point it is important to note that the diffusion coefficient is related to the measured parameters. The input curve for the phosphorylation of the STAT5 molecules (pJAK(t)) must be recalculated in a form of concentration using the number of receptors of the specific cell type. For the CFU-E cell, we have:

$$\text{pJAK}(t,x) = \begin{cases} 1.244 \cdot \text{pJAK}(t) & \text{for } x \in \partial\Omega_{cyt} \\ 0 & \text{elsewhere} \end{cases}$$

and for the NIH3T3 cell:

$$\text{pJAK}(t, x) = \begin{cases} 5.65 \cdot \text{pJAK}(t) & \text{for } x \in \partial\Omega_{cyt} \\ 0 & \text{elsewhere.} \end{cases}$$

In this input the number of receptor molecules on the cell surface is hidden. For CFU-E cells the maximum number of ligand binding sites, i.e. available receptor dimers at the cell surface, has been reported to be approximately 1,000 [9] yielding 2,000 receptor molecules at the surface of a CFU-E cell. In NIH3T3 cells we overexpress the receptor, presumably generating a higher density of receptor at the cell surface. As the number of cell surface receptors is difficult to measure we have assumed that the receptor density is similar to that in another cell line with overexpressed receptor (BaF3-EpoR, V. Becker, personal communication) and have assumed 28,000 receptor molecules at the surface of a NIH3T3 cell for our model.

For the reaction diffusion system (16)–(27), we have to introduce the corresponding functional spaces, which will assure the solvability:

Definition 1. Let X denote a real Banach space with Norm $|| \; ||$. Ω is an open and bounded subset in $I R^n$. The space $L^p((0, T), X)$ consists of all strongly measurable functions $u : [0, T] \to X$ with

$$||u||_{L^p((0,T),X)} := \left(\int_0^T ||u(t)||^p dt \right)^{\frac{1}{p}} < \infty$$

for $1 \le p < \infty$, and

$$||u||_{L^\infty((0,T),X)} := ess \; sup||u(t)|| < \infty.$$

For our system $X = H_0^1(\Omega)$, H^{-1} is the dual space of X, and we have

$$H_0^1(\Omega) \subset L^2(\Omega) \subset H^{-1}(\Omega).$$

The well-posedness is shown by following theorem:

Theorem 2. *For the given set of data the initial-boundary value problem* (16)–(27) *is well-posed. There is a unique global solution* $u = (u_0, ..., u_7)$, *with* $u_0, u_1 \in L^2((0, T); H^1(\Omega))$, $\partial_t u_0, \partial_t u_1 \in L^2((0, T); H^{-1}(\Omega))$, *and* $u_2, ..., u_7 \in C^1[0, T]$ *on any time interval* $[0.T]$.

The proof of this theorem can be done in a standard way by decoupling the equations for u_i, $i = 0, 1, 3, ..., 7$ and applying the Banach fixpoint theorem to u_2. It can be found in [2] and will be addressed in a forthcoming paper together with some additional properties of the solutions for generalized nonlinear mixed systems resulting from signaling.

Interaction Between Experiment, Modeling and Simulation of Spatial Aspects

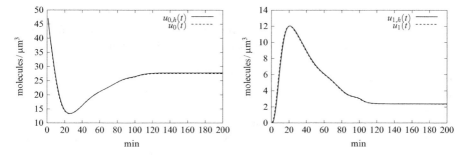

Fig. 5 The solution $u_0(t)$ (unphosphorylated STAT5) of system (1)–(13) in comparison to the solution $u_{0,h}(t) = \int_{\Omega_{cyt}} u_0(t,x)dx$ of system (16)–(27) (*left*) and the solution $u_1(t)$ (phosphorylated STAT5) of system (1)–(13) in comparison to the solution $u_{1,h}(t) = \int_{\Omega_{cyt}} u_1(t,x)dx$ of system (16)–(27) (*right*) in a CFU-E cell

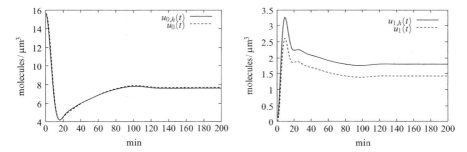

Fig. 6 The solution $u_0(t)$ (unphosphorylated STAT5) of system (1)–(13) in comparison to the solution $u_{0,h}(t) = \int_{\Omega_{cyt}} u_0(t,x)dx$ of system (16)–(27) (*left*) and the solution $u_1(t)$ (phosphorylated STAT5) of system (1)–(13) in comparison to the solution $u_{1,h}(t) = \int_{\Omega_{cyt}} u_1(t,x)dx$ of system (16)–(27) (*right*) in a NIH3T3-cell

The numerical simulations use the in house Finite Element software Gascoigne [10], which provides tools for grid generation, discretization by bilinear Finite Elements (Q_1), timestepping by the Crank–Nicolson, and solution of linear systems by a multigrid algorithm.

4 Results

The results of the simulations of system (1)–(13), dotted line, and system (16)–(27), solid line, are plotted for CFU-E and NIH3T3 in Figs. 5 and 6, respectively. There, we visualized the differences between the time distribution of the concentration of unphosphorylated STAT5 and phosphorylated STAT5 from the system without diffusion with $u_{i,h}(t) = \int_{\Omega_{cyt}} u_i(t,x)dx$, $i = 0,1$, where $u_0(t,x), u_1(t,x)$ are solutions of the system with diffusion. In CFU-E cells no effect of diffusion is observed on the time distribution of phosphorylated STAT5 concentration. There, the distance from the cell membrane to the nucleus is very short (Fig. 1). Therefore, addition of diffusion of cytoplasmatic STAT5 does not alter the results. The model (1)–(13) describes the considered pathway in the CFU-E cells very well (Fig. 5). In NIH3T3 cells adding diffusion of cytoplasmatic STAT5 cause a higher time distribution of the concentration of phosphorylated STAT5 in the cytoplasm ($u_1(t)$) as well as a different steady state (Fig. 6). This is due to the long distance, which the molecules have to travel, to get from the membrane to the nucleus. The higher concentration of phosphorylated STAT5 in the cytoplasm is compensated by a lower concentration of phosphorylated STAT5 in the nucleus, which indicates that the addition of diffusion does not change the amount of phosphorylated STAT5 in the whole cell.

For NIH3T3 cells the discrepancy between the model with diffusion and without diffusion is small (smaller than 1 molecule/μm^3, Fig. 6). To answer the biological question, we simulated various transport processes considering the same parameters in each model (data not shown). Each transport process shows a distinct concentration development during time but the same dynamics due to the linearity of the system. Only because of fast diffusion the differences are very small. Also, the cell geometry influences the temporal development of the phosphorylated STAT5 molecules concentration in time. Here, the differences are more visible, even with fast diffusion (Fig. 6).

5 Perspectives

A meaningful interpretation of the simulation results can be obtained only after analyzing the reliability of the parameters. Sensitivity analysis is used to determine how sensitive a mathematical model is with respect to changes in its parameters and structure. We perform a series of tests, in which we choose different parameter values from our fit, to see how changes in parameters cause changes in the dynamical behavior of the model. The model responds to changes in the parameters is shown in Fig. 7. There, the *absolute* deviations are presented of the concentrations of STAT5 from those of pSTAT5 calculated with the specific parameters. The calculation of the *relative* deviation of the concentrations lets us conclude that their effect on the concentration parameters is of the same order or even larger than the effect of the diffusion. The level of accuracy of our parameters is not sufficient. Thus the question, which transport process is used by the STAT5 molecules to shuttle to the nucleus, remains open.

Additionally, the limitations of the experimental techniques make the interpretation of the results difficult. Simulation results and experimental results are not always directly comparable as illustrated by the biochemical experiments.

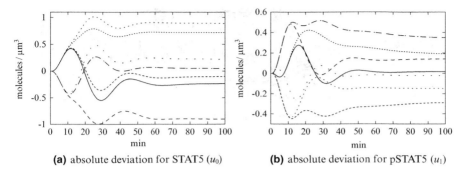

Fig. 7 Sensitivity analysis as a series of tests in which we set different parameter values to see how a change in the parameter causes a change in the dynamic behavior of our model. In this figure we see the absolute deviation between the concentrations of the un- and phosphorylated STAT5 in the cytoplasm of a NIH3T3 cell with rates from Table 2

From the simulation results, we obtain the concentrations of the involved pathway components in a single cell whereas the biochemical data is generated from a large cell population and is affected by losses of cellular components due to the biochemical purification. This yields only averaged, relative concentrations that are difficult to convert to absolute numbers. A direct comparison of the observables from experiments with the ones from simulations is therefore not possible here.

The detailed study of the two models give us some indications under which conditions an answer of the biological question might be possible. For similar linear models in signal transduction, we can conclude for which geometry and diffusion coefficient diffusion may play an important role in the dynamics of the observed pathway. First steps towards such experimental design have been done here: For the considered pathway, we have to reduce the relative deviation of the model parameters ϑ in order to better answer the biological question. One possibility is to add the observable $y(t, \vartheta)$,

$$y(t, \vartheta) = \frac{u_2(t, \vartheta) + u_3(t, \vartheta)}{u_0(t, \vartheta) + u_1(t, \vartheta)},$$

from live cell imaging experiments (Fig. 8) to the estimation of the parameters. In microscopy, a better time resolution can be easily achieved, which should have a positive effect on the relative deviation of the parameters. Recently, spatially resolved diffusion times have been measured for fluorescently labeled molecules in cells by diffusion imaging microscopy [4]. Similar measurements could be performed to investigate possible anisotropic diffusion of STAT5. Experimental design could be further optimized to determine which measurements should be done in order to obtain better results. Furthermore, new biological questions could arise from the observed model behavior, e.g., is STAT5 phosphorylated on the appendices of fibroblasts and are there active Epo receptors at all?

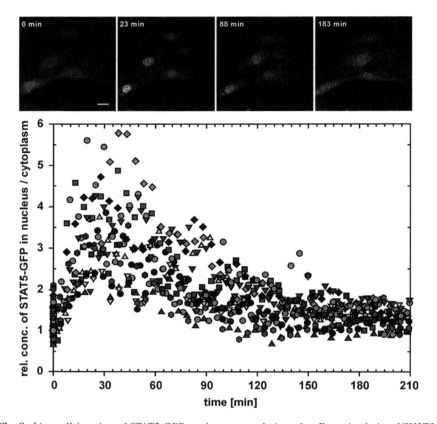

Fig. 8 *Live cell imaging of STAT5-GFP nuclear accumulation after Epo stimulation.* NIH3T3-EpoR cells expressing STAT5-GFP were serum-starved for 5 h and transferred to the microscope. After addition of Epo, cells were imaged on a confocal microscope for a minimum of 4 h at 37°C. The maximum of nuclear accumulation of STAT5 was observed around 23 min. Scale bar, 20 μm. The graph shows the ratio of total nuclear STAT5-GFP to total cytoplasmic STAT-GFP after stimulation with Epo. Different symbols and grey scales represent individual cells

In this paper our aim is to emphasize the importance of interdisciplinary research on the example of the JAK2/STAT5 signaling pathway. High-quality quantitative measurements (Fig. 2) together with mathematical modeling, computational simulations, parameter estimation and experimental design helps to decide whether and which scientific question can be answered based on existing laboratory constraints.

Acknowledgements This work was supported by the Helmholtz Alliance on Systems Biology (SBCancer, Submodule V.7). ACP was supported by the German Federal Ministry of Education and Research (BMBF) grant FORSYS-ViroQuant (#0313923). We thank Julie Bachmann for providing biochemical data for the CFU-E cells and V. Becker for personal communication.

References

1. Maiwald, T. and Timmer, J., *Dynamical modeling and multi-experiment fitting with Potters-Wheel*, Bioinformatics 24(18):2037–43, 2008.
2. Neumann, R., Räumliche Aspekte in der Signaltransduktion, *Diplomarbeit* Ruprecht-Karls-Universität Heidelberg, 2009.
3. Pfeifer A.C., Kaschek D., Bachmann J., Klingmüller U. and Timmer J., *Model-based extension of high-throughput to high-content data*, BMC Syst Biol., 4:106 2010.
4. Roth C. M., Heinlein P. I., Heilemann M., and Herten D.-P., *Imaging Diffusion in Living Cells Using Time-Correlated Single-Photon Counting*, Anal. Chem., 79 (19), pp 7340–7345, 2007.
5. Schilling M., Maiwald T., Bohl S., Kollmann M., Kreutz C., Timmer J. and Klingmüller U., *Computational processing and error reduction strategies for standardized quantitative data in biological networks*, FEBS J. 272(24):6400–11, 2005.
6. Schilling M., Maiwald T., Bohl S., Kollmann M., Kreutz C., Timmer J. and Klingmüller U., *Quantitative data generation for systems biology: the impact of randomisation, calibrators and normalisers*, Syst Biol 2005, 152(4):193–200, 2005.
7. Schilling M., Pfeifer A.C., Bohl S. and Klingmüller U., *Standardizing experimental protocols*, Curr Opin Biotechnol. 2008.
8. Schwille P. and Haustein E., *Fluorescence Correlation Spectroscopy. An Introduction to its Concepts and Applications*, Annu Rev Biophys Biomol Struct. 36:151–69, 2007.
9. Youssoufian H., Longmore G., Neumann D., Yoshimura A. and Lodish H.F., *Structure, function, and activation of the erythropoietin receptor*, Blood. May 1;81(9):2223–36, 1993.
10. *High Perfomance Adaptive Finite Element Toolkit Gascoigne*, http://gascoigne.uni-hd.de.

The Importance and Challenges of Bayesian Parameter Learning in Systems Biology

Johanna Mazur and Lars Kaderali

Abstract In the last decade a new research field has emerged: *Systems Biology*. Based on experimental data and using mathematical and computational methods, systems biology attempts to describe biological behavior in a quantitative dynamic way. Biological data contains a lot of noise and there is only a limited amount available due to high experimental effort and cost. Thus, for parameter estimation from this kind of data, their stochasticity and the problem of non-identifiability of model parameters has to be taken into account. One way to do this is using a Bayesian framework, where one obtains distributions over possible parameter values, and these are then further analyzed. In this article we describe the potential impact of Bayesian parameter learning on systems biology, and discuss challenges arising from a Bayesian approach.

1 Modeling and Parameter Estimation in Systems Biology

Systems biology deals with the description of biological aspects by mathematical and computational approaches to obtain a systems point of view. Traditionally, a lot of qualitative data and knowledge was collected for single molecules of a cell or an organism. Systems biology now deals with the task to describe, in quantitative terms, the dynamic system's behavior resulting from the interplay between these single parts. The paradigm of systems biology approaches is that new properties emerge as the system is considered as a whole, which are not apparent at the level of individual components.

J. Mazur (✉) · L. Kaderali
Viroquant Research Group Modeling, University of Heidelberg, Bioquant BQ26, INF 267,
69120 Heidelberg, Germany
e-mail: johanna.mazur@bioquant.uni-heidelberg.de; lars.kaderali@bioquant.uni-heidelberg.de

H.G. Bock et al. (eds.), *Model Based Parameter Estimation*, Contributions
in Mathematical and Computational Sciences 4, DOI 10.1007/978-3-642-30367-8_6,
© Springer-Verlag Berlin Heidelberg 2013

Fig. 1 Schematic view of gene expression in a eukaryotic cell. First, in the nucleus the DNA is transcribed into messenger RNA (mRNA). In the next step, mRNA is exported into the cytoplasm. There it is read and translated into a protein

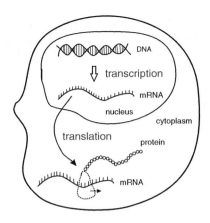

For example, let us consider the case of *gene expression*. In Fig. 1, a scheme is provided for the basic mechanisms occurring in gene expression in a eukaryotic cell. First, a gene encoded on the DNA is read and a complementary copy, *messenger RNA* (mRNA), is produced. This process is called *transcription*. This mRNA is then exported into the cytoplasm, where it is read and a *protein* is formed. This process is called *translation*. The obtained protein is further used for one or several special cellular functions. One might now ask, which gene is responsible for which cellular function, but biology is not that easy, e.g., a special cellular function may only be maintained if several proteins are present at the same time, or a protein may be used to start an increase or decrease of the transcription of another gene. Thus, in systems biology, we look at *gene regulatory networks* (GRNs), which are used to analyze the mechanisms occurring in a cell from a systems point of view. GRNs represent a collection of gene products, e.g., mRNAs or proteins, and their interactions. They have been modeled in several ways. Basically, there are four classes of models [12,13]:

1. *Logical models* discretize gene expression data to describe GRNs qualitatively.
2. *Correlation based models* connect similarly expressed genes to describe GRNs qualitatively.
3. *Continuous models* are deterministic models based on ordinary differential equations (ODEs) which describe GRNs quantitatively.
4. *Single-molecule models* are stochastic models, i.e., starting with an initial point, the underlying model will produce different trajectories according to a provided probability, describing GRNs quantitatively.

All of these models contain parameters, which have to be specified either by searching the literature, e.g., for reaction and/or degradation rates of molecules, or by fitting the model to data, e.g., for abstract parameters not directly corresponding to biology or for reaction rates not yet known.

In the next section we will describe the importance of Bayesian parameter learning in systems biology, motivated by the example of modeling GRNs. Furthermore, we give the challenges that arise when using a Bayesian framework. In Sect. 3, we

The Importance and Challenges of Bayesian Parameter Learning in Systems Biology 147

motivate Bayesian experimental design and give a promising approach of how it can be performed efficiently.

2 Bayesian Parameter Estimation in Biological Systems

In the last years, Bayesian learning has become more and more important in the life sciences. Here we only want to name a few recently published methods and software tools available for systems biological purposes. As a first example, Wilkinson recently reviewed Bayesian methods for heterogeneous biological systems [26]. Further, he summarized his work on stochastic kinetic models by using Bayesian learning [25]. Girolami gave an introduction to Bayesian methodology applied to models based on ODEs [7]. Furthermore, he and his colleague offer a software tool called *BioBayes*, where one can import SBML descriptions of biochemical models together with experimental data to perform Bayesian parameter learning and compare models by calculating Bayes factors [23, 24]. For reverse engineering of GRNs, we recently embedded a system of non-linear ODEs into a Bayesian framework [16].

In Bayesian parameter estimation, one starts with data $D = \{d_1, \ldots, d_n\}$ and a model $M(\omega)$ with parameters $\omega = (\omega_1, \ldots, \omega_k)$. The idea is now to obtain a *posterior distribution* of the model parameters ω given the data D. This is obtained by using Bayes' theorem:

$$p(\omega \mid D) = \frac{p(D \mid \omega)p(\omega)}{p(D)}, \tag{1}$$

where $p(D \mid \omega)$ is the *likelihood* function giving a probability density of the data given model parameters ω, i.e., a density relating the model $M(\omega)$ with its parameters to the given data. Further, $p(\omega)$ denotes a *prior distribution* on the model parameters and $p(D)$ is a normalizing factor independent of the model parameters. Thus, the posterior distribution according to (1) gives a trade-off between the available prior knowledge and the knowledge about the model parameters being inherent in the data. In other words:

The posterior distribution updates the knowledge of the model parameters ω according to the measured data D.

For deterministic models, we can give a likelihood, which coincides with the least squares approach usually used for parameter estimation problems. This is done by assuming that every data point d_i, $i \in \{1, \ldots, n\}$, is corrupted by mean zero Gaussian noise ε_i with variance σ_i^2, i.e., $\varepsilon_i \sim \mathcal{N}(0, \sigma_i^2)$. Let us further denote by σ^2 the vector containing all noise parameters σ_i^2, i.e., $\sigma^2 = \{\sigma_1^2, \ldots, \sigma_n^2\}$. With this we obtain a likelihood:

$$p(D \mid \omega, \sigma^2) = \prod_{i=1}^{n} \frac{e^{-\frac{1}{2\sigma_i^2}\left(d_i - M(\omega)\right)^2}}{\sqrt{2\pi\sigma_i^2}}, \tag{2}$$

which is the probability density of the data given some parameters ω and σ^2. By using a maximum likelihood approach one can now obtain the parameters which are most likely to describe the data. This corresponds to the parameters obtained with a least squares approach, as can be seen when taking the logarithm of $p(D \mid \omega, \sigma^2)$ and neglecting terms independent of ω. However, a likelihood can also be created for other noise models and can have, in general, any desired form which relates the model parameters to the data.

2.1 Importance

This subsection deals with the question, why Bayesian parameter learning is important in systems biology. Let us again look at the case of gene expression. The rough sketch described in Sect. 1 only tells half of the story. The processes described there correspond to biochemical reactions which, of course, can only occur if the desired molecules are present at the same time and at the same place. Since this cannot be guaranteed, reaction events are not predictable. We refer to this as *intrinsic noise* in biological systems. In the gene expression case, intrinsic noise may arise, for example because of the stochasticity of mRNA transcription or degradation, i.e., mRNA is decomposed into smaller chemical substances.

Another important noise type for parameter estimation is the measurement noise, here called *extrinsic noise*. To measure gene expression, there are two common methods: microarray and qPCR. The first one is a high-throughput method able to give DNA abundance for the whole genome of a species. But it also provides a lot of extrinsic noise. qPCR on the other hand, produces data with lower experimental noise but is expensive to perform. Thus, the data available is either limited, or contains a lot of (extrinsic and intrinsic) noise or, as in most cases, both.

To obtain now the desired quantitative models of a biological system, one has to take into account the different kinds of noise present in the data. Single-molecule models, as presented in the previous section, are suitable to capture the intrinsic noise for the case where only a small number of cells are considered. Although parameter estimation given stochastic models is difficult to perform and an ongoing research topic far from being solved, Boys et al. [3] showed that a Bayesian parameter estimation framework is capable of accomplishing this task. Bear in mind, that for stochastic models, one cannot define a distance function to use in optimization in an obvious way, as in the deterministic case described above. Regarding a larger number of cells, the importance of the intrinsic noise decreases, since the experimental data can be considered as the mean value over the gene expression values for all the cells. In that case, deterministic models provide a suitable approach.

Parameter estimation problems in systems biology have to deal with limited data which contain a lot of noise. Thus, these parameter estimation problems pose challenging tasks and are usually non-identifiable, i.e., different parameter sets may

The Importance and Challenges of Bayesian Parameter Learning in Systems Biology   149

describe the given data in a similar way. The non-identifiability problem is of special importance for gene regulatory networks, where different network structures represent the same data. In that case, one wants to know all underlying networks to investigate them further. Another difficulty arising in parameter estimation problems in systems biology is *sloppiness*, which was recently argued to be present in systems biological models [9]. Sloppiness means that the eigenvalues of the covariance matrix around a point estimate span many decades and their eigenvectors are skewed from their underlying parameter axes. Thus, even if we have only one mode in the posterior distribution, a wide range of parameter values may describe the experimental data in a similar way.

So we have seen that we have to deal with a lot of problems in parameter learning for systems biological models. Bayesian parameter estimation is able to deal with these difficulties and desires. In (2) we saw how noise inherent in the experimental data is taken into account to obtain a probability density of the experimental data, given the parameters of the underlying model. By using Bayes' theorem, we incorporate prior knowledge into the parameter estimation task to bound the search space for the parameters which helps with the non-identifiability problem. Such prior knowledge can be extracted from one of the numerous databases that exist for qualitative descriptions of biological behavior. In cases where no prior knowledge is available, one can use the uniform distribution over model parameters. The most important advantage of Bayesian parameter learning is its ability to provide a complete distribution over model parameters. Thus, one is able to look into, first, if there are different parameter sets consistent with the data and, second, if so, look into more detail at every one of them and rework the model, or measure additional data if necessary. The distributions obtained in Bayesian approaches are highly useful to design such additional experiments.

2.2   Challenges

Although we have seen, that Bayesian parameter learning is highly desirable and important, we also have to deal with some problems in the efficient treatment of Bayesian parameter learning algorithms, which has to be kept in mind when using them.

First, one has to consider that since the posterior distribution (1) is analytically tractable only in simple cases, e.g., for linear models, one has to approximate it. Methods which can find out of a family of given distributions the one which can best approximate the posterior distribution are: *Laplace approximation* (fits a Gaussian distribution) or *variational Bayesian methods* (fit all kinds of distributions). Since one usually does not know in advance, which distribution, or even if a standard distribution will fit the desired posterior distribution, we will focus on Markov chain Monte Carlo (MCMC) algorithms to obtain a set of samples from the posterior. In MCMC algorithms one starts with a starting value for the parameters in the model, and then creates a Markov chain which reaches the desired distribution, also called the *stationary distribution*, after a burn-in phase independent of the starting value. Such a Markov chain is also called *ergodic*. Thus, in theory we will obtain with infinitely many samples an excellent approximation of the desired distribution. Since in practice we have to deal with finite samples, we need an efficient algorithm which:

1. Converges fast to the stationary distribution.
2. Has a good *mixing rate*, i.e., the algorithm samples from the whole distribution in an efficient way and does not get stuck in modes.

The Metropolis–Hastings (MH) algorithm [10, 17], one of the simpler MCMC schemes, samples from the desired distribution $\pi(\cdot)$. It requires a *proposal distribution* $q(x, \cdot)$, also called *transition kernel*, to propose a new sample y which is accepted with the probability

$$\alpha(x, y) = \min\left\{1, \frac{\pi(y)q(y, x)}{\pi(x)q(x, y)}\right\}. \tag{3}$$

Otherwise the chain stays at state x and a new sample is proposed. The MH algorithm has the drawback that it samples only locally, and thus requires a large number of steps to sample the full support of a distribution. Furthermore, it shows a random walk behavior, such that the mixing rate is poor. This random walk behavior can be circumvented by performing the steps in a more global way. To give one example, the *hybrid Monte Carlo* algorithm [6] introduces momentum variables ρ with associated energy $K(\rho)$, and then iteratively samples ρ from $K(\rho)$ and follows the *Hamiltonian dynamics* of the system $H(\cdot, \rho) := -\ln \pi(\cdot) + K(\rho)$ to explore the whole state space. This algorithm is based on the physical fact that a state is completely described by its position and its velocity, and the energy of an unperturbed system is always constant and is the sum of the energy of the position and the energy of the velocity. The Bayesian inference problem for model parameters is then represented by an imaginary physical system in which the state corresponds to the set of unknown parameters ω. The energy of the positions, the *potential energy*, is described by the distribution $\pi(\cdot)$ we are interested in. The energy of the momentum variables $K(\rho)$, the *kinetic energy*, which describes the velocity of the system, follows a multivariate Gaussian distribution, and thus is easily sampled. A detailed analysis of the hybrid Monte Carlo algorithm can be found in [18]. As a last remark, one has to mention that this algorithm is only usable, provided derivatives of $\pi(\cdot)$ according to the parameters can be calculated.

Another approach is by using *adaptive MCMC* [1], where a combination of parameterized proposal distributions $q_{k,\theta}$ are used and the parameters θ are updated during the sampling procedure to obtain an "optimal" (often a criteria expressed by expectations of the stationary distributions reached with the considered Markov chains) transition kernel which converges fast to the stationary distribution and has a good mixing rate. The problem with adaptive MCMC is, that the ergodicity constraint may be violated. This can be bypassed by stopping adaptation after one

The Importance and Challenges of Bayesian Parameter Learning in Systems Biology 151

is sure to have reached an optimal value of θ. However, then one is going in circles like Andrieu and Thoms [1] state:

> ... most criteria of interest [on the parameters θ] depend explicitly on features of π, which can only be evaluated with... MCMC algorithms.

A further approach are population-based MCMC algorithms [11]. The idea is to start several Markov chains with different densities π_n, $n \in \{1, \ldots, N\}$, and to construct a valid MCMC algorithm which has

$$\pi^* := \prod_{n=1}^{N} \pi_n$$

as its invariant distribution where at least for one n, we have $\pi_n = \pi$. The sequence of distributions π_n can be selected in such a way that they are easier to simulate than π. Jasra et al. [11] suggest to use a combination of tempered and stratified densities for π^*. The first one introduces temperature variables $\zeta_n \in (0, 1]$ and uses $\pi_n \propto [\pi]^{\zeta_n}$. The idea is that for high temperatures, i.e., ζ_n close to zero, the distributions are easily sampled and help to enlarge the mixing rate of the algorithm. The objective of the second one is to take a partition of the parameter space and provide an optimal kernel on each subpart. The chains in population-based MCMC algorithms are started with different starting points, chosen from a distribution which is over-dispersed compared to the desired distribution. On the chains then one of the following three "genetic operators" is applied: *mutation*, *crossover* or *exchange*. The mutation operator deals with the local exploration of the state space. The crossover and the exchange operator deal with the global exploration of the state space to exchange information between chains. To ensure that the ergodicity constraint is fulfilled, the genetic operators are accepted according to (3).

Which of these MCMC algorithms is suitable for a special problem, cannot be answered in general. This is a problem specific question. However, since models in systems biology are usually nonlinear, as they should be able to describe complex biological behavior, like oscillations or switch-like behavior, the underlying posterior distribution may contain a lot of ridges and is multimodal. This is nicely illustrated in [7] for a simple model of a Circadian Oscillator. In our opinion, population-based MCMC algorithms are a good choice for complex biological models.

Another problem often seen is that the posterior is flat in a particular coordinate direction, indicating a model which is not identifiable. Recently, Gustafson [8] and Xie and Carlin [27] gave some proposals as to how the informativeness of data on Bayesian parameter estimates can be measured in non-identified models. This may be even more relevant in the case where only limited amounts of and vague biological prior knowledge is known. To obtain more informative data, an experimental design framework for the Bayesian context is described in the next section.

3 Bayesian Experimental Design in Systems Biology

As described in the previous section, we usually have only a few data points to use to reverse engineer the parameters of the model, which leads to identifiability problems of the parameters. The incorporation of prior knowledge helps with this shortcoming, but still the search space is typically very large and several regions in the parameter space may describe the data and the biological system equally well.

In Fig. 2, a typical framework for the development of a mathematical model is shown. First, from biological knowledge, a mathematical model is designed, which is used to predict some biological behavior. If this prediction does not adequately describe the biological behavior, experimental design has to be performed to determine new wetlab experiments which will improve the mathematical model and/or the parameter estimates of this model the most. If now the prediction of this newly obtained model describes biology accurately, we have found a model of interest. Since the uniqueness of such a model cannot be guaranteed, we can end up with several indistinguishable models. To stick to one model at the end, Occam's razor can be used to find the "simplest" model which describes the biological behavior of interest. However, since biological experiments take a lot of effort and are very expensive, the experimental design step is a crucial one.

For a review of experimental design methods used in systems biology see [15]. Most often one "minimizes" the inverse of the *Fisher information* matrix F, which is defined as the covariance matrix of the estimated parameters. Common scalar values for the "minimization" of F^{-1} are the minimization of the determinant (D-optimal design) or the minimization of the sum of eigenvalues (A-optimal design). These classical *alphabetic experimental designs* were introduced by Kiefer in 1959 [14]. An application of an alphabetic optimal experimental design for the parameter estimation of a cell signaling model is given in [2].

The whole motivation for Bayesian parameter estimation deals with the fact that different parameter values may describe the given data in a similar way. Thus classical alphabetical experimental designs are not appropriate, since they assume that the estimated parameters are near to the true parameters, and therefore only the covariance matrix around this point is considered. To deal with several possible parameter values, one can use *Bayesian experimental design* (BED). For a detailed introduction into BED, see [5].

Already in 1978, O'Hagan [19] stated concerning BED:

> ... I cannot see how one can realistically approach optimal design in any other way.

Now, we want to give the main idea of BED. As a first ingredient we need the *predictive distribution* for future data d for experiments e of interest, e.g., concentrations of mRNA at different time points. With a model $M(\omega)$ at hand and prior knowledge on the model's parameter values ω, we obtain as predictive distribution

$$p(d \mid M) = \int p(d \mid \omega, M) p(\omega) \, d\omega. \tag{4}$$

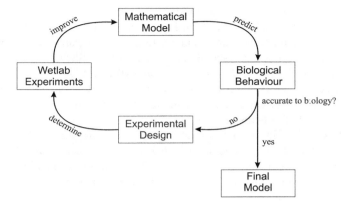

Fig. 2 General cycle of creating mathematical models in systems biology. Starting with a mathematical model, some biological behavior is predicted. If this does not describe the biology, experimental design is performed to determine optimal new wetlab experiments, which are then used to improve the mathematical model. Once we obtained a model which predicts biological behavior accurately, we are done

The next step now is to use *Bayesian decision theory* to decide which experiment to perform next. For this purpose one needs a *utility function* $U(d, e)$ which depends on the experiments we can perform and on the data these experiments will give us. Since we do not know how the data will look like before the experiment is made, the experiment of interest is where the *expected utility*

$$EU(e) = \int U(d,e) p(d \mid M) \, dd \qquad (5)$$

is maximal. Most often, the *information* of the new posterior distribution is used as the utility function for BED. For a continuous random variable X, with corresponding density function f, the information $I(X)$ is defined as

$$I(X) = \int f(x) \ln(f(x)) \, dx = -\text{Ent}(X), \qquad (6)$$

where $\text{Ent}(X)$ denotes the *entropy* of X. The entropy of a random variable gives a measure of the uncertainty of this random variable, i.e., high entropy contained in X means low information contained in X, and vice versa. Lately, Busetto et al. [4] showed that BED outperforms alphabetic experimental designs for models for the TOR pathway of *Saccharomyces cerevisiæ*.

However, although there is a rich theory concerning BED, analytical solutions are only available for linear models and assuming Gaussian distributions. Recently, Steinke et al. [21] used Bayesian experimental design to identify gene regulatory networks out of a linearized ODEs model together with perturbations of the steady states. In the case of realistic biological models, i.e., non-linear models, BED has to be solved with sampling methods, i.e., with MCMC algorithms. This is computationally intractable without further considerations, since triple integrals over prior model parameters, data and posterior model parameters have to be solved (for illustration simply set (4) and (6) into (5)). Shewry and Wynn [20] introduced in 1987 *Maximum Entropy Sampling* (MES) for Bayesian experimental design purposes. Recently, MES was applied by van den Berg et al. [22] on seismic amplitude versus offset experiments. Thus far, to our knowledge, MES was not applied in the field of systems biology.

The idea of MES in BED is to reformulate the computationally intractable problem of finding the experiment, such that the new posterior distribution over model parameters has maximal information. This is accomplished by finding the experiment such that the distribution of the data generated with this experiment has maximal entropy. In other words:

Perform the experiment where you know the least!

This strategy is numerically much simpler since we only need the predictive distribution of the data generated from some specific experiment. Thus, we just have to simulate data of the underlying model according to the distribution of the model parameters obtained in a first run of Bayesian parameter estimation. Hence, we only have to solve one integral over prior model parameters (see (4)).

4 Summary and Outlook

In this article we argue that, since there is only a small amount of biological data available, which additionally contains a lot of noise, Bayesian parameter estimation is important in the field of systems biology. This is due to the fact that Bayesian parameter estimation procedures give probability distributions over models and model parameters, and such provide the possibility to account for different possible explanations for the underlying data. Additionally, prior biological knowledge can be used to reduce the search space.

We also discussed the challenges concerning Bayesian parameter learning. Since the desired posterior distributions are usually not of a simple form and contain a lot of ridges and modes, we need an efficient MCMC algorithm to get the desired samples. Several possibilities exist, such as the hybrid Monte Carlo algorithm, adaptive MCMC algorithms or, as a very promising approach for complex biological models, population-based MCMC algorithms.

In our opinion, a very important focus for future research will be the development and implementation of efficient algorithms for Bayesian experimental design. Maximum entropy sampling offers a potent possibility to make this task computationally feasible.

Acknowledgements We are grateful for funding to the German Ministry of Education and Research (BMBF), grant number 0313923 (FORSYS/Viroquant).

References

1. C. Andrieu and J. Thoms. A tutorial on adaptive MCMC. *Stat. Comput.*, 18:343–373, 2008.
2. S. Bandara, J. P. Schlöder, R. Eils, H. G. Bock, and T. Meyer. Optimal experimental design for parameter estimation of a cell signaling model. *PLoS Comput. Biol.*, 5(11):e1000558, 2009.
3. R. J. Boys, D. J. Wilkinson, and T. B. L. Kirkwood. Bayesian inference for a discretely observed stochastic kinetic model. *Stat. Comput.*, 18(2):125–135, 2008.
4. A. G. Busetto, C. S. Ong, and J. M. Buhmann. Optimized expected information gain for nonlinear dynamical systems. In A. P. Danyluk, L. Bottou, and M. L. Littman, editors, *ICML*, volume 382 of *ACM International Conference Proceeding Series*, page 13. ACM, 2009.
5. K. Chaloner and I. Verdinelli. Bayesian experimental design: A review. *Statist. Sci.*, 10(3):273–304, 1995.
6. D. Duane, A. D. Kennedy, B. J. Pendleton, and D. Roweth. Hybrid Monte Carlo. *Phys. Lett. B*, 195:216–222, 1987.
7. M. Girolami. Bayesian inference for differential equations. *Theor. Comp. Sci.*, 408:4–16, 2008.
8. P. Gustafson. What are the limits of posterior distributions arising from nonidentified models, and why should we care? *J. Am. Stat. Assoc.*, 104(488):1682–1695, 2009.
9. R. N. Gutenkunst, J. J. Waterfall, F. P. Casey, K. S. Brown, C. R. Myers, and J. P. Sethna. Universally sloppy parameter sensitivities in systems biology models. *PLoS Comput. Biol.*, 3(10):e189, 2007.
10. W. K. Hastings. Monte Carlo sampling methods using Markov chains and their applications. *Biometrika*, 57(1):97–109, 1970.
11. A. Jasra, D. A. Stephens, and C. C. Holmes. On population-based simulation for static inference. *Stat. Comput.*, 17:263–279, 2007.
12. L. Kaderali and N. Radde. Inferring gene regulatory networks from expression data. In A. Kelemen, A. Abraham, and Y. Chen, editors, *Computational Intelligence in Bioinformatics*, volume 94 of *Studies in Computational Intelligence*, pages 33–74. Springer, Berlin, 2008.
13. G. Karlebach and R. Shamir. Modelling and analysis of gene regulatory networks. *Nat. Rev. Mol. Cell Biol.*, 9:770–780, 2008.
14. J. Kiefer. Optimum experimental designs. *J. R. Stat. Soc. Series B Methodol.*, 21(2):272–319, 1959.
15. C. Kreutz and J. Timmer. Systems biology: experimental design. *FEBS J.*, 276:923–942, 2009.
16. J. Mazur, D. Ritter, G. Reinelt, and L. Kaderali. Reconstructing nonlinear dynamic models of gene regulation using stochastic sampling. *BMC Bioinformatics*, 10:448, 2009.
17. N. Metropolis, A. W. Rosenbluth, M. N. Rosenbluth, and A. H. Teller. Equations of state calculations by fast computing machines. *J. Chem. Phys.*, 21(6):1087–1092, 1953.
18. R. M. Neal. Probabilistic inference using Markov chain Monte Carlo methods. Technical report, Department of Computer Science, University of Toronto, 1993.
19. A. O'Hagan and J. F. C. Kingman. Curve fitting and optimal design for prediction. *J. R. Stat. Soc. Series B Methodol.*, 40(1):1–42, 1978.
20. M. C. Shewry and H. P. Wynn. Maximum entropy sampling. *J. Appl. Stat.*, 14(2):165–170, 1987.
21. F. Steinke, M. Seeger, and K. Tsuda. Experimental design for efficient identification of gene regulatory networks using sparse Bayesian models. *BMC Syst. Biol.*, 1:51, 2007.
22. J. van den Berg, A. Curtis, and J. Trampert. Optimal nonlinear Bayesian experimental design: an application to amplitude versus offset experiments. *Geophys. J. Int.*, 155(2):411–421, 2003.
23. V. Vyshemirsky and M. Girolami. Bayesian ranking of biochemical system models. *Bioinformatics*, 24(6):833–839, 2008.
24. V. Vyshemirsky and M. Girolami. BioBayes: A software package for Bayesian inference in systems biology. *Bioinformatics*, 24(17):1933–1934, 2008.
25. D. J. Wilkinson. *Stochastic Modelling for Systems Biology*. Chapman & Hall/CRC, Boca Raton, FL, USA, 2006.

26. D. J. Wilkinson. Stochastic modelling for quantitative description of heterogeneous biological systems. *Nat. Rev. Genet.*, 10:122–133, 2009.
27. Y. Xie and B. P. Carlin. Measures of Bayesian learning and identifiability in hierarchical models. *J. Stat. Plan. Inference*, 136:3458–3477, 2006.

Experiment Setups and Parameter Estimation in Fluorescence Recovery After Photobleaching Experiments: A Review of Current Practice

J. Beaudouin, Mario S. Mommer, Hans Georg Bock, and Roland Eils

Abstract Fluorescence Recovery After Photobleaching (FRAP) is a popular and versatile family of methods used to estimate mobility and reaction parameters in cellular systems. Part of an area containing a fluorescently labeled species is bleached using a laser, and the effect of the perturbation of the spatial concentration profile of the fluorescent species is monitored. Subsequently, the collected data is reconciled with a model of the dynamics, thus yielding estimates for the parameters of interest. While originally devised to elucidate transport parameters, it was soon extended to the estimation of reaction rates, and is also used nowadays to answer a variety of questions on the organization of cellular systems.

In this chapter, we review a variety of different approaches, classifying them according to the sources of uncertainty that are addressed or ignored, and the type of parameter that they attempt to estimate. We would like to highlight the importance of the general methodology as a tool that can be widely applied to a large number of situations.

J. Beaudouin (✉)
Division of Theoretical Bioinformatics, German Cancer Research Center (DKFZ),
Im Neuenheimer Feld 280, 69120 Heidelberg, Germany
e-mail: j.beaudouin@dkfz.de

M.S. Mommer (✉) · H.G. Bock
Interdisciplinary Center for Scientific Computing, Im Neuenheimer Feld 368,
69123 Heidelberg, Germany
e-mail: mario.mommer@iwr.uni-heidelberg.de; bock@iwr.uni-heidelberg.de

R. Eils
Division of Theoretical Bioinformatics, German Cancer Research Center (DKFZ),
Im Neuenheimer Feld 280, 69120 Heidelberg, Germany

Department for Bioinformatics and Functional Genomics, Institute for Pharmacy and
Molecular Biotechnology (IPMB), and Bioquant, Heidelberg University, Heidelberg, Germany
e-mail: r.eils@dkfz.de

H.G. Bock et al. (eds.), *Model Based Parameter Estimation*, Contributions
in Mathematical and Computational Sciences 4, DOI 10.1007/978-3-642-30367-8_7,
© Springer-Verlag Berlin Heidelberg 2013

1 Introduction

Fluorescence recovery after photobleaching (FRAP) has become a method of choice to investigate protein mobility in different media including living cells. The method relies on the introduction of a fluorescent dye on a molecule of interest, on the local perturbation of fluorescence distribution of the molecule population and on the monitoring of fluorescence redistribution over time. The quantitative interpretation of the kinetics of fluorescence redistribution can eventually lead to the identification of the main processes that drive molecule mobility, and of their kinetic parameters, and thus ultimately to a better understanding of intracellular processes.

The mathematical description of the general FRAP methodology can be stated as follows. We start by considering a domain Ω, usually representing the spatial extent of a cell or an organelle.

Within this domain Ω, we consider a fluid which includes the fluorescently labelled species of interest. Transport within this fluid obeys, in the general case, a reaction–diffusion–convection law (Anomalous diffusion can also be modelled using reaction diffusion equations, see [15]). This law is parameterized by the quantities of interest (reaction, diffusion, or convection coefficients) whose estimation is the immediate objective of the FRAP experiment.

To achieve this objective, the experimenter perturbs the concentration profile of the fluorescent component, and records the response of the system to that perturbation. Finally, a model of the process is used to obtain the parameters, using either optimization methods, or closed form expressions that have been derived for some cases.

Essentially, we have the following setting. The transport of the set of species of interest is given by a law of the general form

$$\frac{\partial u}{\partial t} = D(p)\triangle u + a(p) \cdot \nabla u + f(u, p, v) \qquad \text{on } \Omega \qquad (1)$$

together with boundary conditions

$$0 = F\left(u, \frac{\partial u}{\partial \mathbf{n}}\right) \qquad \text{on } \partial\Omega \qquad (2)$$

where p captures the relevant parameters, and v is a time-dependent control representing the bleaching process. The domain Ω is in $\mathbb{R}^3$, but is often considered for modelling purposes to be in $\mathbb{R}^2$, as discussed later.

To obtain measurement values $y = J_q(u)$ for the estimation of parameters, there is an a measurement function J_q involved, that has its own parameters q, which may or may not have to be estimated separately or together with the quantities of interest, depending on the situation. Such additional parameters may be physiological properties of the cell, as well as performance characteristics of the microscope. In a FRAP experiment, the measurement function is always defined in terms of spatial distribution of the fluorescence of the sample.

A Review of FRAP Experiments 159

It is important to note that the setting of (1) is not in general a feasible model for parameter estimation, and we have to regard it as an idealized setting from where to derive more practical models. The actual, physical domain Ω in a cell under study has a very complex geometry whose details remain hidden to the experimenter under normal conditions. Additionally, including this complex geometry for the analysis may not be trivial (see e.g. [19]). Likewise, (1) has a very complex form, if considered in full, that includes many different types of interactions and additional species.

Thus, to obtain the parameters of interest, the investigator must use a simplified model for the estimation of the parameters. In it, a simplified geometry for Ω is chosen, and some of the terms in (1) are simplified using additional information on what is relevant for the system. The FRAP family of techniques is notable from a parameter estimation and experiment design perspective due to the richness of possibilities in the choices of the experimental setup, measurement, modelling, simulation, and parameter estimation techniques involved. The choices are driven by the uncertainty present in the data, the particularities of the subject of study, as well as by the limitations and capabilities of the wide range of instruments that can be used.

In the following pages we will give an overview on how this setting occurs in actual applications. Different quantitative analysis approaches proposed in the literature over the last three decades will be presented, with a main focus on the analysis of protein mobility in living cells. We will also discuss the assumptions and the limitations of the methods, related to potential experimental artefacts, complex geometries and parameter identifiability. We will observe that the freedom of choice of ingredients mentioned in the previous paragraph is indeed an important practical aspect of the methodology, and not simply a theoretical possibility. We will also be able to appreciate the importance of the technique by noticing that many instances of the methodology are developed and published alongside a study of a relevant biological problem.

2 The FRAP Experiment

To perform a FRAP experiment in living cells, one must first fluorescently label molecules, most of the time proteins, of interest. In the context of protein, this step has been greatly facilitated by the introduction of fluorescent proteins as a standard tool of molecular biology in the 1990s [10]. Many different variants of fluorescent proteins have been developed, covering the whole range of visible light. These fluorescent proteins can easily be fused to proteins of interest using standard molecular biology approaches to clone genes that encode for such protein fusions and transfection to bring these genes in living cells where they are eventually expressed.

The experiment itself is performed by perturbing the fluorescence distribution of the tagged proteins [14]. The standard experiment consists of using photobleaching.

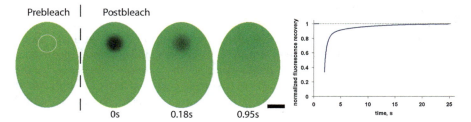

Fig. 1 FRAP experiment simulation. A population of green molecules (Prebleach) diffusing with a diffusion coefficient of 6 μm^2/s was locally bleached within the *white circle* for 1 s. Fluorescence redistribution was simulated over time (Postbleach), and fluorescence recovery inside the bleached region was measured and normalized with the total intensity of the molecule population. Scale bar: 5 mm

Briefly, when a fluorescent molecule is brought from its ground to its first electronic state by a photon of the right energy, it can emit a photon with a lower energy, a process called fluorescence. Once in the excited state, there is also a probability that the fluorescence molecule switches to states where it is not fluorescent anymore, a process that we summarize as photobleaching. While this process occurs when fluorescent molecules are observed, it can be enhanced by applying a strong exciting light to pump as many molecules as possible in the excited states and to bring them in the photobleached state.

To achieve the local perturbation of fluorescence distribution required for FRAP, one needs to photobleach a pool of the fluorescent protein population. This is typically achieved by using a laser at a wavelength that can excite the fluorescent protein. Such an experimental technique is now available in any laboratory equipped with a confocal laser scanning microscope (CLSM). Such an instrument is producing images of fluorescent samples by scanning them pixel-per-pixel and measuring the fluorescence intensity of each pixel. The instrument is typically equipped with an acousto-optical tunable filter that permits an almost instantaneous switch from 0 to 100% of laser intensity from one pixel to the other during the scanning process. This allows to perform the photobleaching in any desired region within the fluorescent sample. Other microscopes used in biological laboratories like wide-field or spinning disc microscopes may also be equipped for FRAP.

The typical setup of a FRAP experiment using a CLSM is as follows. Images are acquired at a specific set of time points, not necessarily evenly spaced. During the acquisition of one of those images, a local pool of fluorescent molecules is bleached through an increase of intensity of the scanning laser. Images acquired after this point are used to monitor the redistribution of fluorescence (see Fig. 1). Photobleaching is typically kept as short as possible but easily requires hundreds of milliseconds to achieve a good contrast between bleached and non-bleached regions. The resulting perturbed concentration of fluorescently labeled proteins, also known as the *bleach profile*, is used as an initial condition in the simulations needed for parameter estimation.

Due to the time constrains imposed by the fluorescent recovery, and technical constraints of the equipment, in many cases images are acquired in two dimensions, even if the sample is three-dimensional. For all the methods, we assume that the chemical properties of molecules are not changed by bleaching and that the FRAP is performed on population of molecules that remains in steady state over the whole experiment.

3 Quantitative FRAP Analysis

The quantitative analysis of a FRAP experiment can be seen as a two-dimensional process: it depends not only of the design of the experiment, but also on the type of process that is being investigated. In the biological context, these processes are limited to diffusion, chemical interactions and flow. In this section, we will mostly focus on the type of analysis depending on the design of the experiment and cover two processes that have raised interest in cell biology in the past years: diffusion and interactions.

3.1 Spot Bleaching

The historically first FRAP approach to be used and quantitatively analyzed was the so-called spot bleaching. It consists of bleaching a small 2D-isotropic area of the fluorescent sample and monitoring the fluorescence recovery within this region. The data that is used for estimating the parameters consists in the percentage of prebleaching fluorescence intensity observed in the bleached area over time, and is called the *recovery curve*. Most analysis proposed in the literature assume instantaneous bleaching, or that molecules do not significantly move during the time interval when the laser is bleaching. Axelrod et al. [1] proposed the first closed solutions for the fluorescence recovery in the context of diffusion combined with flow. The analysis covered two types of bleaching profile: Gaussian and circular. Diffusion is mathematically modelled by the following differential equation,

$$\frac{\partial C}{\partial t}(\mathbf{r}, t) = D \triangle C(\mathbf{r}, t),$$

where C denotes the concentration of the species of interest and D is a number representing the diffusion coefficient. In this case, for the Gaussian bleach profile, the closed form solution for the recovery curve is

$$F_k(t) = F_k \nu K^{-\nu} \Gamma(\nu) P(2K|2\nu),$$

where F_k is the intensity before bleaching, ν is $(1 + 8Dt/\omega^2)$, $\Gamma(\nu)$ is the gamma function and $P(2K|2\nu)$ the χ^2-probability distribution. The number ω is the half-width of the Gaussian at e^{-2} height, and K corresponds to the amount of bleaching and can be determined from the intensity ratio of intensities before, F_k, and just after bleaching, $F_k(0)$:

$$\frac{F_k(0)}{F_k} = \frac{1 - e^{-K}}{K}.$$

The recovery solution has been extended to the case when only a fraction α of the fluorescent molecules is mobile [27],

$$\frac{F_k(t)}{F_k} = \alpha v K^{-v} \Gamma(v) P(2K|2v) + (1 - \alpha)\frac{F_k(0)}{F_k}.$$

For the circular bleaching profile, Soumpasis [23] gave the following closed solution that is independent of K,

$$f_k(t) = e^{-2\tau_D/t} [I_0(2\tau_D/t) + I_1(2\tau_D/t)],$$

where $f_k(t)$ is the fractional recovery curve $(F_k(t) - F_k(0))/(F_k - F_k(0))$ and where I_0 and I_1 are modified Bessel functions of the first kind.

The spot bleaching analysis with a circular profile has been extended to the case of reaction-diffusion models. Sprague et al. [25] analyzed the case of a diffusing fluorescent molecule interacting with fixed binding sites homogeneously distributed. A closed solution was given in the form of a Laplace transform,

$$\overline{frap}(p) = \frac{1}{p} - \frac{F_{eq}}{p}(1 - 2K_1(qw)I_1(qw)) \\ \times \left(1 + \frac{k_{on}^*}{p + k_{off}}\right) - \frac{C_{eq}}{p + k_{off}}, \tag{3}$$

where I_1 and K_1 are modified Bessel functions of the first and second kind, p is the Laplace variable, k_{off} and k_{on}^* are the off-rate and apparent on-rate of the interaction. Also,

$$F_{eq} = \frac{k_{off}}{k_{on}^* + k_{off}} \quad \text{and} \quad C_{eq} = \frac{k_{on}^*}{k_{on}^* + k_{off}}.$$

Equation (3) is inverted numerically to yield the average intensity on the bleach spot as a function of time. This curve can then be fitted to data.

It is also possible to obtain fastly converging series expansions of the fluorescence in a bleach spot in terms of Fourier series. In Kumar et al. [13] this approach is implemented for estimating the mobility of proteins in E. coli bacteria, yielding a series expansion of the recovery of fluorescence over time,

$$R(t; D, L) = \sum_{j=0}^{\infty} h_j^m u_j^0 \exp\left\{-t\left(\frac{j\pi}{L}\right)^2 D\right\}.$$

A Review of FRAP Experiments

This form includes the Fourier coefficients of the characteristic function of the measurement spot and idealized initial conditions, $\{h_j^m\}$ and $\{u_j^0\}$, as well as the length L of the cell. In this approach, only a few terms in the series are needed to compute R to machine precision.

An interesting alternative to spot FRAP is the continuous bleaching approach first proposed by Peters [16]. Instead of bleaching a spot and monitoring the recovery of fluorescence, one can continuously bleach the spot and analyze the fluorescence decay kinetics. The decay given by fixed fluorescent molecules corresponds to their bleaching kinetics, but mobile molecules moving inside the volume induce a slower decay that can be quantitatively interpreted. Wachsmuth et al. [28] proposed solutions for reaction–diffusion systems with fixed binding sites, which followed the following differential equation, which is a specialization of (1).

$$\frac{\partial c_{\text{diff}}(\mathbf{r}, t)}{\partial t} = D \triangle c_{\text{diff}}(\mathbf{r}, t) + Q(\mathbf{r}) c_{\text{diff}}(\mathbf{r}, t)$$

$$- k_{\text{on}} c_B(\mathbf{r}) c_{\text{diff}}(\mathbf{r}, t) + k_{\text{off}} c_{\text{immo}}(\mathbf{r}, t)$$

where c_{diff} and c_{immo} correspond to diffusing and bound protein concentrations, c_B is the binding site distribution. $Q(\mathbf{r})$ is the bleaching profile, assumed to induce a first-order fluorescence decay, and D, k_{on} and k_{off} are the diffusion coefficient, and the on- and off-rates of the interactions.

3.2 Analysis on Raw Image Data

The spot bleaching approach has been extensively developed for different models and bleaching profiles. While the approach did match the experimental setup when the method was developed (e.g. [1]), which meant that image analysis could not be performed reliably, most experiments are nowadays performed on laser scanning microscopes, which provide not only temporal but also spatial information. Spot bleaching analysis average this spatial information over the bleach region, while it can actually be combined with the temporal information to fit models. The approaches in this subsection use spatially resolved information directly to estimate the parameters of interest.

FRAP experiments performed by bleaching a spot can be interpreted by measuring the profile of fluorescence across the spot over time. If the fluorescence profile is Gaussian and the system can be considered as infinite in size and if molecules diffuse, then the profile remains Gaussian over time and the variance increases as $2Dt$, D being the diffusion coefficient, independently of the initial conditions [7],

$$C(x, y, t) = C_0(t) + [C_U - C_0(t)]$$

$$\left[1 - \exp\left(-\frac{\{[x - x_0(t)]^2 + [y - y_0(t)]^2\}}{R_0^2(0) + 8Dt} \right) \right]$$

where $x_0(t)$ and $y_0(t)$ are the locations of the center of the Gaussian, $R_0(0)$ is the Gaussian width at time 0, $C_0(t)$ and C_U are the intensity at position 0 and infinitely far from the center. This profile analysis can be extended to other bleach profiles [12].

For fluorescent samples that are large enough, another approach to analyze diffusion consists of bleaching a periodic pattern [22]. As high frequencies decay faster than low ones, the decay of the amplitude of the bleached pattern rapidly tends towards a single exponential with a characteristic time of $P^2/(4D\pi^2)$, D being the diffusion coefficient and P the period of the pattern.

Extending this method, one can also easily analyze post-bleach images in the Fourier space, as each spatial frequency q gives an exponential decay over time with a characteristic time of $1/(4D\pi^2q^2)$ [3, 26].

Finally computers are now powerful enough to allow direct simulations of fluorescence distribution over time and space. This was performed in Beaudouin et al. to study nuclear protein dynamics in mammalian cells [2]. In this case, a reaction–diffusion system was simulated using a finite difference approach taking into account the real geometry of the nucleus.

4 Limitations Introduced by the Assumptions

4.1 Bleaching Duration and Initial Conditions

Original studies on spot bleaching approaches assumed instantaneous bleaching, meaning that molecules do not significantly move during the bleaching. In practice this means that bleach duration has to be much shorter than the characteristic time of fluorescence recovery. The commonly used CLSM has the disadvantage of being slow compared to a system like in Axelrod [1] where the laser is continuously illuminating the same spot. This can lead to a first postbleach image spot that is larger than expected due to movement of fluorescent molecules inside the spot and photobleached molecules outside the spot during the bleach. Applying the spot bleaching fitting functions can lead to a significant underestimation of diffusion coefficients [5, 29]. This would notably happen for soluble proteins with diffusion coefficients in the range of $10\text{--}100\,\mu\text{m}^2/\text{s}$ [11]. In this context, analysis approaches that do not require information about the bleaching itself are more appropriate.

The problem of molecule mobility during the bleaching is more complex in the context of reaction–diffusion systems. If the fluorescent molecule interacts with another molecule, the observed intensity corresponds to two populations: the free and the bound molecules. As their mobility is different, their distribution on the first postbleach image is also different. So far analysis methods, including the ones independent of bleaching parameters, have neglected this effect and may lead to wrong parameter estimates.

4.2 Sample Geometry

Most analysis methods assume a cell of infinite spatial extent, which is reasonable when the size of the bleach spot is much smaller than the cell under study. As a consequence, when this assumption is necessary, the bleach region is typically kept as small as possible in comparison to the size of the sample, which can limit the signal-to-noise ratio and thus degrade the accuracy of the parameter estimates.

This can also lead to overestimation of immobile fraction when the normalization is not performed accurately: bleaching leads to the depletion of a fluorescent pool, therefore the recovery intensity cannot reach the initial intensity. One way to compensate for this is to normalize with the total fluorescence intensity of the sample over time (e.g. [2, 17]).

Models can also be designed to take into account the finite size of the sample. This has been performed in the context of bacteria, where the elongated geometry of the cell can be modelled as a one-dimensional interval [8, 13]. In both studies, methods were based on the analysis of the Fourier transform of the images. This was also applied for mammalian cell nucleus, where the system was assumed to be a disk and where parameters were estimated numerically using a Gaussian spot bleaching approach [6]. Numerical simulations of the partial differential equations derived from reaction–diffusion models used in [2] take the real closed geometry of the system into account, discretized in the finite difference grid.

While performing FRAP with CLSM is almost always limited to two dimensions, living cells are three-dimensional samples. Nevertheless live cell imaging experiments are often performed with adherent cells, so that samples are typically thin. Therefore one strategy to circumvent the third dimension problem is to use objectives of low numerical aperture that generate quasi cylindrical vertical illumination at the scale of the sample [2], and consider the two dimensional images as projections of the three-dimensional fluorescence distribution. One can also consider some aspects of the third dimension, notably the point-spread function (PSF) and the confocality of the CLSM, to build recovery models. This was performed for the case of diffusion and circular bleaching by Braeckmans et al. [4]. Finally, as images are the convolution product of the original object with the PSF of the instrument, the easiest approach for this problem may be to use the Fourier space: the Fourier transform of the fluorescence recovery images are simply the product of the PSF Fourier transform and the one of the spatial fluorescence distribution. Therefore the exponential decays described above remain valid [3].

These approaches require samples that have constant or infinite depth. This assumption was tested in the case of mammalian cell nucleus by simulating FRAP with the real 3D geometry of the nucleus and simulating CLSM imaging from this simulation [2]. In this case the quality of the fits was not affected, but a parameter deviation of around 10% was observed between 3D and CLSM simulations.

4.3 Reversibility of Photobleaching

The use of fluorescent proteins considerably facilitates FRAP experiments in living cells. One assumption of all the analysis methods is that photobleaching is irreversible. This is nevertheless only an approximation as there is a probability that fluorescent proteins in the excited states transiently switch to a dark state, on top of the irreversible bleached state, before coming back to the ground state. As this is a probabilistic process, the dark state is enriched during bleaching and is depleted again during the postbleach acquisition when excitation is low. For the GFP mutant S65T, the residency time in the dark state is 1.6s and the depletion of the dark state during the first postbleach images can therefore interfere with fluorescence recovery due to protein mobility [9]. This reversibility was also observed for other fluorescent proteins [21]. This effect would affect all the analysis methods; its correction is non-trivial and has not been performed so far. It would require a good knowledge of the reversible process and its incorporation in the spatio-temporal model of bleaching and recovery. Therefore it may be easier to implement it in methods that already incorporate bleach duration and amplitude. An experimental alternative would consist in using techniques that generate fluorescence, by photoactivation/conversion/switching, instead of photobleaching it. Even if the process is reversible or not instantaneous, one can easily normalize it by dividing local intensities with the total one [2]. Such an approach can be improved nowadays thanks to the development of such fluorescent proteins [20].

4.4 Model Choice and Parameter Identifiability

One of the difficulties in the field has been the generation of models to fit to the data. Diffusion models have often been used even in cases where diffusion was not the only process driving or limiting molecule mobility. On the other hand, other studies have used interaction models neglecting diffusion while this assumption was not justified (see [24]). Such studies were notably performed in the nucleus of mammalian cells and the argument to justify this assumption was that the time of fluorescence redistribution was long compared to the expected one if the protein of interest was not interacting. The mistake in that case was that binding sites fill the whole sample and can therefore generate fluorescence recoveries that are slow but still limited by diffusion.

Inaccurate estimation of parameters may not only come from a wrong choice of model but also from limited parameter identifiability. This is a typical problem in reaction–diffusion systems. In the ideal case, one can estimate the diffusion coefficient of the different species, an off-rate and an apparent on-rate of the interaction. But often the system may reach the limit when the diffusion or the interaction kinetics do not show any contribution [2, 6, 25]. In the first case where diffusion is not contributing, one can use simple interaction models that lead to exponential fluorescence recovery [18]. In the second case where the reaction is on faster timescales than diffusion, the system follows a diffusive model where the diffusion coefficient is a linear combination of the one of the free molecule and the one of the molecule complexed with its interacting partner. The coefficients of the linear combination correspond to the percentage of free and bound molecules. Determining which model is correct is not a trivial task, as a direct comparison of fits of different models is not enough. A complex model typically gives better fits than a simpler one containing less parameters, even if a simpler one is also correct. To discriminate between the complete reaction-diffusion model and the simpler cases, one can compare their capacity to fit the data using statistical tests that takes into account the different amount of parameters of the different models.

5 Discussion and Conclusions

We believe that with this article we have made a case for considering FRAP as an important and interesting experimental technique where many open fundamental questions remain. We have seen that while the basic idea is simple, it presents a great range of opportunities to make creative use of the physical and mathematical characteristics of the underlying processes.

Thus, the choice of bleach profile is motivated by the properties of the laser, the microscope used, and the mathematical description of the properties of diffusion–convection–reaction phenomena. It can be of different shapes, and have different temporal profiles, and thus lead to a variety of different parameter estimation problems.

We have also demonstrated that the flexibility of the method is not just a theoretical possibility, but a fundamental property of the methodology. It is this adaptability that makes it such a great asset in practice. We also believe that there is still a lot of potential in the methodology, as few studies have really looked at the methodology as a whole and analyzed the popular choices from the point of view of parameter estimation and experimental design as established disciplines. We hope that by raising awareness on this issue, and through future work, we will be able to contribute to change this state of affairs, and bring the FRAP approach closer to realizing its full potential.

References

1. D. Axelrod, D. E. Koppel, J. Schlessinger, E. Elson, and W. W. Webb. Mobility measurement by analysis of fluorescence photobleaching recovery kinetics. *Biophys J*, 16(9):1055–69, 1976.
2. J. Beaudouin, F. Mora-Bermudez, T. Klee, N. Daigle, and J. Ellenberg. Dissecting the contribution of diffusion and interactions to the mobility of nuclear proteins. *Biophys J*, 90(6):1878–94, 2006.

3. D. A. Berk, F. Yuan, M. Leunig, and R. K. Jain. Fluorescence photobleaching with spatial fourier analysis: measurement of diffusion in light-scattering media. *Biophys J*, 65(6):2428–36, 1993.
4. K. Braeckmans, L. Peeters, N. N. Sanders, S. C. De Smedt, and J. Demeester. Three-dimensional fluorescence recovery after photobleaching with the confocal scanning laser microscope. *Biophys J*, 85(4):2240–52, 2003.
5. J. Braga, J. M. Desterro, and M. Carmo-Fonseca. Intracellular macromolecular mobility measured by fluorescence recovery after photobleaching with confocal laser scanning microscopes. *Mol Biol Cell*, 15(10):4749–60, 2004.
6. J. Braga, J. G. McNally, and M. Carmo-Fonseca. A reaction-diffusion model to study rna motion by quantitative fluorescence recovery after photobleaching. *Biophys J*, 92(8):2694–703, 2007.
7. S. R. Chary and R. K. Jain. Direct measurement of interstitial convection and diffusion of albumin in normal and neoplastic tissues by fluorescence photobleaching. *Proc Natl Acad Sci U S A*, 86(14):5385–9, 1989.
8. M. B. Elowitz, M. G. Surette, P. E. Wolf, J. B. Stock, and S. Leibler. Protein mobility in the cytoplasm of escherichia coli. *J Bacteriol*, 181(1):197–203, 1999.
9. M. F. Garcia-Parajo, G. M. Segers-Nolten, J. A. Veerman, J. Greve, and N. F. van Hulst. Real-time light-driven dynamics of the fluorescence emission in single green fluorescent protein molecules. *Proc Natl Acad Sci U S A*, 97(13):7237–42, 2000.
10. B. N. Giepmans, S. R. Adams, M. H. Ellisman, and R. Y. Tsien. The fluorescent toolbox for assessing protein location and function. *Science*, 312(5771):217–24, 2006.
11. N. Klonis, M. Rug, I. Harper, M. Wickham, A. Cowman, and L. Tilley. Fluorescence photobleaching analysis for the study of cellular dynamics. *Eur Biophys J*, 31(1):36–51, 2002.
12. U. Kubitscheck, P. Wedekind, and R. Peters. Lateral diffusion measurement at high spatial resolution by scanning microphotolysis in a confocal microscope. *Biophys J*, 67(3):948–56, 1994.
13. M. Kumar, M. S. Mommer, and V. Sourjik. Mobility of cytoplasmic, membrane, and dna-binding proteins in escherichia coli. *Biophys J*, 98(4):552–9, 2010.
14. J. Lippincott-Schwartz, E. Snapp, and A. Kenworthy. Studying protein dynamics in living cells. *Nat Rev Mol Cell Biol*, 2(6):444–56, 2001.
15. M. S. Mommer and D. Lebiedz. Modelling subdiffusion using reaction diffusion systems. *SIAM J. on Appl. Math.*, 70(1):112–132, 2009.
16. R. Peters, A. Brunger, and K. Schulten. Continuous fluorescence microphotolysis: A sensitive method for study of diffusion processes in single cells. *Proc Natl Acad Sci U S A*, 78(2):962–966, 1981.
17. R. D. Phair and T. Misteli. High mobility of proteins in the mammalian cell nucleus. *Nature*, 404(6778):604–9, 2000.
18. G. Rabut, V. Doye, and J. Ellenberg. Mapping the dynamic organization of the nuclear pore complex inside single living cells. *Nat Cell Biol*, 6(11):1114–21, 2004.
19. I. F. Sbalzarini, A. Mezzacasa, A. Helenius, and P. Koumoutsakos. Effects of organelle shape on fluorescence recovery after photobleaching. *Biophys J*, 89(3):1482–92, 2005.
20. N. C. Shaner, G. H. Patterson, and M. W. Davidson. Advances in fluorescent protein technology. *J Cell Sci*, 120(Pt 24):4247–60, 2007.
21. D. Sinnecker, P. Voigt, N. Hellwig, and M. Schaefer. Reversible photobleaching of enhanced green fluorescent proteins. *Biochemistry*, 44(18):7085–94, 2005.
22. B. A. Smith and H. M. McConnell. Determination of molecular motion in membranes using periodic pattern photobleaching. *Proc Natl Acad Sci U S A*, 75(6):2759–63, 1978.
23. D. M. Soumpasis. Theoretical analysis of fluorescence photobleaching recovery experiments. *Biophys J*, 41(1):95–7, 1983.
24. B. L. Sprague and J. G. McNally. Frap analysis of binding: proper and fitting. *Trends Cell Biol*, 15(2):84–91, 2005.
25. B. L. Sprague, R. L. Pego, D. A. Stavreva, and J. G. McNally. Analysis of binding reactions by fluorescence recovery after photobleaching. *Biophys J*, 86(6):3473–95, 2004.

A Review of FRAP Experiments

26. T. T. Tsay and K. A. Jacobson. Spatial fourier analysis of video photobleaching measurements. principles and optimization. *Biophys J*, 60(2):360–8, 1991.

27. A. Tsuji and S. Ohnishi. Restriction of the lateral motion of band 3 in the erythrocyte membrane by the cytoskeletal network: dependence on spectrin association state. *Biochemistry*, 25(20):6133–9, 1986.

28. M. Wachsmuth, T. Weidemann, G. Muller, U. W. Hoffmann-Rohrer, T. A. Knoch, W. Waldeck, and J. Langowski. Analyzing intracellular binding and diffusion with continuous fluorescence photobleaching. *Biophys J*, 84(5):3353–63, 2003.

29. M. Weiss. Challenges and artifacts in quantitative photobleaching experiments. *Traffic*, 5(9):662–71, 2004.

Drug Resistance in Infectious Diseases: Modeling, Parameter Estimation and Numerical Simulation

Le Thi Thanh An and Willi Jäger

Abstract Infectious diseases are drawing our attention again. In recent years, we have been confronted with SARS, bird-flu, swine-flu and many other severe diseases. In addition, pathogens are getting resistant to drugs controlling them. In this paper we establish a new population dynamical system of infectious diseases including drug treatment. We take into account both sensitive and resistant parasites. The unknown model parameters are fitted based on a set of data for malaria from Cisse, Burkina Faso. The fitted model is used to investigate the influence of drug treatment on drug resistance. Based on these investigations, treatment strategies to reduce drug resistance can be elaborated.

Keywords Differential equation models • Drug resistance • Infectious diseases • Vector-borne diseases

1 Introduction

This paper is organized as follows. In the first section we give a short overview on the drug resistance models existing so far. In Section 2, we present a new model describing the dynamics of pathogen, transmitter and human populations including drug treatment. Section 3 is devoted to doing parameter estimation and simulation. Estimation results and simulations with controllable factors are performed here. The last section is reserved for discussion of the simulation results and conclusions.

As we know, infectious diseases were and are still being a big burden to every country. For a long time a lot of scientists have been studying them intensively [1, 8, 12]. Although being considered as classical field of mathematical modeling,

---

L.T.T. An (⊠) · W. Jäger
Interdisciplinary Center for Scientific Computing (IWR), Heidelberg, Germany
e-mail: an.lethithanh@iwr.uni-heidelberg.de; jaeger@iwr.uni-heidelberg.de

H.G. Bock et al. (eds.), *Model Based Parameter Estimation*, Contributions
in Mathematical and Computational Sciences 4, DOI 10.1007/978-3-642-30367-8_8,
© Springer-Verlag Berlin Heidelberg 2013

it is now attracting a lot of attention again. Pathogens are getting more resistant to drugs, infectious diseases can spread very fast and become a danger for the populations. Including this effect in the model equations is an important condition to meet reality in epidemics.

We are going to set up a mathematical model under general aspects, but we select malaria as the central case, where we have a chance to obtain data to calibrate and validate our model.

For the rest of this section, we mainly discuss modeling of drug resistance with central focus on malaria. Historically, there has been a lot of research on this disease, see [5, 7, 13] and references therein. However, up to now it is very difficult to obtain satisfactory malaria models. In general, before modeling authors have their own questions. They would then call a new model to be "good enough" if it is able to answer their questions. One of the challenges with malaria is that it comes together with Anopheles mosquitoes and parasites—*Plasmodium falciparum*, *Plasmodium vivax*, *Plasmodium ovale* and *Plasmodium malariae*. Their organisms are much more complex compared to other pathogens with single-cell or even simpler- such as bacteria, archaea, viruses, etc. This can also partly explain why there are only few mathematical models concerning drug treatment in malaria.

What may influence drug resistance?

- In most diseases, it takes long time to find out which factors may characterize and reduce drug efficiency. Usually, there are many individuals of the pathogen population which already have "resistant" mechanisms. They are naturally not eliminated by drugs. When infected people take medicine, it reduces sensitive parasites and makes good environment for resistance growth.
- In malaria, parasite organisms are very complex, they would either mutate or adapt variously. They have a better chance to survive after natural selection and maintain their best genetic system. Hence, their new generations are more "fit" in a drug environment.
- Another important reason for drug resistance is that malaria is more widespread in tropical countries, like in Africa, where it is difficult to have a global policy to treat all patients in a proper way.

For the rest of this paper, we are more interested in mathematical malaria modeling. In history, there were many valuable contributions by Ross, MacDonald, May, Dietz, see [1, 5] and references therein. However, resistance was not well investigated. Mathematicians have only recently started to pay attention to this field. Two recent malaria models concerning drug resistance came from Aneke [2] and Esteva et al. [6]. Here the authors described a model of hosts and vectors dealing with both resistant and sensitive strains. In these papers and also some other cases, authors focus on analyzing the models to establish positive equilibrium and studying their stability under certain assumptions. Although in [6] the authors tried numerical simulation, they said that the model needed to be validated using realistic epidemiological data. So far there was very few data available to calibrate malaria and other vector-borne disease models.

Drug Resistance in Infectious Diseases 173

In this paper, we derive a model which consists of differential equations including all host and pathogen populations. Such models are complex to perform analysis, and thus we are going to study them numerically. Since from experiments not all parameters are known directly, we have to use parameter estimation techniques to determine their values. After this, the model is used for computer experiments. We obtain results on treatment strategies, some of which have been recently confirmed by experimentalists after a lot of works in hospitals and laboratories [7, 13].

2 Modeling of Population Dynamics Including Drug Treatment

We divide this part into two sub-sections: the first one introduces all model compartments and their network, the second one presents model formulation.

2.1 The Network of Compartments

There are two main groups of infectious diseases: the first group (such as HIV/AIDS, tuberculosis) can be transmitted directly from one person to another and the second group (such as malaria, dengue hemorrhagic fever) is transmitted only via an intermediate host. From the perspective of infectious diseases, vectors are the transmitters of disease-causing organisms that carry the pathogens from one host to another [14]. That's why a disease belonging to the second group is called a vector-borne disease.

As we discussed before, we would like to model dynamics of pathogen, vector and human populations. We consider two hosts: number of humans denoted by H_1 and vectors denoted by H_2. The pathogens within human hosts are denoted by P. Like in the classical case, there are three compartments for human hosts which we call susceptible H_{1s}, infected H_{1i}, recovered H_{1r}. We have two compartments for vector hosts H_{2s}, H_{2i}—susceptible and infected. Since drugs are employed, there are two compartments for pathogens, non-resistant and resistant P_n, P_r, respectively. All transmissions among the seven groups are described in Fig. 1.

2.2 Model Formulation

Now we give an overview of the modeled processes, as schematically indicated in Fig. 1. Since our focus is drug treatment for humans, so the model may not include some interference applying on vectors directly. We include the influence of drugs on the host population and the subsequent effects on pathogen population.

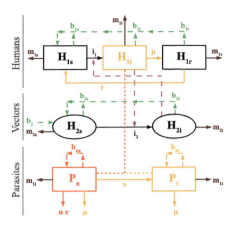

Fig. 1 Network for the population dynamics with drug treatment

Remark. During the course of our study, we discover that some parts of the extracted data are not precise enough and can have some effect on parameter estimation and simulation. We revise them and exclude the imprecise values. The model is improved in parallel with this process. Despite the fact that the usable data is less, we can obtain better fit compared to the first estimation, see Fig. 3.

2.2.1 Susceptible Human Population H_{1s}

In vector-borne diseases, human hosts become infected only after coming in contact with vectors, so there is often no vertical transmission, e.g. from parents to children. Every new offspring goes to the susceptible class. Like other epidemic models, we use common denotations for the natural birth rates of susceptible, infected, recovered human classes b_{1s}, b_{1i}, b_{1r}. Also as convention, mortality rate of H_{1s} is denoted by m_{1s}, infection rate by i_1 and re-susceptibility rate by r (individuals from recovered class H_{1r} after a certain time come back to susceptible class H_{1s}). Values of m_{1s}, i_1 are given in the next section and r is an unknown parameter which can be found after fitting the data. The dynamic equation of the susceptible human population is the following:

$$\frac{dH_{1s}}{dt} = b_{1s}H_{1s} + b_{1i}H_{1i} + b_{1r}H_{1r} - i_1 \frac{H_{2i}}{H_{2i} + H_{2s}} H_{1s} + rH_{1r} - m_{1s}H_{1s}. \quad (1)$$

Notice that the infection term depends on $\frac{H_{2i}}{H_{2i} + H_{2s}}$, which implies that there are too many vectors so that they have to compete with each other to get successful contact to humans (e.g., obtain a blood meal in case of malaria). This is a quite usual situation in Africa, where the tropical climate is suitable for vector (mosquito) development. Although mainly infected female Anopheles mosquitoes can transmit

Drug Resistance in Infectious Diseases 175

malaria, they still have to compete with other un-infected mosquitoes to get blood meals. On the other hand, the size of the mosquito population induces a corresponding protection force from human side.

Concerning additional details for parameter r, we are aware that different drugs have different half-lives. They partly remain inside human bodies after treatment and provide the same protection as immunity. Medically, we consider five times of drug half-life equal to wash-out time. When drug is washed out and immunity is no longer active, the person returns to the susceptible class. This process is modeled by formula $r = k_r r_0$, where $r_0 = [\max(T, d)]^{-1}$, T is natural immune duration, d is drug wash-out duration; k_r is a scalar factor subject to estimation. After fitting, we use this formula for control simulation, with the controllable parameter d.

2.2.2   Infected Human Population H_{1i}

After susceptible individuals get infected with rate i_1, they go to the infected class. Here are two possibilities how they can move out of this class: either they recover after treatment or they can not recover and die. The natural and disease-induced mortality rate is denoted by m_{1i} with value based on literature [11], and the recovery rate μ is unknown parameter.

$$\frac{dH_{1i}}{dt} = i_1 \frac{H_{2i}}{H_{2i} + H_{2s}} H_{1s} - \mu H_{1i} - m_{1t} H_{1i}. \tag{2}$$

Moreover, we foresee that the recovery rate μ depends on the individual immunity, the employed medical treatment and the way how the drug is administrated. We are going to take the data from Cisse in Burkina Faso, a small and poor village. We should notice that most of the local population does not have access to any treatment easily. With support from a project between Heidelberg researchers and local people, classical Chloroquine regimen was given to all patients and showed good effect. That is why in simulation we can assume that drugs start with a good efficacy. When one patient takes the full dose as recommended, all his/her sensitive pathogens are killed at rate e. The proportion of people that follows proper treatment is expressed by α.

In general, most of the patients carry both sensitive and resistant parasites. Drug treatment strengthens the immunity and shortens the recovery process. In other words, it helps to increase the recovery rate. We combine immune and treatment effects in formula

$$\mu = k_\beta e \beta \alpha + \beta(1 - \alpha),$$

where β is the rate of parasites which are eliminated by natural host immunity without treatment, $k_\beta e$ is the factor which expresses how drugs strengthen the immunity in treated patients. The value of k_β is subject to estimation.

2.2.3 Recovered Human Population H_{1r}

All recovered individuals from H_{1i} go to the recovered class H_{1r}. For a certain while they have protection due to the retained drug and natural immunity. Depending on different diseases they can or can not be reinfected again. With malaria, especially in highly endemic areas, all recovered individuals eventually lose their immunity and return to susceptible class. The process can be fast or slow depending on drug wash-out duration and individual health. We have already explained the meaning of r in the Sect. 2.2.1 above. We also call m_{1r} mortality rate of recovered class H_{1r}, the value of m_{1r} is given in Table 1.

$$\frac{dH_{1r}}{dt} = \mu H_{1i} - rH_{1r} - m_{1r}H_{1r}. \tag{3}$$

2.2.4 Susceptible Vector Population H_{2s}

This compartment is similar to the susceptible human compartment. However, since parasites do not influence vectors in the same way as they influence humans, there is no recovery or re-susceptibility effect.

Let b_{2s}, b_{2i} be the unknown birth-rates of susceptible and infected vectors. In addition, eggs from vectors (mosquitoes) are supposed to be able to hibernate over unfavorable time spans and hatch to become new vectors when the season provides better conditions. This process has an effect close to immigration and emigration, modeled by parameter b_2. Vectors become infected on contacting infected humans in class H_{1i}, this is modeled by rate i_2. All of these four parameters are subject to estimation. Mortality rate of vectors m_{2s} is taken as in [11].

$$\frac{dH_{2s}}{dt} = b_{2s} H_{2s} + b_{2i} H_{2i} + b_2 - i_2 \left(\frac{H_{1i}}{H_{1s} + H_{1i} + H_{1r}} \right) H_{2s} - m_{2s} H_{2s}. \tag{4}$$

2.2.5 Infected Vector Population H_{2i}

Expectedly, the infected vector population and the infected human population have similar dynamics. All infected individuals from susceptible class H_{2s} go to H_{2i}. Since the life-cycle of vectors is not so long, in most cases parasites stay in vectors' abdomen until they die. The mortality rate of vectors is denoted by m_{2i}, its value is given in Table 1.

$$\frac{dH_{2i}}{dt} = i_2 \left(\frac{H_{1i}}{H_{1s} + H_{1i} + H_{1r}} \right) H_{2s} - m_{2i} H_{2i}. \tag{5}$$

Drug Resistance in Infectious Diseases

2.2.6 Non-resistant Parasite Population P_n

Now we pay attention to parasite population in humans. They are supposed to be inside the infected human class H_{1i}, so naturally depend on H_{1i}. Since this dependence is likely reduced when the number of infected human becomes large, we use Michaelis–Menten relation, involving $\dfrac{H_{1i}}{C + H_{1i}}$. We also assume that parasite population growth follows the logistic rule governed by rate b_{pn}, b_{pr} and a maximal capacity K.

Unlike the hosts, parasites usually clone themselves, so in some sense there is no "death." It practically applies for malaria parasites in humans, they only multiply asexually inside human red blood cells. On the other hand, all parasites inside infected hosts also die out due to the mortality of the infected hosts.

For a given treatment, there is the proportion α of the infected human population H_{1i}, who received full doses. Immunity and drug can eliminate pathogens with rate $(\mu + \alpha e)$, e presents the rate at which (sensitive) parasites are eliminated by drug.

Also due to drug presence, some sensitive pathogens need to change themselves to be able to increase their chances of survival, such as by mutation or adaptation. This makes a certain ratio of sensitive pathogens become resistant, the process is occurring with rate

$$s = s_0 + k_d \frac{\alpha \min(d, d_0)}{e}.$$

The rate s_0 describes the mutation and the part $k_d \alpha \min(d, d_0)/e$ describes the adaption rate. We assume that the adaption rate is directly increased in the same time with the proportion α of treatment and drug wash-out duration d, but inversely proportional to the efficacy e of the given treatment. There is also a parameter d_0, an upper threshold of drug wash out duration d, which implies that above a certain limit parasite adaptation rate would not increase considerably but rather stay the same. Factor k_d is an unknown subject to estimation.

$$\frac{dP_n}{dt} = b_{pn} \frac{H_{1i}}{C + H_{1i}} \left(1 - \frac{P_n + P_r}{K}\right) P_n - m_{1i} P_n - (\mu + \alpha e) P_n - s P_n. \quad (6)$$

2.2.7 Resistant Parasite Population P_r

As we mentioned before, the resistant parasite population P_r grows with the rate b_{pr}. It decreases due to H_{1i}-death rate m_{1i} or is eliminated by host immunity μ. Since resistant parasite population is fitter in drug presence, it gains some new ones from sensitive population $s P_n$.

$$\frac{dP_r}{dt} = b_{pr} \frac{H_{1i}}{C + H_{1i}} \left(1 - \frac{P_n + P_r}{K}\right) P_r - m_{1i} P_r - \mu P_r + s P_n. \quad (7)$$

Now we put together all the equations (1)–(7). We obtain one system which is related to our network in Fig. 1. This shows clearly what is included in the dynamical system:

$$\frac{dH_{1s}}{dt} = b_{1s} H_{1s} + b_{1i} H_{1i} + b_{1r} H_{1r} - i_1 \frac{H_{2i}}{H_{2i} + H_{2s}} H_{1s} + r H_{1r} - m_{1s} H_{1s},$$

$$\frac{dH_{1i}}{dt} = i_1 \frac{H_{2i}}{H_{2i} + H_{2s}} H_{1s} - \mu H_{1i} - m_{1i} H_{1i},$$

$$\frac{dH_{1r}}{dt} = \mu H_{1i} - r H_{1r} - m_{1r} H_{1r},$$

$$\frac{dH_{2s}}{dt} = b_{2s} H_{2s} + b_{2i} H_{2i} + b_2 - i_2 \frac{H_{1i}}{H_{1s} + H_{1i} + H_{1r}} H_{2s} - m_{2s} H_{2s}, \qquad (8)$$

$$\frac{dH_{2i}}{dt} = i_2 \frac{H_{1i}}{H_{1s} + H_{1i} + H_{1r}} H_{2s} - m_{2i} H_{2i},$$

$$\frac{dP_n}{dt} = b_{pn} \frac{H_{1i}}{C + H_{1i}} \left(1 - \frac{P_n + P_r}{K}\right) P_n - m_{1i} P_n - (\mu + \alpha e) P_n - s P_n,$$

$$\frac{dP_r}{dt} = b_{pr} \frac{H_{1i}}{C + H_{1i}} \left(1 - \frac{P_n + P_r}{K}\right) P_r - m_{1i} P_r - \mu P_r + s P_n$$

where

$$r = k_r r_0 = k_r \max^{-1}(T, d),$$
$$\mu = k_\beta e \beta \alpha + \beta(1 - \alpha),$$
$$s = s_0 + k_d \alpha e^{-1} \min(d, d_0).$$

All initial conditions are given:

$$(H_{1s}, H_{1i}, H_{1r}, H_{2s}, H_{2i}, P_n, P_r)(t^0) = (H_{1s}^0, H_{1i}^0, H_{1r}^0, H_{2s}^0, H_{2i}^0, P_n^0, P_r^0) \geq 0,$$

especially the susceptible classes $H_{1s}^0, H_{2s}^0 > 0$.

This is a non-linear differential equation system. The system is designed for vector-borne diseases. Compared to the already established models concerning drug resistance of vector-borne diseases, such as in [2,6], we add two new compartments of parasites. They are either sensitive or resistant to the given drug. Drug treatment can take effect on infected human hosts H_{1i}, sensitive parasites P_n and partly on resistant parasites P_r inside the infected humans.

3 Parameter Estimation and Numerical Simulation

For numerical study, we often use finite interval. For short time periods, parameters related to human hosts do not change very much in general. That is why we are going to specify some factors, in order to reduce the complexity of the problem.

This section contains data extraction, all known and unknown parameters, establish a parameter estimation problem and a simulation problem. Using the software package VPLAN [9], we can solve the two problems. All the results are presented in detail.

3.1 Data Extraction

Concerning observation data, they are mainly extracted from [11], a project which have been done in Burkina Faso. We make the following assumptions:

- The study focused on children and all observations were generalized for the whole town.
- In the period of study, most of the clinical malaria cases were treated so most of the infected people recovered after treatment. Whole population was only slightly decreased due to malaria.
- Given a certain drug, all parasites are either sensitive or resistant.

In addition, we assume that the observation errors are normally distributed and can be approximated by 20% of the peak values of each variable measurement.

3.2 Parameter Values

Before simulation in the case of malaria in Burkina Faso, we have to specify all parameters to meet the specific situation. We present all parameters in two tables: the first one for known parameters and the second for unknown parameters. Our time unit is 5 days.

After several discussions with experts in epidemiology, also based on the data from literature, we take some factors as constants over the year of study (2004): birth rate and mortality rate of human classes, mortality rate of vectors (mosquitoes), etc. We also take into account the properties of the drugs that were given (mainly Chloroquine) and its efficacy. We then obtain a list of the known parameters as given in Table 1. Infection rate i_1 is a piecewise linear function as in Fig. 2.

In Table 2 we present all unknown parameters. We plan to estimate them by piecewise functions, each interval corresponds to 1 month. We state also some constraints needed to analyze estimation results.

Table 1 All known parameters

Parameter	Meaning	Value (unit)	References
b_{1s}	Birth rate of susceptible humans	0.00063 (time^{-1})	[11]
b_{1i}	Birth rate of infected humans	0.0005 (time^{-1})	[11]
b_{1r}	Birth rate of recovered humans	0.00063 (time^{-1})	[11]
T	Natural immune duration in humans	12 (time)	[15]
d	Drug wash-out duration	40 (time)	[4]
m_{1s}	Mortality rate of susceptible humans	0.000285 (time^{-1})	[16]
α	Proportion of full treatments	0.9 (dimensionless)	[11]
β	Rate of parasites eliminated by immunity	0.025 (time^{-1})	[10]
e	Elimination rate of sensitive parasites by drug	0.7 (time^{-1})	Assumed, [11]
m_{1i}	Mortality rate of infected humans	0.000613 (time^{-1})	[11, 16]
m_{1r}	Mortality rate of recovered humans	0.000285 (time^{-1})	[16]
m_{2s}	Mortality rate of susceptible vectors	0.75 (time^{-1})	[11]
m_{2i}	Mortality rate of infected vectors	0.75 (time^{-1})	[11]
C	Constant in Michaelis–Menten formula	5 (dimensionless)	Assumed,
K	Parasite-holding capacity of human hosts	25×10^6 (dimensionless)	[11], scaled
s_0	Parasite mutation rate (become resistant)	7×10^{-7} (time^{-1})	[7]
d_0	Effective threshold of drug-wash-out durations	73 (time)	[4]

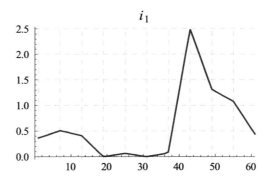

Fig. 2 Approximated function of human infection rate i_1, based on [11] and information about the proportion of infected human class

3.3 Setup of Parameter Estimation Problem and Simulation Problem

In this part, we recall a classical problem concerning parameter estimation in differential equations and specify our problems to be solved later. Interested readers are referred to [3] and references therein for related information.

Drug Resistance in Infectious Diseases 181

Table 2 All unknown parameters for estimation

Parameter	Meaning	Unit	Remark
k_r	Unknown factor, needed in re-susceptibility rate in humans	dimensionless	$k_r \geq 0$
k_β	Unknown factor describing treatment effect	time	$k_\beta \geq 0$
b_{2s}	Birth rate of susceptible vectors	time^{-1}	$b_{2s} \geq 0$
b_{2i}	Birth rate of infected vectors	time^{-1}	$b_{2i} \geq 0$
b_2	Hibernated eggs contributing to susceptible vectors	dimensionless	$b_2 \geq 0$
i_2	Infection rate of susceptible vectors	time^{-1}	$i_2 \geq 0$
b_{pn}	Relative birth rate of sensitive parasites	time^{-1}	$b_{pn} \geq 0$
b_{pr}	Relative birth rate of resistant parasites	time^{-1}	$b_{pr} \geq 0$
k_d	Unknown factor, needed in selection force	time^{-3}	$k_d \geq 0$

For the rest of this section, without further notice, all variables are elements of $\mathbb{R}^n$. After the modeling section, we have established a population dynamics in form of a differential equation system (8). Generalizing the system, we denote time by t, state variable by $x(t)$, unknown parameters by p, control parameters by q, control functions by $u(t)$, the right hand side of system (8) by f, the constraints by g. We have a problem:

$$\dot{x}(t) = f(t, x(t), p, q, u(t)), \quad t \in [t^0, t^f],$$
$$0 = g(x(t^0), p, q). \tag{9}$$

For all parameter estimation problems, we need data. It is assumed that experiments $i, i = 1, \dots, N$ have been carried out at the given times $t^j, j = 1, \dots, M$, yielding measurements η_{ij}. On the other hand, measurement errors are ε_{ij} and the "true" model response corresponding to these measurements are b_{ij}:

$$\eta_{ij} = b_{ij}(t^j, x(t^j), p) + \varepsilon_{ij}.$$

Parameter p is found by minimizing the deviation between data and model response. Due to statistical reasons, some weights σ_{ij}^{-1} can be introduced, see details in [3]. In addition, if we know in advance certain information about some approximate value p_0 of parameter p then we can add a regularization term $\delta(p - p_0)^2$ to the objective function. Vector δ has the same dimension as vector p and all components δ^l are nonnegative.

Summing up, a general parameter estimation problem can be formulated as:

$$\min_{(x,p)} \left[\sum_{i,j} \left(\frac{\eta_{ij} - b_{ij}(t^j, x(t^j), p)}{\sigma_{ij}} \right)^2 + \sum_l \delta^l (p^l - p_0^l)^2 \right], \tag{10}$$

s.t. (x, p) solves equation (9).

In our case, we have one experiment ($i = 1$) with all observation data η_j about the state variables $x(t_j)$ and the weights σ_j. The unknown parameters are given in Table 2

$$p = (k_r, k_\beta, b_{2s}, b_{2i}, b_2, i_2, b_{pn}, b_{pr}, k_d).$$

The initial values for the state variables, which are implied by g, are given at $t = t^0$. In the parameter estimation problem, all the control factors are given, we are interested in finding p.

After parameter estimation, values for the parameters are determined. Now we can consider simulations. Unlike before, the control parameters (q) play the key roles. They are the proportion of full treatment (α), the drug wash-out duration (d) and the treatment efficacy e. To each simulation, there is a fixed set of values. With different values of control parameters, we need to solve the differential equation (8) to compute the simulation.

For parameter estimation and simulations, we use software package VPLAN, see [9] and references therein. VPLAN is a software package developed at the Interdisciplinary Center for Scientific Computing (IWR), University of Heidelberg.

3.4 Result of Parameter Estimation

In this part we are going to present the estimation results. As mentioned before, to minimize the residual, we take into account all parameters in form of piecewise constant functions. We divide the domain into an appropriate number of intervals and find parameter values in each interval. To be more specific, we solve problem (10) in the first interval, then pass the last state variable values as the initial values of the next intervals. This assures the continuity of the solutions. The results of p in all intervals form piecewise constant functions, see Fig. 4.

In addition, we use multiple shooting [3] and maximize the usage of data information to deliver a good fit, see Fig. 3. Since parasite populations are very large compared to the host populations, we scale them by $1/10^{10}$.

As we can see in the Fig. 3, all the predicted populations (the dashed curves) are in good agreement with data (the points) from Cisse, Burkina Faso. The different scales between hosts and parasites were taken into account by using a weighted least squares function. Due to technical reasons, in some place we use small regularization factors (see the problem setup 3.3). When summing up, the overall residual is very low.

For clear visibility, we show only half of the data points. The study was carried out from the end of 2003 to the end of 2004. Since data at the beginning and closing periods were not very good, so we only take the part from middle of January to November 2004, around 300 days or 60 units of time. There are the peaks in mosquitoes and parasites around August since the season becomes more suitable for vector development. Notice that the eggs can survive through dry season and hatch in rainy season to become larvae, pupae and then mosquitoes.

Drug Resistance in Infectious Diseases

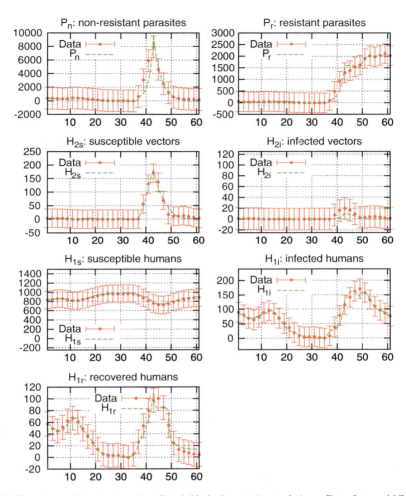

Fig. 3 The observed (points) and predicted (dashed curves) populations. Data from middle of January to November 2004, [11]. All together we use 427 measurement data. These data points are plotted together with their error bars (20% of each peak measurements). Time unit is 5 days

All the parameters are presented in Fig. 4. It shows that most of the parameters are not constant during the year, they vary a lot due to seasonal conditions, host environments, interactions among different populations.

As expected, the rainy season creates the peaks of mosquito growth rates. There are many more susceptible mosquitoes available compared to dry season. Through the contact with infected humans, they also become infected by parasites. Due to the weather condition, mosquitoes are much more active than in dry months. With Human Land Capture method, volunteers caught many more mosquitoes in rainy months while just a few in the other months [11].

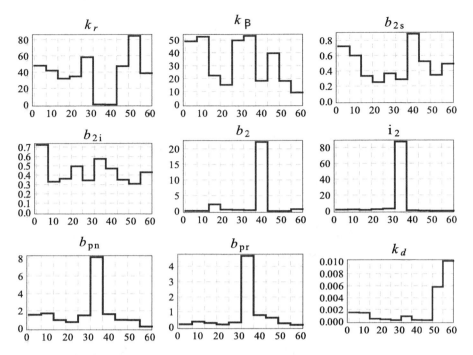

Fig. 4 The fitted parameters at the study time (2004). The horizontal axes are time axes, unit 5 days

In the same time, the value of k_r has a big drop, indicating that there are almost no recovered people that come back to susceptible class. In malaria, no recovery or complete clearance is expected in a short period of time. We should also mention that patients with low parasite densities sometime are not detected and can be considered healthy. In fact, the parasite density develops slowly and symptoms appear in patients after few weeks or much later than the exposure to mosquitoes. This explains why the peak of selection force of resistance also comes later than the intensive treatment period, this is expressed in the last parameter k_d in Fig. 4.

3.5 Result of Simulation with Controllable Parameters

In this section, we simulate the model (8) using the set of fitted parameters. Here we can control the proportion of full treatment α, the drug regimen corresponding to drug wash-out duration d and the treatment efficacy e. Values of (α, d, e) are varied within certain ranges. We also use VPLAN for solving the differential equation system under different controls.

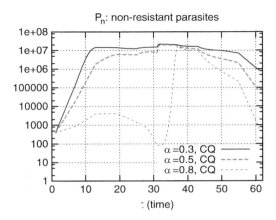

Fig. 5 Simulations of sensitive parasite populations for different proportions of treatments. Increasing α leads to rapid decrease in sensitive parasite population to both treatments. Time unit is 5 days

3.5.1 Proportion of Infected Human Class with Full Treatment

In this part α is our control parameter. This is the proportion of infected human class with full treatment. In this case, two drugs are chosen with treatment efficacy $e = 0.7(\text{time}^{-1})$, Chloroquine (CQ) with wash-out duration $d = 40.0$ and Sulfadoxine-pyrimethamine (SP) with $d = 6.0$. Their results turn out to be quite similar. As result of the treatment, the sensitive parasite population is changed much faster than the resistant one.

Simulation in Fig. 5 shows that the proportion of infected humans receiving full medical treatment is very essential in disease control. In addition, the figure indicates that *at least a certain proportion* of the population needs to be treated properly in order to avoid deadly clinical malaria in the case of very high density of parasites. Recall that we have shown parasite populations with scale $1/10^{10}$ and Cisse is only a small town with less than one thousand habitants. We can calculate the average density of parasites in each patient accordingly.

Beside that, different levels of treatment strongly influence the fitness between sensitive and resistant parasite populations. As we see in Fig. 6, full treatments give resistant parasites a chance to better compete with sensitive parasites to invade the environment.

3.5.2 Drug Regimens with Different Half-Lives

Using drug regimens with different half-life time would effect the picture of sensitive and resistant parasite populations. This effects the initial ratios of sensitive and resistant populations and also the number of parasites which become adapted to the drug environment. In Cisse–Burkina Faso we do not have any data about different drug usages and how parasites would resist to certain treatment. It is necessary to run virtual simulations with different possible values of parameters d.

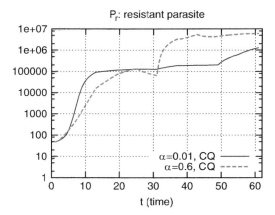

Fig. 6 Simulations of resistant parasite population for different proportions of treatment. High proportion of full treatment makes resistant parasite population increase faster than low proportion of treatment in the later period. The *dashed line* represents 60% full treatment with Chloroquine while the *solid line* represents only 1% Chloroquine full treatment. Time unit is 5 days

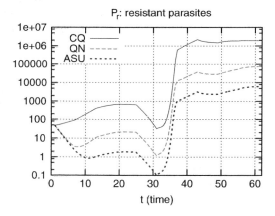

Fig. 7 Simulations of resistant parasites for different drug half-lives. Switching between different therapies leads to noticeable changes in the number of resistant parasites. Simulations are done for Artesunate (ASU), Quinine (QN), Chloroquine (CQ) with approximated wash-out durations $d_{ASU} = 0.04$, $d_{QN} = 0.5$, $d_{CQ} = 40.0$, based on their average half-lives given in [4]

For clear comparison, we use the same initial values for the seven populations and keep the same treatment efficacy in all simulations. Assuming that all drug regimens have the same efficacy $e = 0.7(\text{time}^{-1})$, 80% of infected human receiving full treatment $\alpha = 0.8$, different therapies lead to noticeable changes in the resistant parasite population, see Fig. 7.

According to our simulation, for long treatment periods using drugs with shorter half-life gives better performance. However, in the first period of treatment, the

Drug Resistance in Infectious Diseases

Fig. 8 Simulations of resistant parasites and infected humans with different drug treatment policies. Time unit is 5 days

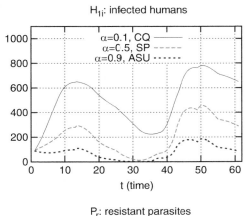
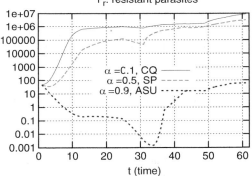

different is not much due to the fact that a large proportion of parasites is sensitive to drugs. That is why in this period it does not matter which type of antimalarial drugs are used. The effect shows later, indicated by the fast increase starting from $t = 30$ in Fig. 7.

We also simulate the case when first Chloroquine is used and afterward switched to Artesunate in the last 3 months of study. As expected, the resistant population was reduced considerably fast, this opens a good chance for alternative treatments.

3.5.3 Different Treatment Policies: Combined Control of α and d

With the setting similar to Burkina Faso, we begin with high efficacies for most of the antimalarial drugs. We simulate here the case when we combine the two mentioned controls, the proportion of full treatment α and the drug type expressed by d. Shortly speaking, we have combined advantages. We can keep both the sensitive and resistant parasite populations under control. The resistant parasite population is relatively small, as we see with Artesunate treatment in the lower panel of Fig. 8. On the upper panel, combined control leads to noticeable change in the infected human population.

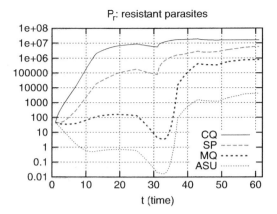

Fig. 9 Simulations of resistant parasites for different treatment efficacies. Simulations are done for Chloroquine (CQ), Sunfadioxine-pyrimethamine (SP), Mefloquine (MQ), Artesunate (ASU) with the values of e are $0.2, 0.4, 0.6, 0.6$ respectively. The wash-out duration d of all drugs is based on [4]: $d_{ASU} = 0.04$, $d_{QN} = 0.5$, $d_{MQ} = 16.0$, $d_{CQ} = 40.0$. The time unit is 5 days

3.5.4 Extended Setting: Different Efficacies of Drug Treatments

In this part we consider the case in which all the infected people can afford their necessary medication or the health care systems are good enough to cover all the cost. So all patients can be treated and $\alpha = 1.0$. The treatment efficacy e and drug d are our control parameters. Simulations are done for Chloroquine (CQ), Sunfadioxine-pyrimethamine (SP), Mefloquine (MQ), Artesunate (ASU). The treatment efficacies are close to the values of antimalarial drugs taken from Asia, especially Vietnam. They are based on a report by WHO in [17]. Note that in the Burkina Faso setting, most of the antimalarial medication would begin with high efficacy—since most of the people living in the rural regions have had no access to them before.

Figure 9 using logarithmic scale, lets us easily see that there are big differences between Chloroquine, Sunfadioxine-pyrimethamine, Mefloquine and Artesunate. In the first three drugs, parasite resistance levels in the last period are very high. In contrast, with Artesunate or the similar Artemisinin Combination Therapy, the regimens are still effective, they keep parasite resistance low.

4 Conclusions

We discuss the simulation results and summarize the works in this paper.

Drug Resistance in Infectious Diseases

4.1 Interpretation of the Simulation Results

The simulations which have been done with the fitted model lead to the following interpretations:

- From our result, the use of medication do speed up the resistant parasite populations. However, we need them avoid the high (sensitive) parasite density in infected humans, e.g. to keep the average parasite density in human blood below a specific level. Here our model can serve as the potential basis for the control problems, providing an optimal strategy of treatment. We can optimize the proportion (or number) of infected human population which need to be treated with full doses.
- In case of malaria, our model simulations suggest that parasite mutation and drug adaptation both play big roles in resistance. In quantitative sense, the simulation shows that: when drug with long half-life is employed, drug adaption is dominant. This is not the case with short half-life drugs. So the shorter the drug half-life is, the fewer resistant parasites are newly created.
- For the first time of treatment, our simulations show that the outcome of different drug treatments are quite similar. It is explained by the fact that in Burkina Faso, not many people have access to treatment. Most of the drugs are too expensive for the region, especially in a village like Cisse, where our data come from.

  That is why in Burkina Faso or similar regions in Africa, treatment can start with any drug and later switch to a new therapy when the resistant pathogens to the old drugs become dominant. This shows clearly efficacy since most of the parasites which resist to the old drug are sensitive to the new one.
- We also consider different regions with different drug usage. In Asia, antimalarial drugs are sold openly and any person can buy them without prescription from doctors. For most patients, treatment is not their first time. We observe in our simulations that the drug therapies and their efficacies strongly influence the overall treatment results.

By using a quantitative model, we can simulate a lot of scenarios in advance. Depending on preferred criteria, such as keeping the total parasite density below certain threshold or reducing the resistant parasite population, clinicians should be able to choose the suitable therapies. In general, the model results are valid not only for malaria but also for other infectious diseases whose biology are similar to malaria. Based on these conclusions, treatment strategies to reduce drug resistance can be elaborated.

4.2 Summary

To sum up, in this paper we have built a model describing the population dynamics as they appear in vector-borne diseases. In comparison to most of the mathematical models before, our model has two new compartments for parasite populations and includes parameters describing drug treatment.

With numerical study, we have solved parameter estimation and simulation problems. We have obtained fitted parameters and a good agreement with the data in Burkina Faso. We have also simulated the fitted model with controllable parameters. Using the VPLAN package, we were able to solve the systems and to see the influence of drug treatment not only on parasites but also on the host populations.

Acknowledgements We would like to thank professor Klaus Dietz, Dr. Yazoumé Yé, Dr. Johannes Schlöder and the anonymous referee for a lot of valuable comments, discussions and corrections; Dr. Stefan Körkel and Cetin Sert for their great helps in computation. Many thanks to Dr. Maria Neuss-Radu, who encouraged me (LTTA) a lot to complete the final manuscript.

This work was financially supported by the Vietnamese Government, the DFG (International Graduiertenkolleg 710 and Heidelberg Graduate School MathComp) and ABB company.

References

1. Anderson R.M., May R.M.: Infectious Diseases of Humans: Dynamics and Control. Oxford University Press, (1991).
2. Aneke S.J.: Mathematical modeling of drug resistant malaria parasites and vector populations. Math. Meth. Appl. Sci. **25**, 335–346, (2002).
3. Bock H.G., Kostina E., Schlöder J.P.: Numerical Methods for Parameter Estimation in Nonlinear Differential Algebraic Equations. GAMM-Mitt. **30**, no. 2, 376–408, (2007).
4. Bloland P.B.: Drug resistance in malaria, World Health Organization, (2001).
5. Dietz K.: Mathematical models for transmissions and control of malaria. In: Malaria: Principles and Practice of Malariology. W. Wernsdorfer, I. McGregor (Eds.), Churchill Livingstone, Edinburgh, 1091–1133, (1988).
6. Esteva L., Gumel A.B., de León C.V.: Qualitative study of transmission dynamics of drug-resistant malaria. Mathematical and Computer Modelling **50** (3–4), 611–630, (2009).
7. Hastings I.M., D'Alessandro U.: Modelling a Predictable Disaster: The Rise and Spread of Drug-resistant Malaria. Parasitology Today, **16**, no. 8, 340–347, (2000).
8. Keeling M.J., Rohani P.: Modeling Infectious Diseases in Humans and Animals. Princeton University Press, (2008).
9. Körkel S.: Numerische Methoden für Optimale Versuchsplanungsprobleme bei nichtlinearen DAE-Modellen, doctoral thesis, University of Heidelberg, (2002).
10. Molineaux L., Diebner H. H., Eichner M., Collins W. E., Jeffery G. M., Dietz K.: Plasmodium falciparum parasitaemia described by a new mathematical model. Parasitology, **122**, 379–391, (2001).
11. Yé Y.: Incorporating environmental factors in modelling malaria transmission in under five children in rural Burkina Faso, doctoral thesis, University of Heidelberg, (2005).
12. Vynnycky E., White R.G.: An introduction to Infectious Disease Modelling, Oxford University Press, (2010).
13. White N.J.: The role of anti-malarial drugs in eliminating malaria (review), Malaria Journal, 7(Suppl 1):S8. doi:10.1186/1475-2875-7-S1-S8, (2008).
14. http://www.enotes.com/public-health-encyclopedia/vector-borne-diseases.
15. Bulletin of the World Health Organization, 79 (10), (2001). http://www.scielosp.org/pdf/bwho/v79n10/79n10a21.pdf.
16. Mortality country fact sheet 2006, Burkina Faso. http://www.who.int/whosis/mort/profiles/mort_afro_bfa_burkinafaso.pdf.
17. Malaria drug resistance, World Health Organization, "http://www.wpro.who.int/sites/mvp/Malaria+drug+resistance.htm."

Mathematical Models of Hematopoietic Reconstitution After Stem Cell Transplantation

Anna Marciniak-Czochra and Thomas Stiehl

Abstract Transplantation of bone marrow stem cells is a widely used option to treat leukemias and other diseases. Nevertheless this intervention is linked to life-threatening complications. Numerous clinical trials have been performed to evaluate various treatment options. Since there exist strong interindividual variations in patients' responses, results of clinical trials are hardly applicable to individual patients. In this paper a mathematical model of hematopoiesis introduced by us in (Marciniak-Czochra et al.: Stem Cells Dev. 18:377–85, 2009) is calibrated based on clinical data and applied to study several aspects of short term reconstitution after bone marrow transplantation. Parameter estimation is performed based on the data of time evolution of leukocyte counts after chemotherapy and bone marrow transplantation obtained for individual patients. The model allows to simulate various treatment options for large groups of individual patients, to compare the effects of the treatments on individual patients and to evaluate how the properties of the transplant and cytokine treatment affect the time of reconstitution.

Keywords Differential equations • Hematopoiesis • Mathematical model • Stem cell transplantation

1 Introduction

Stem cell research is an important field of life sciences with promising clinical impacts. During last years an enormous amount of information on specific factors, genes and cellular interactions involved in tissue homeostasis, cell differentiation

---

A. Marciniak-Czochra (✉) · T. Stiehl
Interdisciplinary Center for Scientific Computing, Im Neuenheimer Feld 368,
69120 Heidelberg, Germany
e-mail: anna.marciniak@iwr.uni-heidelberg.de; Thomas.Stiehl@iwr.uni-heidelberg.de

H.G. Bock et al. (eds.), *Model Based Parameter Estimation*, Contributions
in Mathematical and Computational Sciences 4, DOI 10.1007/978-3-642-30367-8_9,
© Springer-Verlag Berlin Heidelberg 2013

and stem cell function has been collected. However, the precise mechanisms regulating stem cell self-renewal, maintenance and differentiation in clinically important scenarios are still poorly understood.

Mathematical modelling has evolved as an important tool in describing and understanding of bio-medical processes. Such processes are characterised by a high level of complexity. In most cases important sub-processes are poorly defined or not known. This fact and the mathematical or computational complexity of resulting models make it impossible and useless to include all involved processes into a mathematical model. Nevertheless mathematical modelling is a powerful tool in addressing specific biological questions or in comparing different well defined hypotheses.

In higher organisms, a steady supply of somatic cells is accomplished by proliferation of corresponding stem cells, which retain the capability for almost indefinite self-renewal. Driven by hormonal signals from the organism, some stem cells commit to differentiation and maturation into specialised lineages. Organs and tissues of complex organisms, such as blood, consist of a large number of different specialised (differentiated) cell types. Unspecialised cells that are able to give rise to cells with different biological properties are called stem cells. Most tissues contain small populations of specific tissue stem cells (adult stem cells) accounting for homeostasis and tissue repair. The hematopoietic system is maintained by a population of hematopoietic stem cells (HSC) that is able to give rise to blood cells of all lineages (erythrocytes, leukocytes, platelets).

Understanding of the mechanisms governing hematopoiesis is of central interest for stem cell biology, especially because of its clinical impact [1]. High regenerative properties of hematopoietic stem cells are used to reconstitute blood structure of patients after treatment with high-dose chemotherapy, which results in a rapid decline of blood cell counts. Several diseases of blood and bone marrow, e.g., leukemias and other cancers, can be treated and sometimes cured using stem cell transplantation, formerly known as bone marrow transplantation. Intravenous administration of growth factors allows to mobilize HSC and early progenitor cells into blood stream. From blood stream cells can be collected due to expression of surface markers.

A transplantation is called autologous if donor and recipient are the same person, otherwise it is called allogeneic. Stem cell transplantations allow application of aggressive chemo-therapies with high marrow toxicity. After chemotherapy infusion of the cryopreserved stem/progenitor cell graft allows repopulation of bone marrow and recovery of blood cell counts. Until occurrence of blood cell recovery patients are susceptible to infections due to impairment of immune function. For this reason it is an important clinical goal to keep the period of aplasia as short as possible.

A major advantage of autologous transplantation is the absence of mismatch reactions. The major disadvantage of this approach is possible contamination of the transplant by malignant cells. In case of allogeneic transplantations, malignant cells are attacked and eliminated by graft cells. The major problems of allogeneic

Mathematical Models of Hematopoiesis 193

transplantation are the search of an adequate donator and control of possible acute or chronic graft versus host reaction.

Because of relative facility in sampling blood or bone marrow, the hematopoietic system is well suited for modelling and validation. Hematopoiesis involves complex processes that can be examined at the level of genes, signal transduction proteins, or the populations of diverse cell types. By viewing hematopoiesis as a dynamical system and stress and disease as a perturbation of the system, the system biology approach may contribute to a better understanding of physiological and pathological states of hematopoiesis.

Depending on the specific scientific questions, different mathematical models were developed to study hematopoiesis. One established method of modelling of hierarchical cell systems is to use a discrete collection of ordinary differential equations, each of which describes dynamics of cells at a single maturation stage. In such framework, a range of mathematical results have been obtained, such as stability and oscillation criteria, some of which are applicable to modelling of the underlying biological systems, e.g., [2]. Another group of models addresses the effects of stochasticity in the cell fate decisions [3] or in the regulation of cell quiescence [4]. These models were designed to simulate the time dynamics of blood reconstitution [5], spread of cancer stem cells [6] or the perturbations in the blood dynamics in diseases such as e.g., myleoid leukemia [2, 6]. They provided explanations of many observations and the model of erythroid production of Lasota and Wazewska (as cited in [7]) was successfully validated in patients already in the early 1980s.

In the following we describe a new approach to mathematical modelling of the hematopoietic (blood-forming) system and possible applications to clinical scenarios of stem cell transplantation. The starting point of our study are the models of stem cell differentiation, which we have recently developed in a series of papers [8–11]. The aim is to provide insights how different cellular properties influence hematopoietic recovery after stem cell transplantation. In particular, the role of the asymmetry of cell division is being investigated motivated by the recent experimental observations [12].

In this paper we focus on the application of proposed models to clinical data. The equations are calibrated based on the data obtained from patients with multiple myeloma after high-dose chemotherapy and stem cells transplantation. As a practical application, the models are used to investigate the dependence of the reconstitution time on the size of transplant and its properties. With our approach it is possible to compare the effect of various medical treatments on individual patients.

The paper is organised as follows. In Sect. 2 we recall the assumptions and formulation of the basic model and present a specific choice of nonlinearities. In Sect. 3 a link to a continuous maturation model is presented. Section 4 is devoted to the clinical application of the proposed model. It includes model calibration, application to large patient groups to enable comparison to clinical trials performed on such groups, and application to some treatment scenarios. We conclude in Sect. 5 with some final comments and suggestions for further investigation.

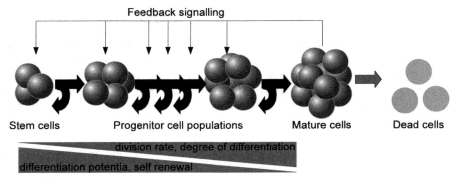

Fig. 1 Schematic illustration of the model assumptions concerning the increasing rate of proliferation and decreasing fraction of self-renewal versus differentiation along the differentiation tree

2 Model of Hematopoiesis with Discrete Maturation

2.1 Assumptions

Our model is based on the traditional assumption that in each lineage of blood cell precursors, there exists a discrete chain of maturation stages, which are sequentially traversed [13]. Cell behaviour is characterised by the proliferative activity, the probability to differentiate and the probability to die. We assume that the cell properties change during the maturation process. For simplicity we restrict our model to one cell lineage. This restriction is also motivated from a clinical point of view. Most complications occurring shortly after stem cell transplantation are caused by impaired immune function due to a lack of mature leukocytes [14]. For this reason, in a clinical applications of the model it is sufficient to consider leukocytes.

We assume that cell behaviour is regulated by feedback signalling. This assumption is based on experimental data, see, e.g., [15]. As an approximation of reality and motivated by clinical data, we assume that feedback signalling depends only on the level of mature cells [16]. For simplicity we focus on one kind of feedback signal, e.g., G-CSF, which is the major regulator of granulopoiesis [15]. Model assumptions and general stem cell properties are summarised in Fig. 1.

Equations for the cell numbers follow from an accepted model of cell cycle that is treated as a well-mixed tank, from which cells may either enter division or death and the length of cell cycle is equally distributed among individuals. Since the hematopoietic system consists of large numbers of cells (of the order of 10^9 leukocytes per litre blood) differential equations are a suitable tool to describe the processes of interest.

Mathematical Models of Hematopoiesis

2.2 Equations

Denote by $c_i(t)$ the population density of cell type i at time t. The density of the stem cell population at time t is denoted as $c_1(t)$ and the concentration of signalling molecules as $s(t)$. Time evolution of cell system is described by the following system of ordinary differential equations,

$$
\frac{dc_1(t)}{dt} = f_1(s(t), c_1(t)),
$$

$$
\frac{dc_2(t)}{dt} = f_2(s(t), c_2(t)) + g_1(s(t), c_1(t)),
$$

$$
\vdots \quad \vdots \quad \vdots
$$

$$
\frac{dc_n(t)}{dt} = f_n(s(t), c_n(t)) + g_{n-1}(s(t), c_{n-1}(t)). \tag{1}
$$

$g_i(s(t), c_i(t))$ denotes a flux of cells from the maturation stage i to the maturation stage $i + 1$ due to differentiation. Since we neglect any de-differentiation events we assume that $g_i(s(t), c_i(t))$ is nonnegative. The term $f_i(s(t), c_i(t))$ denotes a change of $c_i(t)$ that is caused by processes taking place at the ith stage of maturation. If the gain of cells caused by proliferation and self-renewal is stronger than the loss caused by differentiation or death, then $f_i(s(t), c_i(t))$ is positive. Otherwise $f_i(s(t), c_i(t))$ is negative. Since mature cells are postmitotic, i.e., they cannot proliferate, the term $f_n(s(t), c_n(t))$ accounts only for cell death and is, therefore, negative. The whole process is regulated by a single feedback mechanism based on the assumption that there exist signalling molecules (cytokines) which regulate the differentiation or proliferation process. The intensity of the signal is denoted by $s(t)$. Following [8] we assume that it depends on the level of mature cells, $s(t) = s(c_n(t))$. A specific choice of the function s is presented in the next section. A mathematical analysis of this model and biological implications on the definitions of stem cells are described in [9].

2.3 A Special Case

For understanding of clinical processes it is helpful to replace the general terms f_i and g_i in the above model by explicit functions depending on quantitative cell properties that are, at least in principle, accessible to experiments.

To model the differentiation process quantitatively we introduce the fraction of self-renewal that describes which fraction of the progeny cells is identical to the mother cells (i.e. does not differentiate). This parameter can be interpreted as the probability that a daughter does not differentiate after cell division. The fraction of self-renewal is a property defined on the scale of a whole population.

Denote the proliferation rate of the subpopulation of type i at time t by $p_i(t)$, the fraction of self-renewal by $a_i(t)$ and the death rate by $d_i(t)$. Then, the flux to mitosis at time t is given by $p_i(t)c_i(t)$. The fraction $a_i(t)$ of daughter cells stays undifferentiated. Therefore, the influx to cell population i after cell division is given by $2a_i(t)p_i(t)c_i(t)$ and the flux to the next cell compartment is given by $2(1 - a_i(t))p_i(t)c_i(t)$. The flux to death at time t is given by $d_i(t)c_i(t)$. Choice of

$$f_i(t) = (2a_i(t)-1)p_i(t)c_i(t) - d_i(t)c_i(t), \quad \text{for} \quad i < n,$$
$$g_i(t) = 2(1-a_{i-1}(t))p_{i-1}(t)c_{i-1}(t), \quad \text{for} \quad 1 < i < n,$$
$$f_n(t) = -d_n(t)c_n(t).$$

leads to the model described in [8].

2.3.1 Signal Feedback

In this section we present a derivation of a specific form of the signalling feedback, motivated by the observation that dynamics of signalling molecules takes place on a faster time scale than the process of cell proliferation and differentiation [17, 18].

We assume that $a_i(t)$ and $p_i(t)$ depend solely on the feedback signal at time t, i.e., $a_i(t) \equiv a_i(s(t))$ and $p_i(t) \equiv p_i(s(t))$.

Assuming that signal molecules are secreted by specialised cells at a constant rate α and degraded proportional to the level of mature cells c_n and at a constant rate μ, we obtain the following differential equation for the dynamics of the concentration of signalling molecules, denoted by $S(t)$:

$$\frac{d}{dt}S(t) = \alpha - \mu S(T) - \beta S(t)c_n(t).$$

As discussed above a secretion of cytokines is very fast in comparison to cell proliferation and differentiation and the level of cytokines approaches a quasi-steady state. Applying a quasi-steady state approximation and substitution of $s(t) := \frac{\mu}{\alpha}S(t)$ and $k := \frac{\beta}{\mu}$ enables us to calculate signal intensity s as

$$s := s(c_n(t)) = \frac{1}{1 + kc_n(t)},$$

which is between 0 and 1. Considering different plausible regulatory feedback mechanisms leads to different types of nonlinearities in the model equations. In particular, in [8] two hypotheses were tested. The Hypothesis 1 assumes that differentiation is governed by enhancing the rate of proliferation only, while in the Hypothesis 2 it is the ratio of the rate of self-renewal to the rate of differentiation

Mathematical Models of Hematopoiesis

that is regulated by external signals. Consequently, two different regulatory modes were proposed:

(M1) constant p_i and $a_i(s) = \frac{a_{i,max}}{1+kc_n}$,

(M2) $p_i(s) = \frac{p_i}{1+kc_n}$ and constant a_i.

Different approaches to model signal feedback are also possible. We checked numerically that, for example, an exponential dependence of signal intensity on mature cell counts, $s(t) = e^{-const \cdot c_n(t)}$, similar to that of the Lasota–Wazewska model, leads to similar qualitative results.

Numerical simulations and analysis of the model solutions under both hypotheses (each leading to a different mathematical model) demonstrate that the regulation of self-renewal fractions is more efficient and can be achieved in the clinically relevant time scale [8]. Regulation of the rates of proliferation (parameters p_i) is not sufficient for that purpose.

Therefore, in the reminder of this paper we assume the proliferation rates to be constant in time and self-renewal fractions to be controlled by the signal feedback, $a_i(t) \equiv a_i(s(c_n(t))) = a_{i,max} \cdot s(c_n(t))$, where $a_{i,max}$ is the maximal fraction of self-renewal. Furthermore, death rates are assumed to be constant in time. This results in the following set of equations,

$$\frac{dc_1}{dt} = (2a_{1,max}s - 1)p_1c_1 - d_1c_1,$$
$$\ldots$$
$$\frac{dc_i}{dt} = (2a_{i,max}s - 1)p_ic_i + 2(1 - a_{i-1,max}s)p_{i-1}c_{i-1} - d_ic_i, \qquad (2)$$
$$\ldots$$
$$\frac{dc_n}{dt} = 2(1 - a_{n-1,max}s)p_{n-1}c_{n-1} - d_nc_n,$$
$$s = \frac{1}{(1 + kc_n)}$$

The following, biologically relevant, assumptions are made for the model parameters and initial data:

$$
\begin{aligned}
t &\in [0, \infty), \\
c_i(0) &\geq 0, \text{for } i = 1\ldots n, \\
d_i &\geq 0, \text{for } i = 1\ldots n-1, \\
d_n &> 0, \\
p_i &> 0, \text{for } i = 1\ldots n-1, \\
a_{i,max} &\in [0, 1) \text{ for } i = 1\ldots n-1. \\
k &> 0.
\end{aligned}
$$

3 Continuous Model of Hematopoiesis

One of the fundamental biological questions is whether cell differentiation is a discrete or a continuous process and what is the measure of cell maturation? Is the pace of maturation (commitment) dictated by successive divisions, or is maturation a continuous process decoupled from proliferation that may take place also between cell divisions? Consequently we approach the question how to choose an appropriate model. To address these questions and to investigate the impact of possible continuous transformations on the differentiation process, a new model based on partial differential equations of transport type was introduced [11]. The model consists of three differential equations of the following form,

$$\frac{dc_1(t)}{dt} = f_1(s(t), c_1(t)),$$

$$\partial_t c(x,t) + \partial_x [g(x,s)c(x,t)] = f(s(t), c(x,t)),$$

$$g(0,s)c(0,t) = g_1(s(t), c_1(t)), \quad t > 0,$$

$$\frac{dc_2(t)}{dt} = f_2(s(t), c_2(t)) + g(s(t), c(x^*, t)). \tag{3}$$

Here $c_1(t)$ denotes the number of stem cells, $c_2(t)$—the number of mature cells and $c(x, t)$—the distribution density of progenitor cells structured with respect to the maturity level x, so that $\int_{x_1}^{x_2} c(x,t)dx$ is equal to the number of progenitors with maturity between x_1 and x_2. This includes maturity stages between stem cells and differentiated cells. We assume that $x = x^*$ denotes the last maturity level of immature cells. Moreover, functions f_1, f, f_2, g_1 and g are analogous to those in the system (3). The specific form of this model corresponding to our choice of kinetic functions was studied analytically and numerically in [11]. Among others, it was shown that the discrete and continuous differentiation models are not equivalent. However, it was also demonstrated that the models can exhibit exactly the same dynamics for a suitable scaling of the maturation rate function $g(x,s)$ and under assumptions on parameters which provide asymptotic convergence of solutions to a unique positive stationary solution. The latest result indicates that having a calibration of the discrete differentiation model based on the clinical data we can easily fit the continuous model. Therefore, in the reminder of this paper we focus on the discrete maturation model (2).

4 Simulation of Stem Cell Transplantation

In this paragraph we compare model simulations to clinical data obtained from patients with multiple myeloma after high-dose chemotherapy and stem cell transplantation, and discuss possible applications.

4.1 Comparison of Simulations to Clinical Data

For simulations we choose 8 compartments, corresponding to the following maturity stages of hematopoiesis: HSC, LTC-IC, CFU-GM, CFU-G, Myeloblast, Promyelocyte, Myelocyte, Neutrophil Granulocyte [13]. We focused on neutrophils, since they are the major part of leukocytes (about 50–75%)[13]. Parameter values of proliferation rates and death rates as well as initial conditions have been assumed based on the literature. Self-renewal fractions and the coefficient k, which cannot be measured, were estimated in such way that steady state cell counts are in accordance with literature. We assumed that self-renewal decreases from primitive to more specialised cell types and that stem cells divide about two orders of magnitude slower than most mature progenitors. In two parameter sets proliferation increases strictly monotonous with maturity. These assumptions are in agreement with the classical understanding of stem cell systems. The choice of three parameter sets were done manually based on the patients data.

We simulate the model with three different parameter sets to investigate the influence of interindividual differences on reconstitution kinetics. Steady state population sizes increase from stem cells to most mature progenitors: stem cells of order 10^2 per kg to 10^3 per kg, LTC-IC of order 10^5 per kg, CFU-GM of order 10^6 per kg, CFU-G of order 10^7 per kg, Myeloblasts, Promyelocytes and Myelocytes of order 10^8 per kg. For initial conditions we assume that 0.1% of transplanted cells are stem cells. The ratio of HSC:LTC-IC is assumed to be 1:10, the ratios LTC-IC:CFU-GM and CFU-GM:CFU:G are assumed to be 1:3. Furthermore it is assumed that the transplant does not contain any myeloblasts, promyelocytes nor myelocytes.

Figure 2 compares evolution of leukocyte counts to clinical data. Figure 3 shows post-transplant time evolution of all cell populations. Clinical data concerning hematopoietic reconstitution on HSC transplantation are collected from a clinical trial performed by Klaus and colleagues [19] on patients with advanced multiple myeloma. Median age is 44 (37–66 years). Treatment was performed at the Department of Medicine V, University of Heidelberg using a myeloablative high-dose chemotherapy (Melphalan $100 \, mg/m^2$ body surface) at days -3 and -2 followed by autografting of a unmanipulated HSC transplant (median of 3.6 (2.0–10.3) $\cdot 10^6$ CD34+ cells/kg body weight) at day 0. The transplant has been harvested in advance by leukapheresis from granulocyte colony-stimulating factor mobilised peripheral blood.

Simulations show that our model is able to cover the observed range of reconstitution patterns by slight variation of the model parameters. Since simulated blood leukocyte counts are in reasonable agreement with reality we use the model to investigate the dependence of time needed for recovery on transplant size.

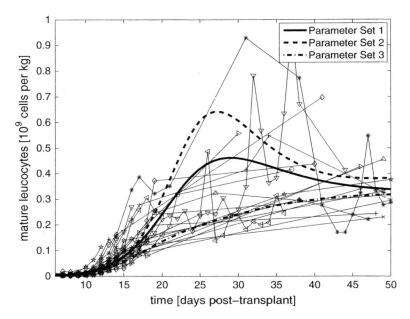

Fig. 2 Clinical data of blood reconstitution following a bone marrow transplant, and simulated solutions of the mathematical model for three different parameter sets. To calculate cells per kg body weight, it was assumed that body weight is 70 kg and blood volume 5 l. The following parameter values were used. Parameter Set 1: $p_1 = 0.0060$, $p_2 = 0.0300$, $p_3 = 0.1500$, $p_4 = 0.6000$, $p_5 = 0.6500$, $p_6 = 1.000$, $p_7 = 1.5000$, $a_{1,max} = 0.7350$, $a_{2,max} = 0.7298$, $a_{3,max} = 0.7245$, $a_{4,max} = 0.7140$, $a_{5,max} = 0.5775$, $a_{6,max} = 0.4725$, $a_{7,max} = 0.3675$, $d_1 = \cdots = d_7 = 0$, $d_8 = 2.7700$, $k = 1.2800 \cdot 10-9$. Parameter Set 2: $p1 = 0.0060$, $p2 = 0.0300$, $p3 = 0.1500$, $p4 = 0.6000$, $p5 = 0.6500$, $p6 = 1.000$, $p7 = 1.5000$, $a_{1,max} = 0.7700$, $a_{2,max} = 0.7645$, $a_{3,max} = 0.7590$, $a_{4,max} = 0.7590$, $a_{5,max} = 0.5500$, $a_{6,max} = 0.4400$, $a_{7,max} = 0.3300$, $d_1 = \cdots = d_7 = 0$, $d_8 = 2.7700$, $k = 1.2800 \cdot 10-9$. Parameter Set 3: $p1 = 0.0057$, $p2 = 0.0297$, $p3 = 0.03$, $p4 = 0.4200$, $p5 = 0.4800$, $p6 = 1.0500$, $p7 = 1.5000$, $a_{1,max} = 0.7000$, $a_{2,max} = 0.6990$, $a_{3,max} = 0.6690$, $a_{4,max} = 0.6980$, $a_{5,max} = 0.5000$, $a_{6,max} = 0.5000$, $a_{7,max} = 0.5500$, $d_1 = \cdots = d_7 = 0$, $d_8 = 2.7700$, $k = 1.2800 \cdot 10^{-9}$

4.2 Application to Large Patient Groups

The proposed model can be applied to large patient groups. To account for interindividual variability we choose the fractions of self-renewal a_i randomly between 0 and 1 with the additional assumptions $a_1 > 0.5$ and $a_1 > a_i$ for all $i > 1$. The last conditions are necessary for existence of a positive steady state, for details see [9]. To exclude unrealistic parameter sets only choices with steady state leukocyte values greater or equal $2 \cdot 10^9/l$ are taken into account. In Fig. 6 simulation results are compared to clinical data collected by Lowenthal and colleagues [28]. Lowenthal defined hematopoietic recovery as exceedance of $5 \cdot 10^8$

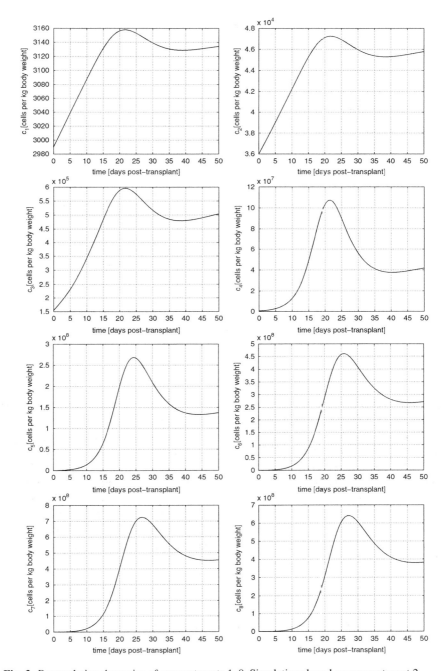

Fig. 3 Repopulation dynamics of compartments 1–8. Simulations based on parameter set 2

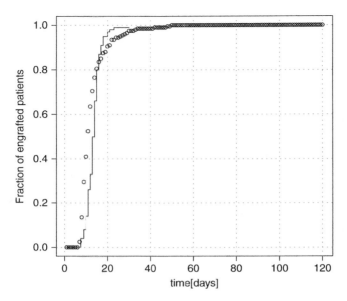

Fig. 4 Simulation of blood cell recovery of a large population. The plot shows the percentage of patients reaching $5 \cdot 10^8$ granulocytes (corresponding to 10^9 leukocytes) per litre blood at different time points. Model results calculated in cells per kg body weight are compared to the clinical data assuming using average body weight of 70 kg and blood volume of 5 l. Experimental data is taken from [28]

neutrophil granulocytes per liter blood. Since neutrophil granulocytes account for about 50% of peripheral leukocytes, we take a threshold of 10^9 leukocytes per litre blood as the criterion for recovery. For simplicity all parameters besides a_i are chosen as in Parameter Set 2. Perturbing randomly other parameter values leads to similar results. For the presented simulation we assume a transplant size of $9.5 \cdot 10^6$ CD34 positive cells per kg body weight, which corresponds to the median transplant reported by Lowenthal [28]. Simulations are run for 200 parameter sets obtained by a random choice of self-renewal fractions as described above. Simulation results are in good agreement with clinical data. This demonstrates that our model is, in principle, able to describe behaviour of large patient groups as they are assessed in clinical trials (Fig. 4).

4.3 Application to Clinical Scenarios

Stem cell transplantation is an important clinical treatment option for hematologic and non hematologic diseases. The clinical scenario of autologous transplantation is summarised in Fig. 5. Reduction of the period of immune impairment is an

Mathematical Models of Hematopoiesis

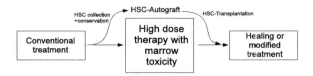

Fig. 5 Autologous stem cell transplantation: collection, conservation and reinfusion of stem and progenitor cells allows application of marrow toxic therapies

important clinical goal and has been addressed by different studies, e.g., [19]. Clinical results indicate that augmented transplant sizes lead to faster recovery. Nevertheless, the qualitative shape of relationship between transplant size and time needed for recovery has not yet been described. Due to occurring infections and other complications it is difficult to establish such relationship based on clinical data alone.

On the other hand it has been reported that cells with leukemia associated mutations can be found in the marrow of healthy subjects [20,21] and it is speculated that the host's marrow can be supportive to proliferation of such cells after transplantation [22]. To a certain extent bone marrow seems to serve as a threshold of adult stem cells of various types that migrate to appropriate sites when there is need [23–26]. A similar mechanism for cancer stem cells has not yet been evaluated and cannot be excluded. If cancer stem cells behaved similar to their benign counterparts, bone marrow could be a storage of cancer cells of different types. If this was the case, excessive mobilisation of bone marrow derived cells could be linked to a higher risk of donor derived diseases. For this reason it is necessary to understand better the dependence of transplant size on reconstitution time.

We use the model to investigate qualitatively the dependence of time needed for recovery on transplant size. We defined hematopoietic recovery as blood leukocyte counts exceeding 10^9/l. Figure 6 shows the dependence of recovery time on the number of transplanted cells for three different parameter sets. Simulation results show that the benefit of additionally transplanted cells decreases with increasing transplant size. For this reason it cannot be recommended to excessively increase transplant sizes in case of autologous transplantations.

Since our model covers behaviour of large patient groups these results might be considered as representative for qualitative dependencies in larger populations.

5 Discussion

In this paper we showed that the models of stem cell differentiation with cell self-renewal regulated by a nonlinear signalling feedback can be applied to describe the reconstitution pattern of patients after autologous stem cell transplantation. We calibrated the model using proliferation rates and death rates reported in the literature. The unknown parameters were estimated manually based on patients data

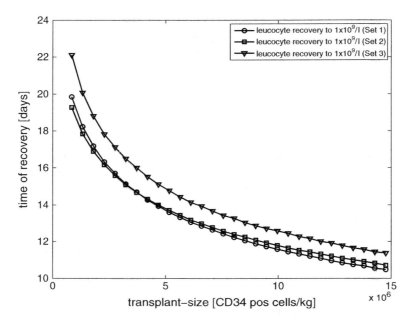

Fig. 6 Reconstitution time depending on transplant size for three representative parameter sets. The values of the parameters are given in Fig. 2

on the time dynamics of leucocyte counts, steady state values and the transplant data. Model calibration using inverse numerical methods could help to obtain better insight into dependence of model outcome on the specific parameters and accelerate the modelling process. Numerical parameter estimation can be applied to identify the parameters influencing the process of blood reconstitution critically.

Among others, the models were applied to get insights into the qualitative dependence of recovery process after stem cell transplantation on transplant size. Our models show that interindividual differences in cell properties (model parameters) may account for interindividual differences observed in clinical practice.

The proposed models contain different simplifications, such as restriction to one cell lineage, exclusive dependence of feedback on mature cells and incorporation of only one feedback signal. These simplifications were adjusted based on clinical data. The main problem of clinical applications is the lack of knowledge of in vivo model parameters and of exact regulatory modes. The crosstalk between different cytokines seems to be very complex and is not yet well understood from a biological point of view [27]. Furthermore our model does not include immune reactions after allogeneic transplantations for which underlying processes are not well defined. Our approach focuses on modelling the impact of cellular properties on clinical measurable results. The models are able to provide insights which parameters are important to obtain fast clinical recovery. It may, therefore, be helpful for further optimisation of transplantation strategies in areas where clinical experiments cannot

Mathematical Models of Hematopoiesis

be performed due to ethical reasons. Furthermore, since our model covers the behaviour of large patient groups, it allows to perform clinical trials in silico and enables to better compare simulation results with clinical data.

Acknowledgements The authors were supported by the WIN Kolleg of Heidelberg Academy of Sciences and Humanities "A man is so old as his stem cells?" and Collaborative Research Center, SFB 873, "Maintenance and Differentiation of Stem Cells in Development and Disease." AM-C was supported by ERC Starting Grant "Biostruct" and Emmy Noether Programme of German Research Council (DFG).

References

1. Giebel B., Zhang T., Beckmann J., Spanholtz J., Wernet P., Ho AD, Punzel M.: Primitive human hematopoietic cells give rise to differentially specified daughter cells upon their initial cell division. Blood. **107**, 2146–2152 (2006)
2. Colijn C, Mackey M.C.: A mathematical model of hematopoiesis–I. Periodic chronic myelogenous leukemia. J. Theor Biol.**237**, 117–132 (2005)
3. Till J.E., Siminovitch L., McCulloch E.A.: Stochastic Model of Stem Cell Proliferation Based on Growth of Spleen Colony-Forming Cells. PNAS **51**, 29–49 (1964)
4. Roeder I., Loeffler M.: A novel dynamic model of hematopoietic stem cell organization based on the concept of within-tissue plasticity. Exp Hematol. **30**, 853–861 (2002)
5. Ostby I., Kvalheim G., Rusten L.S., Grottum P.: Mathematical modeling of granulocyte reconstitution after high-dose chemotherapy with stem cell support: effect of post-transplant G-CSF treatment. J Theor Biol. **231**, 69–83 (2004)
6. Michor F., Hughes T.P., Iwasa Y., Branford S., Shah N.P., Sawyers C.L., Nowak M.A.: Dynamics of chronic myeloid leukaemia. Nature **435**, 1267–1270 (2005)
7. Wazewska-Czyzewska M.: Erythrokinetics radioisotopic methods of investigation and mathematical approach. Foreign Scientific Publications Dept. of the National Center for Scientific, Technical, and Economic Information, Springfield (1984)
8. Marciniak-Czochra A., Stiehl T., Ho A.D., Jäger W., Wagner W.: Modeling of asymmetric cell division in hematopoietic stem cells–regulation of self-renewal is essential for efficient repopulation. Stem Cells Dev. **18**, 377–85 (2009)
9. Stiehl, T., Marciniak-Czochra A.: Characterization of stem cells using mathematical models of multistage cell lineages, Mathematical and Computer Modelling (2010), doi:10.1016/j.mcm.2010.03.057.
10. Marciniak-Czochra A., Stiehl T., Wagner W.: Modeling of replicative senescence in hematopoietic development. Aging (Albany NY) **1**, 723–732 (2009)
11. Doumic, M., Marciniak-Czochra, A., Perthame, B., Zubelli, J.: Structured population model of stem cell differentiation. Preprint available at http://hal.archives-ouvertes.fr/inria-00541860/fr/.
12. Ho A.D., Wagner W.:The beauty of asymmetry: asymmetric divisions and self-renewal in the haematopoietic system. Curr Opin Hematol. **14**,330–336 (2007)
13. Jandl J.H.: Blood cell formation. In: Jandl J.H., ed. Textbook of Hematology. Little, Brown and Company, Boston, MA, 1–69 (1996)
14. Fauci A.S., Braunwald M.D., Kasper D.I., Hauser S.I., Longo D.L., Jameson J.L., Loscalzo J.: Harrison's Principles of Internal Medecine, 17th Edition, Mc GrawHill, New York (2008)
15. Metcalf D.: Hematopoietic cytokines. Blood **111**, 485–91 (2008)
16. Shinjo K., Takeshita A., Ohnishi K., Ohno R.: Granulocyte colony-stimulating factor receptor at various stages of normal and leukemic hematopoietic cells. Leuk Lymphoma. **25**,37–46 (1997)

17. Bogner V., Keil L., Kanz K.G., Kirchhoff C., Leidel B.A., Mutschler, W., Biberthaler P.: Very early posttraumatic serum alterations are signifficantly associated to initial massive rbc substitution, injury severity, multiple organ failure and adverse clinical outcome in multiple injured patients. Eur J Med Res. **14**, 284– 291 (2009)
18. Morgan D., Desai A., Edgar B., Glotzer M., Heald R., Karsenti E., Nasmyth K., Pines J., Sherr, C.: The Cell Cycle. In: Alberts B., Johnson A., Lewis J., Raff M., Roberts K., Walter R. (Eds): Molecular Biology of the Cell, 5th Edition. Garland Science, New York (2007)
19. Klaus J., Herrmann D., Breitkreutz I., Hegenbart U., Mazitschek U., Egerer G., Cremer F.W., Lowenthal R.M., Huesing J., Fruehauf S., Moehler T., Ho A.D. and Goldschmidt H.: Effect of CD34 cell dose on hematopoietic reconstitution and outcome in 508 patients with multiple myeloma undergoing autologous peripheral blood stem cell transplantation. Eur J Haematol. **78**, 21–28 (2007)
20. Janz S., Potter M., Rabkin C.C.: Lymphoma- and leukemia-associated chromosomal translocations in healthy individuals. Genes Chromosomes Cancer **36**, 211–223 (2003)
21. Schueler F.,Hirt C., Doelken G.: Chromosomal translocation t(14;18) in healthy individuals. Semin Cancer Biol. **13** 203–209 (2003)
22. Reichard K.K., Zhang Q.Y., Sanchez L., Hozier J., Viswanat D., Foucar K.: Acute myeloid leukemia of donor origin after allogeneic bone marrow transplantation for precursor t-cell acute lymphoblastic leukemia: case report and review of the literature. Am J Hematol. **81** 178–185 (2006)
23. Ratajczak M.Z., Kucia M., Majka M., Reca R., Ratajczak J.: Heterogeneous populations of bone marrow stem cells-are we spotting on the same cells from the different angles? Folia Histochem Cytobiol. **42** 139–146 (2004)
24. Ratajczak M.Z., Kucia M., Reca R., Majka M., Janowska-Wieczorek A., Ratajczak J.: Stem cell plasticity revisited: Cxcr4-positive cells expressing mrna for early muscle, liver and neural cells 'hide out' in the bone marrow. Leukemia **18**, 29–40 (2004)
25. Kucia M., Ratajczak J., Ratajczak M.Z.: Bone marrow as a source of circulating cxcr4+ tissue-committed stem cells. Biol Cell. **97**, 133–146 (2005)
26. Kucia M., Ratajczak J., Reca R., Janowska-Wieczorek A., Ratajczak M.Z.: Tissuespecific muscle, neural and liver stem/progenitor cells reside in the bone marrow, respond to an sdf-1 gradient and are mobilized into peripheral blood during stress and tissue injury. Blood Cells Mol Dis. **32**, 52–57 (2004)
27. Roeder I., de Haan G., Engel C., Nijhof W., Dontje B., Loeffler M.: Interactions of erythropoietin, granulocyte colony-stimulating factor, stem cell factor and interleukin-11 on murine hematopoiesis during simultaneous administration. Blood, **91**, 3222–3229 (1998)
28. Lowenthal R.M., Faberes C., Marit G., Boiron J.M., Cony-Makhoul P., Pigneux A., Agape P., Vezon G., Bouzgarou R., Dazey B., Fizet D., Bernard P., Lacombe F., Reiffers J.: Factors influencing haemopoietic recovery following chemotherapy-mobilised autologous peripheral blood progenitor cell transplantation for haematological malignancies: A retrospective analysis of a 10-year single institution experience. Bone Marrow Transplant. 1998 **22**, 763–770 (1998)

Combustion Chemistry and Parameter Estimation

Marc Fischer and Uwe Riedel

Abstract Combustion processes in practical systems are marked by an huge complexity stemming from their multi-dimensional character, from the interaction of physical processes (including diffusion, flow dynamics, thermodynamics and heat transfer), and from the extremely complex chemistry potentially involving up to hundreds of species and thousands of elementary reactions. The determination of accurate parameters accounting for the chemical kinetics is an essential step for predicting the behavior of practical combustion systems. This is usually done by carrying out specific experiments in simplified systems whereby many of the physical phenomenons mentioned above can be neglected. The present paper aims at giving an overview over the approaches currently followed to estimate kinetic parameters based on experimental data originating from these simplified systems. The nature and mathematical description of such problems are presented for homogeneous systems where all variables depend on time only. The techniques for the identification of the most significant reactions (and hence parameters) are shown along with methods for mechanism reduction considerably alleviating the computational burden. As an application example, *Mechaeut*, a C++ procedure written by the authors is employed for the reduction of a detailed reaction mechanism aiming at describing the combustion of methane CH_4. In the following section, the estimation of kinetic parameters is formulated as an optimization problem and different approaches found in the current literature are examined.

M. Fischer (✉)
Interdisciplinary Center for Scientific Computing (IWR), University of Heidelberg, Germany
e-mail: marc.fischer_emd@yahoo.fr

U. Riedel
Institute of Combustion Technology, German Aerospace Center (DLR), Stuttgart, Germany
e-mail: uwe.riedel@dlr.de

H.G. Bock et al. (eds.), *Model Based Parameter Estimation*, Contributions
in Mathematical and Computational Sciences 4, DOI 10.1007/978-3-642-30367-8_10,
© Springer-Verlag Berlin Heidelberg 2013

1 Introduction

Realistic simulations of combustion processes in the twenty-first century have become primordial in view of the numerous challenges and problems that have emerged during the past decades: global warming which requires more efficient ways to convert energy while limiting the emissions of gas fostering the green house effect, health concerns with regard to harmful pollutants (including nitric oxides (NOx), photochemical pollutants inducing the formation of smog and particulate matters having hazardous health effects) coupled with the growing influence of emerging countries like China which increasingly exploits its fossil coal resources. The intense research activities aiming at producing sustainable renewable fuels show also that the importance of combustion is not going to be diminished by the disappearance of traditional fossil fuels like oil. For all these reasons, an important research effort has to be done in order to reach an accurate estimation of parameters accounting for the reaction kinetics of combustion problems.

In the next part of this article, the simplified systems used for kinetic studies are described. In the third part, the mathematical description and handling of the associated problems are given and the methods used for solving them are evoked. The fourth part is consecrated to the identification of the most important parameters and to methods used in practice for reducing reaction mechanisms.

In the fifth part as an example, *Mechacut*, a C++ procedure developed recently, is employed for the reduction of the GRI-mechanism aiming at describing the oxidation of methane CH_4. We then go into the practical problem of parameter estimation in the sixth part by formulating it as an optimization problem and examining the different approaches followed in the literature.

2 Description of Simplified Combustion Systems

Practical combustion systems hinge on a very large number of overlapping phenomenons, which not only make realistic predictions hard to attain, but also hinders researchers from deducing the values of parameters out of a class of similar experiments. In effect, the behavior of these systems is extremely over-determined in that many (if not an endless number of) combinations of physical and chemical parameters can lead to the same simulation results (see [34] for an example concerning chemical kinetics alone). To reduce the sources of uncertainty and keeping focused on the chemistry, it is thus necessary to decouple the different processes occurring by devising experiments where only a small part of them is active. Homogeneous systems have turned out to be extremely useful for kinetic studies of combustion problems.

Homogeneous systems are constrained in such a way that all variables (including concentrations, temperature, pressure, and sometimes volume) are only time-dependent. This means that the physical phenomenons may either be neglected or

Combustion Chemistry and Parameter Estimation 209

happen so fast that the evolution of the system is only driven by its chemistry. In this case, a system of ordinary differential equations (see next section) has to be solved and the simulations can be carried out with less resources than for other, more complex types of systems.

Shock tubes have been shown to be appropriate for the investigation of high temperature chemistry (usually for temperatures greater than 1,200 K) [9, 18]. They consist of tubes made up of two sections: a driver section at high pressures and a driven section (containing the experimental mixture) at low pressure separated by a membrane. At the beginning of an experiment, the aluminum membrane is burst provoking a strong shock wave which after having been reflected will heat up the mixture very quickly (in less than $1 \mu s$). Close to the end flanks of the tube, the temperature and pressure remain constant for several milliseconds, during which the experimental measurements are made. The concentration values are estimated through several techniques based on the emission/absorption properties of molecules like mass spectrometry, gas chromatography and laser absorption [31].

Rapid compression machines are currently employed by many researchers all over the world for the study of combustion reactions at lower temperatures (500–950 K) [6, 9]. The gaseous mixture containing the reactants and possibly the inert gas is compressed homogeneously and rapidly so as to reach predetermined reaction conditions (pressure and temperature). Afterwards, the temporal evolution of the system is followed with the same diagnostic techniques as those utilized for shock tube studies.

A third kind of systems consists of plug flow reactors [9, 27] for the investigation of intermediate temperature regimes (900–1,300 K). According to given temperature and possibly pressure profiles the reacting flow is steadily going through the system which is assumed to be homogeneous in the plane perpendicular to the flow direction. In such a way, the variables only depend on the distance from the inlet point, which can be easily converted into the reaction time. The analysis of the mixture can be done at several points of the flow using the methods evoked above.

Ignition delay times are often measured during experiments carried out in either shock tubes or rapid compression machines. They are defined as the period between the beginning of the reactions in the system and the sharp increase in the concentration of free radicals resulting from the onset of chain reactions [10].

In every of the three cases described above, for so-called homogeneous systems output profiles only hinge on initial conditions, kinetic parameters and thermodynamic data.

3 Mathematical Modeling of Homogeneous Combustion System

As mentioned earlier, homogeneous systems are characterized by a relatively simple ensemble of mathematic equations when compared to spatially resolved simulations. The chemistry is accounted for by n_s species and n reactions and is

Table 1 Reaction from the H_2–O_2 system (see (2))

Reactions	A	n	E_a
$O_2+H=OH+O$	2.065E+14	−0.097	62.853
$H_2+O=OH+H$	3.818E+12	0.000	33.256
$H_2+OH=H_2O-H$	2.168E+08	1.520	14.466
$OH+OH=H_2O+O$	3.348E+04	2.420	−8.064

well illustrated by a part of the H_2–O_2 system [10], see Table 1. The following species are involved: H_2, O_2, H, O, OH, and H_2O.

Let us consider a given reaction R

$$\sum_{k=1}^{N_{Rspe}} v_k S_k = 0 \tag{1}$$

where S_k is a participating species (either reactant or product), v_k is the corresponding stoichiometric coefficients being negative for a reactant and positive for a product whereas $N_{R_{spe}} = N_{R_{rea}} + N_{R_{pro}}$ is the total number of participating species, that is the sum of the reactant and the product number.

Since detailed kinetic modeling aims at describing what really happens on the molecular level, the number of products and reactants is usually not greater than two except for pressure dependent reactions where collisions with other molecules are described as a reaction with a third body M, bringing up the number of reactants and products to three [10].

The rate constant of a reaction is given by

$$k_R = AT^n e^{\frac{-E_a}{RT}} \tag{2}$$

where A, n and E_a are respectively the pre-exponential factor, the temperature factor and the activation energy. The rate constant of the backward reaction $k_{R,-1}$ is given by

$$k_{R,-1} = \frac{k_R}{K_c} \tag{3}$$

with K_c being the equilibrium constant. For gaseous reactions, the dependency of K_c (in concentration units) is expressed by the following formula:

$$K_c = K_p \left(\frac{p_0}{RT}\right)^{\sum_{k=1}^{N_{Rspe}} v_k} \tag{4}$$

where K_p is the equilibrium constant in pressure units, p_0 the standard pressure which is equal to 1.013 bar, R the ideal gas constant and T the temperature. The equilibrium constant K_p itself is calculated from the free enthalpies of all species participating in the reaction as follows:

$$K_p = e^{\frac{-\Delta G_0}{RT}} \tag{5}$$

Combustion Chemistry and Parameter Estimation

where ΔG_0 is the free enthalpy of the reaction, obtained by adding the free enthalpies of all products and subtracting those of all reactants.

If the system is considered to be homogeneous, all variables only depend on the reaction time t and chemistry is the only origin of the concentration changes. Their derivatives are therefore given by the formula:

$$\frac{dc_i}{dt} = \Sigma_{R=1}^{n} \nu_{i,R} \, k_R \, \Pi_{j=1}^{j=N_{Rrea}} c_j \tag{6}$$

that is the sum of all stoichiometric coefficients multiplied by the reaction rates, obtained by multiplying the rate constants with the concentrations of all reactants. In this formula, n designates the number of reactions whereas $N_{R_{rea}}$ is the number of reactants which take part in the R-th reaction. A more detailed explanation can be found in [32]. All these reactions form a set of coupled ordinary differential equations which can be written as:

$$\frac{dc_i}{dt} = f_i(t, C, k) \text{ and } C(t = 0) = C_0, \, t \in [0 \; t_e], \, i = 1, 2, ..N_{spe} \tag{7}$$

or

$$\frac{dC}{dt} = f(t, C, k) \text{ and } C(t = 0) = C_0, \, t \in [0 \; t_e] \tag{8}$$

where k represents the rate constants of all reactions and C is the vector of species concentrations.

For solving the system, it is necessary to introduce initial conditions for each species, for the temperature and for the pressure.

One equation characterizing the evolution of the temperature T alongside another one concerning the volume and/or pressure must also be added to uniquely characterize the system.

For solving the system, it is necessary to introduce initial conditions for each species, for the temperature and for the pressure.

The temperature T may either be constant (for isothermal systems) or, in the case of adiabatic systems, evolve according to the heat released by the reactions taking place. The heat itself can be calculated from the thermodynamic data about the species involved in the reactions [32].

In every case, one obtains a system of differential equations which is typically stiff due to many different chemical time scales inherent to the reaction system. This kind of systems implies the use of implicit methods of numerical integrations like DASAC [14] or the Cash–Karp Runge–Kutta and Bader–Deuflhard integrators [11] to deal with it.

After having solved the system of differential equations for a fixed set of k values, one gets more or less a good approximation of the temporal evolution of the species, provided of course that the rate constants are correct from the very beginning. As we will see in the Chap. 6, this is often not the case, so that the values of parameters have to be adapted to the experimental measurements by some fitting procedure.

4 Sensitivity Analysis and Mechanism Reduction

The ultimate goal of detailed reaction kinetics is to give an accurate and complete description of all what can take place on the molecular level. To describe the pyrolysis and rich oxidation of simple fuels such as methane (CH_4), acetylene (C_2H_2) or ethylene (C_2H_4) or the lean oxidation of greater alcanes like iso-octane (i-C_8H_{18}) under all possible pressure, temperature and concentration conditions, several hundreds of species and thousands of reactions are necessary. However, in almost every case, only a smaller number of them will play an important role for the species concentrations or variables (like temperature or ignition delay time) one is interested in. Thus, taking into account all species and reactions from the mechanism for each simulation or iteration of a fitting algorithm requires a large amount of unnecessary computational time since many of them have no significant influence on the variables of interest. It is furthermore highly desirable to identify the most important reactions since they are the ones underlying the whole chemical behavior of the system under given conditions.

Sensitivity analysis are a common way to measure the strength of the influences of parameters on a particular variable which could be either a concentration, the temperature or possibly the ignition delay time. In most cases, sensitivity coefficients of first order are computed according to an OAT (One-At-a-Time) framework whereby the influence of one parameter on the output is evaluated while all other parameters are fixed (see for example [10, 20, 33]). The definition of the sensitivity coefficients s_i is then straightforward and is based upon the derivative of the interesting variable [A] with respect to the parameters p_i:

$$s_i = \frac{\delta[A]}{\delta p_i} . \tag{9}$$

Usually, relative sensitivities are considered in chemistry. These are defined as follows:

$$s_i = \frac{p_i \delta[A]}{[A]\delta p_i} = \frac{\delta ln[A]}{\delta ln p_i} . \tag{10}$$

Sensitivity coefficients for concentrations c_i^{cal} as a function of time can generally be computed simultaneously along the solution of the differential equation system with some additional computational cost [32] according to the formula:

$$\frac{\delta}{\delta t}(\frac{\delta c_i^{cal}}{\delta k_j}) = \frac{\delta}{\delta k_j}(\frac{\delta c_i^{cal}}{\delta t}) = \frac{\delta}{\delta k_j}(f_i(C^{cal}, k_j)) \tag{11}$$

and if one further develops :

$$\frac{\delta}{\delta t}(\frac{\delta c_i^{cal}}{\delta k_j}) = \sum_{s=1}^{s=N_{spe}} \frac{\delta f_i}{\delta c_s^{cal}}(\frac{\delta c_s^{cal}}{\delta k_j}) + \frac{\delta f_i}{\delta k_j} , \tag{12}$$

Combustion Chemistry and Parameter Estimation 213

where f is the right hand-side function characterizing the system of differential equations described in the previous section. $\frac{\delta f_i}{\delta c_s^{cal}}$ form the Jacobian, defined as the matrix of partial derivatives of f with respect to the concentrations. $\frac{\delta c_s^{cal}}{\delta k_j}$ is the sensitivity coefficient of the concentration c_s^{cal} at time t with respect to the j-th kinetic coefficient and $\frac{\delta f_i}{\delta k_j}$ is the derivative of the right hand-side term with respect to the j-th coefficient. It is noteworthy that the differential equation system obtained in such a way is always linear, regardless of the non-linearities characterizing the concentrations.

The reliance upon first derivatives is however criticizable in that the correlated effects of parameter variations is not at all considered although such non-linear effects can often occur for complex reaction mechanisms such as those encountered in combustion systems. To overcome this limitation, methods considering the correlations between the parameters can be employed. Among them, the FAST (Fourier Amplitude Sensitivity Test) [24] has shown promising successes.

In any case, sensitivity analysis allow for a particular situation to make a distinction between the parameters which must be determined accurately (because having a tremendous impact on the simulated variables due to high sensitivities) and those whose knowledge may be less precise. It is nonetheless not sufficient to trust sensitivity analysis alone for the identification of unimportant reactions because it can often occur that fast reactions with low sensitivity coefficients play a critical role. To see how this can happen, one has just to consider the simple (imaginary) chain $C_2H_6 \rightarrow C_2H_5 + H$ and $C_2H_5 \rightarrow C_2H_4 + H$ with rate coefficients k_1 and k_2, if k_1 is much greater than k_2, then the second reaction will be the limiting one whereas the first one will be nearly infinitely fast from this perspective. Thus, to modify the pre-exponential factor of the first reaction slightly (say $0.9\,k_{1,0}$ or $1.1\,k_{1,0}$ instead of $k_{1,0}$) would have very limited impact on the formation of the product so that the sensitivity would be rather small. To exclude it from the reaction mechanism on the ground of this criterion alone would however lead to the rather undesired result that no more C_2H_4 would be produced at all! This thus shows the importance of also considering the flow followed by the element $\mathbf{C}$ and $\mathbf{H}$ between C_2H_6 and C_2H_4.

Generally, for the atom sort a and the reactants i and j, the two following normed flow parameters [29] are introduced: f_{ij}^a, the flow of the atom a during the formation of species i from species j relative to the global formation of i, whereas c_{ij}^a is the flow of the atom a during the consumption of species i for forming species j relative to the global consumption of i:

$$f_{ij}^a = \frac{\sum_{k=1}^{k=n} r_k v'_{jk} v''_{ik} \frac{n_i^a}{\Delta n_k^a}}{\sum_{k=1}^{k=n} v''_{ik} r_k} \tag{13}$$

$$c_{ij}^a = \frac{\sum_{k=1}^{k=n} r_k v'_{jk} v''_{ik} \frac{n_i^a}{\Delta n_k^a}}{\sum_{k=1}^{k=n} v'_{ik} r_k}. \tag{14}$$

n is the set of uni-directional reactions over which it is summed, r_k is the reaction rate of the k-th reaction, n_i^a is the number of atom a in the species i whereas $v_{ik}^{''}$ and $v_{jk}^{'}$ are the stoichiometric coefficients of the species i and j in the k-th reaction.

The number of atoms n_i^a are normalized to the total number of atoms transported in the k-th reaction: $\Delta n_k^a = \sum_{l=1}^{l=N_{Rspe}} v_{lk}^{'} n_l^a$. Both normed flows take on values between 0 and 1 which correspond to the flux part from, respectively, towards species i with respect to species j.

In such a way, flow diagrams showing the relative importance of all formation/consumption pathways for an ensemble of species can be generated. They can be defined either for particular points in time or the flows may be integrated over a defined time interval. If a reaction has small sensitivities according to all variables of interest and if its role in the reaction flow is negligible, then there are good grounds for believing it can be excluded from the mechanism without consequences. The thresholds for f_{ij}^a and c_{ij}^a to be respected depend however on the particular context.

Several methods have been developed for the practical reduction of kinetic differential equation systems for a particular set of conditions. They may be divided in two categories: those which preserve the structural integrity of the mechanism and those which do not. According to Androulakis definition [2], we will say that a reduced mechanism maintains the structural integrity of a detailed mechanism if all species and reactions of the reduced mechanism were already present in the initial one. This implies that the chemical description of the system in terms of reaction flux remains the same over the whole reduction process, only those chemical species and reactions which play no significant role for the desired output are removed.

An intuitive method [25] belonging to this class consists of the successive application of a reaction flow analysis to distinguish between slow and fast reactions, followed by a sensitivity analysis carried out only for the slow reactions to identify those which are crucial for the variables of interest. The rapid reactions and the slow reactions with high sensitivity values are then kept within the reduced mechanism. Other methods imply the stepwise removals of a group of reactions followed by the evaluation of the discrepancies introduced with respect to the original model.

For example, Petzold and Zhu [21] reformulated the mechanism reduction as a non-linear integer-programming problem which was itself transformed into a continuous optimization problem solved with a sequential quadratic programming method. Androulakis [13] also formulated the mechanism reduction as a non-linear integer-programming problem which he solved with the help of a branch and bound algorithm.

The methods belonging to the second group modify the structure of the problem either by lumping reactions and species or by resorting to algebraic simplification of the differential equation system underlying the detailed kinetic scheme.

The most traditional approaches consist of applying the assumption of steady state to very reactive radicals (which means that the temporal derivatives of their concentrations have to remain close to zero over the whole interval) which leads to algebraic relations allowing the simplification of the underlying system of differential equations and the lumping of groups of species and reactions.

Combustion Chemistry and Parameter Estimation

This approach usually involve skillful kineticists and intuitions [21], although Montgomery et al. [19] have proposed an automated procedure for doing this. The system of differential equations can also be simplified by considering very fast bidirectional reactions to be in equilibrium. These methods, which consist in fact of a differentiation between slow and rapid modes of the differential equation system, can be replaced by more general frameworks like the ILDM (Intrinsic Low Dimensional Manifolds) [15] or the CSP (Computational Singular Perturbation) [12] methods.

However, such methods can not be employed for parameter estimation since they do not conserve the mathematical structure upon which the detailed chemistry is based. Thus, only the first class of approaches is of interest for the parameter estimation problem at hand. In the following section, an easily implementable method maintaining the structural integrity of the reaction mechanism developed recently will be presented and demonstrated as one concrete example concerning the oxidation of methane CH_4.

5 Mechacut, a Practical Tool for Simplifying Reaction Mechanisms

For quickly obtaining a simplified mechanism accurately reproducing the variables of interest resulting from the detailed mechanism, the tool Mechacut has been developed by the authors. It relies on the software Homrea for the simulation of homogeneous kinetic systems [10].

As input, it requires the following information:

- The conditions of all experiments considered.
- The reaction mechanism.
- The thermodynamic data.
- The target variables (concentrations or temperature) whose profiles must be reproduced.
- The tolerance in terms of relative differences e (say $e = 2\%$).

Like several methods cited above, it is based upon the fact that typically reactions containing species with small concentrations tend to play a minor role for the whole system.

In a first step, for each species k, $k \in [1, n_S]$, the greatest molar fraction value over all experiments and time points $C_{k,max}$ is determined. The use of molar fractions instead of concentration units allows the simultaneous consideration of experiments with different orders of magnitude of initial reactant and diluent concentrations.

Then, for a given reaction $R : \Sigma_i v_i R_i = \Sigma_j v_j P_j$, the greatest molar fractions $C_{k,max}$ of all participating species are compared, and the species having the lowest (in comparison with the other species of the reaction) value of the greatest molar fractions is identified and the corresponding value is assigned to the reaction.

The reactions are then sorted according to this characteristic value in a decreasing order. Therefore, the reactions situated towards the end of the mechanism will typically contain some species having very low concentrations, whereas the reactions located at the beginning will only have species with greater concentrations.

There may be however situations where small concentrations of some participating species does not necessarily entail the unimportance of the reaction, especially in the case where the reactants or products include free radicals which usually have low concentrations but do play a crucial role for the whole reaction system.

To discriminate important and unimportant reactions, it is therefore necessary to proceed in a stepwise fashion, removing by way of trial a group of reactions and accepting the change if the variations of the target variables are less than a predefined threshold e.

To introduce flexibility, random numbers are employed to choose the first line of the group to be suppressed $n \in [1, n_{reac}]$ and the number of reactions m to be deleted which is included in the interval $[0, m_{max}]$. m, the current number of deletable lines, is generated according to an uniform distribution function, whereas n follows a probability law strongly biased towards the end of the interval $[1, n_{reac}]$, so as to target most of the time the reactions with minor species.

The current maximum number of lines m_{max} is multiplied by a pre-defined factor a ($a > 1$) if the deletion could be accepted, otherwise (the deletion led to some discrepancies greater than the tolerance) it is multiplied by an other pre-defined factor b ($0 < b < 1$). The maximum number of lines m_{max} is however never allowed to exceed the threshold m_{max}^{max}. A species is suppressed from the input file if it no longer appears in the reaction mechanism either as reactant, product or third body.

In this manner, large numbers of reactions may be by way of trial deleted at the beginning of the process, as long as unimportant reactions are considered. From a certain point, the deletion of more influential reactions is tried out and rejected what reduces the length of the interval. If the number of failures exceeds a given threshold, the algorithm leaves the probabilistic mode and tries to delete each reaction one by one.

In what follows, an application example is given to illustrate the algorithm and its efficiency.

As will be presented in detail in the next chapter, the estimation or optimization of kinetic parameters out of experimental data requires solving the differential equations systems underlying the simulated variables for many possible combinations of parameters (several thousands of times for a dozen of adjustable parameters). If several experiments are considered simultaneously for extracting the best set of parameters matching the measurements, each simulation for fixed values of the parameters may cost several minutes if the whole reaction mechanism is employed. This situation is clearly not satisfying for practitioners who often have to consider different combinations of adjustable parameters before finding the one which will lead to acceptable discrepancies with the experiment: the optimization process will then last more than 1 day before either the optimal solution will be found or it may be concluded that the current mechanism is inconsistent with the data.

Reaction mechanism simplification techniques offer here an interesting option to considerably reduce the time of each iteration (corresponding to the simulation of

Combustion Chemistry and Parameter Estimation 217

Table 2 Experimental conditions considered

Experiment	$[CH_4]$ (%)	$[O_2]$ (%)	p	T	t
1	0.4	20	1.48	1,711	6.0e$-$04
2	0.5	10	1.53	1,752	6.0e$-$04
3	0.5	10	1.62	1,843	6.0e$-$04
4	0.4	5	1.57	1,821	6.0e$-$04

all measured variables), thereby leading to significantly shorter CPU time usage for parameter optimization.

The GRI-mechanism 3.0 [28] is a compilation of elementary reactions along with reaction rate values and thermodynamic parameters which aims at describing the oxidation of methane (CH_4). Without the chemistry involving nitrogen N, it contains 38 species and 422 (uni-directional) reactions and is valid over a wide range of conditions for fuel lean to stoichiometric flames. This reaction mechanism has been reduced using Mechacut for a set of experiments concerning the oxidation of CH_4 whereby the temporal profiles of carbon monoxide CO and hydroxyl radicals OH have been measured. These compounds have been considered as target species, that is the reduced mechanism must reproduce the profiles from the detailed mechanism within a given tolerance e. The conditions of the four chosen experiments are given by Table 2. $[CH_4]$ and $[O_2]$ denote the molar fractions of methane and oxygen, T the temperature in Kelvin, p the pressure in bar and t corresponds to the duration of the experiment in seconds.

In what follows, the reaction mechanism has been reduced to reproduce the concentration profiles of CO and OH within a tolerance of 2%. The parameters had the following values: $a = 2$, $b = 0.6$ and $m_{max}^{max} = 150$.

After the end of the procedure, the reduced reaction mechanism contained 172 reactions and 31 species which amounts to a reduction of 59.24% and 18.42% respectively. To illustrate the accuracy of the reduced model, two characteristic profiles were considered and a comparison between detailed and reduced mechanism used in the simulation is given in Figs. 1 and 2.

The differences between the two profiles are extremely small and the results are similar for the six remaining profiles.

A considerable reduction in terms of reactions could thus be reached, showing that many reactions of the GRI-mechanism 3.0 play a negligible role for the evolution of the species CO and OH during the combustion of methane under the conditions considered here.

6 Parameter Estimation for Combustion Chemistry as an Optimization Problem

After having identified the most important parameters and deleting as much irrelevant reactions as possible, the parameter estimation problem for the given set of experimental data can be formulated as an optimization problem whereby the goal is

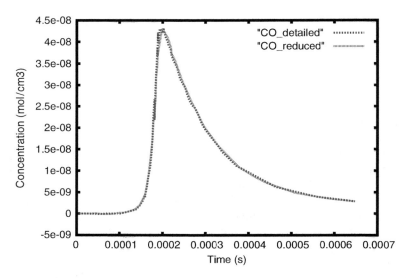

Fig. 1 Comparison between the detailed and reduced mechanisms for the CO profile of the second experiment

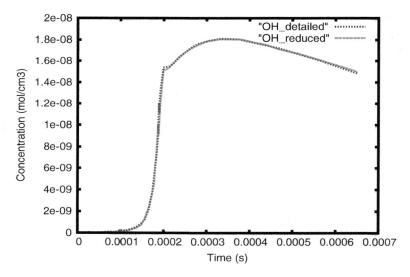

Fig. 2 Comparison between the detailed and reduced mechanisms for the OH profile of the second experiment

to minimize the discrepancies between the results from experiments and those from the model. Let us consider a set of n experiments, the i-th one having n_i measured variables with $i \in [1, n]$. The standard estimators of the differences are the least squares norm and the least absolute values norm which are respectively defined as

$$d = \sum_{i=1}^{n} \sum_{j=1}^{n_i} (c_{i,j} - e_{i,j})^2 \tag{15}$$

and

$$d = \sum_{i=1}^{n} \sum_{j=1}^{n_i} |c_{i,j} - e_{i,j}| \tag{16}$$

where $e_{i,j}$ and $c_{i,j}$ are the variables from the experiments and the model.

The last norm has been proven to be much more insensitive with respect to outliers, that is experimental data lying outside of the expected range due to e.g. measurement errors [1]. While being frequently used in combustion problems for fitting parameters to one unique variable (like the concentration of a species), both norms rapidly show their limits when confronted with variables having different orders of magnitude. It is easy to envisage that by considering the simple case where the temperature T (being more than $1{,}000\,\mathrm{K}$) and the concentration of one species (being around $1e-6$) were measured at the same time: the discrepancies between the simulated and experimental temperatures would completely dominate the discrepancies of the concentration by several orders of magnitude so that the minimization of each of the norm would only adapt the parameters to the temperature profile and ignore the information contained in the concentration profile.

This is the reason why weights must be assigned to all terms so that they may receive the same importance in the evaluation of the total discrepancies. Usually, this is done by minimizing the so-called Chi-squared function, which is defined as:

$$d = \sum_{i=1}^{i=n} \sum_{j=1}^{j=n_i} \frac{(c_{i,j} - e_{i,j})^2}{\sigma_{i,j}^2} \tag{17}$$

whereby $\sigma_{i,j}$ designate the standard deviations of the experimental variable $e_{i,j}$.
In the vast majority of cases, these quantities are unfortunately not determined experimentally, forcing the kineticist to find approximations.

To avoid this problem, the relative least squares and absolute values norms can be considered. They are defined like the norms above except that the terms within the sum are divided by the experimental values of the variable:

$$d = \sum_{i=1}^{i=n} \sum_{j=1}^{j=n_i} \frac{(c_{i,j} - e_{i,j})^2}{e_{i,j}^2} \tag{18}$$

and

$$d = \sum_{i=1}^{i=n} \sum_{j=1}^{j=n_i} \frac{|c_{i,j} - e_{i,j}|}{e_{i,j}}. \tag{19}$$

With these norms, the percentage errors are taken into account in such a way that the relative discrepancies of variables as different as concentrations and temperature have the same weight. Their use is akin to assuming that all standard deviations are proportional to the experimental values. These norms are also particularly useful if concentrations with different orders of magnitude have been measured.

Whatever norm d was chosen, the problem at hand now consists of minimizing the value originating from it, determining in this manner the optimal coefficients. The minimization problem may be generally formulated as :

$$min \ F(P) = d(E, C, P) \tag{20}$$

under the conditions

$$\frac{dC}{dt} = f(t, C, P, Q) \ and \ C(t = 0) = C_0, \ t \in [0 \ t_e] \tag{21}$$

where E is the set of experimental measurements, C the set of modeled values, C_0 the initial values of the concentrations, P the active set of all adjustable parameters and Q the ensemble of fixed parameters. Constraints on the parameter values have almost always to be taken into account, frequently via bound constraints:

$$l_i \leq p_i \leq u_i \tag{22}$$

Often in chemical kinetics, global methods like simulated annealing or genetic algorithms are employed for minimizing the function and they are extremely popular due to their abilities to bypass local minima, which are sub-optimal minima of the function surrounded by points of greater values.

Genetic algorithms were for instance employed by Masoori et al. [17] to determine reaction parameters for the Fischer–Topsch synthesis, Polifke et al. [22] used them for finding optimal values of coefficients for simplified mechanisms whereas Elliott et al. [7] apply them to various chemical kinetic problems including combustion.

Simulated annealing was recently employed by Mani et al. [16] for the determination of activation energies pertaining to the non-isothermal pyrolysis of lignin.

Nevertheless, as Tang et al. [30] pointed out, both global methods can be extremely inefficient in that they ignore problem specific knowledge and rely on randomness for finding better solutions. These authors proposed then a Physically Bounded Gauss–Newton (PGN) method for solving optimization problems in chemical kinetics. The derivatives required by the method are obtained through the sensitivity coefficients, themselves gotten from an additional differential equation solved along the system of differential equations (see Sects. 3 and 4). For a perturbation Φ of the n-parameter set P which is sufficiently small, the new value of the model variable $c_{i,j}$ can be expressed according to a first-order approximation:

$$c_{i,j}(t, P + \Phi) = c_{i,j}(t, P) + (\frac{\delta c_{i,j}}{\delta p_1})\phi_1 + (\frac{\delta c_{i,j}}{\delta p_2})\phi_2 + \ldots + (\frac{\delta c_{i,j}}{\delta p_n})\phi_n. \tag{23}$$

Combustion Chemistry and Parameter Estimation

The set of all concentrations resulting from the model can then be written as:

$$C(t, P + \Phi) = C(t, P) + S(t)\Phi \tag{24}$$

where S is the matrix of sensitivity coefficients. In case of a least square norm, a local Gauss–Newton step can then be taken by solving the following constrained linear least squares problem:

Minimize with respect to the perturbation Φ,

$$F(\Phi) = d(E, C, P + \Phi) \tag{25}$$

with $l_i \leq p_i + \phi_i \leq u_i$. Unfortunately, as the same authors recognize, the successive minimizations according to Gauss-Newton steps can only lead to the next local minimum, which may not be sufficient for the problem at hand. In effect, contrarily to global optimization methods, derivative based local optimization methods like the PGN exploit efficiently the structure of the optimization problem, but the initial values have to be located in a pretty narrow domain to guarantee convergence towards the global minimum of interest.

Aware of this hurdle, Bock proposed a method reaching a compromise between the mathematical efficiency of local methods and the robustness of global methods by applying a multiple shooting method to the optimization problem [4]. Following this approach, the initial value problem

$$\frac{dC}{dt} = f(t, C, P, Q) \ and \ C(t = 0) = C_0, \ t \in [0 \ t_e] \tag{26}$$

is divided into n initial value problems :

$$\frac{dC_j}{dt} = f(t, C_j, P, Q) \ and \ C_j(t_j) = \theta_j, \ t \in [t_j \ t_{j+1}] \tag{27}$$

with $t_1 = 0$, $t_{n+1} = t_e$. The set of optimizable parameters contains not only the adjustable kinetic coefficients P but also the values of the concentrations at the n nodes $\theta_1, \theta_2, \theta_3, \ldots, \theta_n$. The continuity constraints $C_j(t_{j+1}) = C_{j+1}(t_{j+1}) = \theta_{j+1}$ have to be fulfilled by the final solution, ensuring thus its internal continuity. An in-depth review and presentation of the technique may be found in [5]. Bock [4] demonstrated the advantages of the method on a kinetic parameter estimation problem pertaining to the denitrogenation of pyridine: by considering concentrations as variables, a close agreement with the experimental values may be observed from early on and many local minima may be avoided. The method requires however a good initial guess of the concentration values at all nodes for all species, whereas only a small portion of them is measured for combustion problems, thereby making the first estimation very hard. As a consequence, further work needs to be done before this technique may be applied to combustion problems involving large numbers of species and reactions.

Until now, we have focused our attention on parameter fitting based on a given set of experiments without yet considering the purpose of the overall process, namely the construction of a reaction mechanism which has to be consistent with all known data and not just a part of them. According to Frenklach [8, 23] one of the major stumbling block of combustion kinetic lies in the fact that the adjustment of a reaction mechanism to a new experimental set often involves the disappearance of good agreements with previous experiments. This is the case because the dependencies of species profiles measured under different experimental conditions on the reaction parameters is very complex. The traditional cumulative approach in combustion kinetic consists of first considering the simplest possible chemical system (H_2–O_2), determining appropriate coefficients from the experimental data, then *fixing* all the reactions and parameters, and considering the next complex system (CO–O_2) and the corresponding experimental data, only the parameters of the newly introduced reactions involving C-species are allowed to change. In this manner, the good agreement reached for (H_2–O_2) will not be destroyed since the then determined parameters could not be changed and the new reactions of C species have no influence on experiments including the system (H_2–O_2) alone. While this logic remains sound for this particular trivial example, things become much more unwarranted when the same method is applied to the hierarchical construction of large reaction mechanism for species such as CH_4, C_2H_2, C_2H_4, C_2H_6. C_3H_6, C_3H_8 and so on. This stems at least from three reasons:

- For obtaining good agreements with experiments pertaining to a certain compound, it may often occur that changes of the parameters describing the lower chemistry are necessary due to their own imprecision.
- The numbers of experiments concerning one single compound (for instance CH_4) can be extremely large so that it is practically impossible (because of the limitations of current computers) to solve all associated differential equation systems for the optimal determination of kinetic parameters. In reality, only a small subset of the experiments can be used, not considering at all the huge amount of information contained within the other ones.
- The roles of the chemistry of lower and higher compounds may be deeply intertwined and cannot then be seen as a bottom-up dependency only. For instance, reactions involving acetylene C_2H_2 may play an huge role for many pyrolysis or rich oxidation experiments involving methane CH_4 as reactant. This means that after having adjusted methane reactions to fit profiles measured during CH_4 experiments, all the efforts may be spoiled by later modifying C_2H_2 reactions in order to better reproduce experiments involving acetylene.

To overcome these problems, Frenklach focused on the concept of feasible set of parameters [8, 23] which he defined as the interrelated set of all possible parameter values, given theoretical knowledge and information gained from all experiments: the model has to agree with all experimental measurements within the limit of their uncertainties. To allow the reproduction of experimental profiles without having to solve each time a differential equation system, Frenklach et al. employ an HDMR (High Dimensional Model Representation) technique (also called solution mapping) [26] which consists of expressing the modeled species profiles as an algebraic function of the significant parameters. In this way, there is no longer a limit to the quantity of experimental information which may be used to determine kinetic parameters along their uncertainties. As the number of experimental data becomes larger and larger, the unknown parameters have to satisfy an increasing number of constraints (which are implicitly given by the fact that the simulated variables have to reproduce the experimental ones within an interval corresponding to the measurement uncertainties). In this manner, the uncertainties on the parameters become smaller and smaller as new experimental data are taken into account because the intervals of possible values are shrunk.

This approach seems promising since it is potentially able to exploit all kinetic information stemming from simplified combustion systems and not just parts of it. However, it faces considerable problems when applied to the optimization of large reaction mechanisms. If such a model contains, say, 3,000 kinetic parameters, it would be practically impossible to express all variables at every time point as a polynomial function of the 3,000 coefficients. Instead, sensitivity analysis have to be carried out so as to only retain the parameters having an influence on some of the data points (e.g. species concentrations at given time points). As an illustration, let us consider a kinetic model with n parameters p. Let us further suppose that a given variable $m(p)$ can be well approximated by a polynomial function of one part of the parameters p_{act} whereas the sensitivities of the other parameters p_{inac} are low which leads to their exclusion from the function.

The problem now is that modifications of the values of the active parameters p_{act} often lead to changes of the sensitivities of the unconsidered parameters p_{inac} which are then no longer negligible. Indeed, the authors of the present article found this was the case for reduced version of the GRI-mechanism (mentioned above) whereby modifications of the coefficients of certain reactions considerably modified the sensitivities of other for concentration profiles. The results of an optimization using a simplified model (which may be a simplified mechanism or a polynomial function) can therefore be misleading in that significant differences between complete and reduced models may exist for many sets of parameter values. As a consequence, the optimal solution for the reduced model may considerably differ from the optimal solution for the complete model. Caution is therefore mandatory when using the HDMR method or mechanism reduction methods for the optimization of complex reaction mechanisms: the optimizable parameters must be varied within narrow ranges in order that the reduction remains valid.

It is finally worth noting that all parameter estimation approaches aiming at reducing the size of the confidence region of the parameters may take advantage from the method of experimental design developed by Bock et al. [3]. Relying on the exact mathematical description of the chemical kinetic through differential equation systems, experimental settings such as temperature, pressure, initial concentrations and measurement times are optimized in such a way that the uncertainty surrounding targeted kinetics coefficients is reduced as much as possible through the information brought up by these experiments.

7 Conclusion

In this article, an overview has been given on the problem of parameter estimation for combustion chemical kinetic. Homogeneous systems (where variables only depend on the reaction time) are extremely useful to focus on the chemistry and three common techniques, shock tubes, rapid compression machines and plug flow reactors have been described. These systems are governed by a system of differential equations typically exhibiting a huge stiffness which requires implicit numerical methods to tackle it. For the identification of important reactions, sensitivity analysis allow to find the kinetic parameters whose slight variations have important consequences on the variables of interest, whereas reaction flow analysis reveal the most significant pathways in terms of participating molecules. For large reaction mechanisms, reduction of the size of the differential equation system has been traditionally achieved by postulating steady states for free radicals and equilibrium for fast reactions. Recently, methods directly reducing the number of reactions and species have been developed. The program Mechacut presented here is one such approach which has been applied to a problem pertaining to methane oxidation. Afterwards, diverse optimization approaches found in the literature have been presented and the challenges of global optimization and building of reaction mechanisms consistent with all experimental data have been highlighted.

References

1. S. Allende, C. Bouza, and I. Romero. Fitting a linear regression model by combining least squares and least absolute value estimation. *QESTIIO*, 19(1,2,3):107–121, 1985.
2. I.P. Androulakis. Kinetic mechanism reduction based on an integer programming approach. *AIChE Journal*, 46(2):361–371, 2000.
3. I. Bauer, H. G. Bock, S. Körkel, and J. P. Schlöder. Numerical methods for optimum experimental design in DAE systems. *Journal of Computational and Applied Mathematics*, 120:1–25, 2000.
4. H.G. Bock. Modelling of chemical reaction systems. *Proceedings of an International Workshop*, 18:102–125, 1980.
5. H.G. Bock, E. Kostina, and Schloeder J. P. Numerical methods for parameter estimation in non-linear differential algebraic equations. *GAMM-Mitt.*, 30(2):376–408, 2007.
6. M. Carlier, C. Corre, R. Minetti, J-F. Pauwels, M. Ribaucour, and L-R. Sochet. Autoignition of butane: a burner and a rapid compression machine study. *Twenty-Third symposium (International) on combustion / The combustion Institute*, (4):1753–1758, 1990.
7. L. Elliott, D. B. Ingham, A. G. Kyne, N. S. Mera, M. Poukashanian, and C. W. Wilson. Genetic algorithms for optimisation of chemical kinetics reaction mechanisms. *Progress in Energy and Combustion Science*, 30:297–328, 2004.
8. Michael Frenklach. Process informatics for combustion chemistry. *31-th International Symposium on Combustion, Heidelberg*, 2006.
9. R. Q. Gonzales. *Evaluation of a Detailed Reaction Mechanism for Partial and Total Oxidation of C1-C4 Alkanes*. PhD thesis, University of Heidelberg, 2007.
10. C. Heghes. *C1-C4 Hydrocarbon Oxidation Mechanism*. PhD thesis, University of Heidelberg, 2006.

Combustion Chemistry and Parameter Estimation

11. J. C. Ianni. A comparison of the bader-deuflhard and the cash-karp runge-kutta integrators for the gri-mech 3.0 model based on the chemical kinetics code kintecus. *Second MIT Conference on Computational Fluid and Solid Mechanics*, pages 1368–1372, 2003.
12. S.H. Lam. Using csp to understand complex chemical kinetics. *Combustion, Science and Technology*, 89:375, 1993.
13. L. Liang, J.G. Stevens, J.T. Farell, P.T. Huynh, I.P. Androulakis, and M. Ierapetritou. An adaptive approach for coupling detailed chemical kinetics and multidimensional cfd. *5th US Combustion Meeting- Paper C09*, pages 1–15, 2007.
14. A.E. Lutz, R.J. Kee, and J.A. Miller. Senkin: A fortran program for predicting homogeneous gas phase chemical kinetics with sensitivity analysis. *SAND87-8248, Sandia National Laboratories*, pages 4–30, 1988.
15. U. Maas and S.B. Pope. Simplifying chemical kinetics: Intrinsic low-dimensional manifolds in composition space. *Combustion and Flame*, 88:239, 1992.
16. T. Mani, P. Murugan, and N. Mahinpey. Determination of distributed activation energy model kinetic parameters using simulated annealing optimization method for nonisothermal pyrolysis of lignin. *Ind. Eng. Chem. Res.*, 48:1464–1467, 2009.
17. M. Masoori, B. Boozarjomehry, B. Ramin, S.J. Maryam, and N. Reshadi. Application of genetic algorithm in kinetic modeling of fischer-topsch synthesis. *Proceedings of an International Workshop*, 27(1):25–32, 2008.
18. J. V. Michael and K. P. Lim. Shock tube techniques in chemical kinetics. *Annu. Rev. Phys. Chem.*, 44:429–458, 1993.
19. C. J. Montgomery, M.A. Cremer, J.-Y. Chen, C.K. Westbrook, and L.Q. Maurice. Reduced chemical kinetic mechanisms for hydrocarbon fuels. *Journal of Propulsion and Power*, 18(01), 2002.
20. I. I. Naydenova. *Soot Formation Modeling during Hydrocarbon Pyrolysis and Oxidation behind Shock Waves*. PhD thesis, University of Heidelberg, 2007.
21. L. Petzold and W. Zhu. Model reduction for chemical kinetics: An optimization approach. *AIChE Journal*, 45(4):869–886, 1999.
22. W. Polifke, W. Geng, and K. Doebbeling. Optimisation of reaction rate coefficients for simplified reaction mechanisms with genetic algorithms. *Combustion and Flame*, 113:119–135, 1998.
23. T. Russi, A. Packard, R. Feeley, and M. Frenklach. Sensitivity analysis of uncertainty in model prediction. *J. Phys. Chem. A.*, 112(12):2579–2588, 2008.
24. A. Saltelli, M. Ratto, S. Tarantola, and F. Campolongo. Sensitivity analysis for chemical models. *Chemical Review C, American Chemical Society*, Published on Web:A–P, 2004.
25. A. Saylam, M. Ribaucour, M. Carlier, and R. Minetti. Reduction of detailed kinetic mechanisms using analyses of rate and sensitivity of reactions: Application to iso-octane and n-heptane. *Proceedings of the European Combustion Meeting 2005*, pages 1–5, 2005.
26. N. Shenvi, J. M. Geremia, and H. Rabitz. Nonlinear kinetic parameter identification through map inversion. *J. Phys. Chem. A*, 106:12315–12323, 2002.
27. M. S. Skjoeth-Rasmussen, P. Glarborg, M. Oestberg, J. T. Johannessen, H. Livbjerg, A. D. Jensen, and T.S. Christensen. Detailed modeling of pah and soot formation in a laminar premixed benzene/oxygen/argon low-pressure flame. *Combustion and Flame*, 136:91–128, 2004.
28. G.P. Smith et.al. www.me.berkeley.edu/gri-mech.
29. H.S. Soyhan, F. Mauss, and C. Sorusbay. Chemical kinetic modeling of combustion in internal combustion engines using reduced chemistry. *Combustion Science and Technology*, 174(11,12):73–91, 2002.
30. Weiyong Tang, Libin Zhang, Andreas.A. Linninger, Robert S. Tranter, and Brezinsky. Solving kinetic inversion problems via a physically bounded gauss-mewton (pgn) method. *Industrial and Engineering Chemistry Research*, 44(10):3626–3637, 2005.
31. V. Vasudevan, D. F. Davidson, and R. K. Hanson. Shock tube measurements of toluene ignition times and oh concentrations time histories. *Proceedings of the Combustion Institute*, 30:1155–1163, 2005.

32. J. Warnatz, U. Maas, and R.W. Dibble. *Combustion: Physical and Chemical Fundamentals, Modeling and Simulation, Experiments, Pollutant Formation.* Springer-Editions, Berlin, third edition, 2001.

33. T. Zeuch. *Reaktionskinetik von Verbrennungsprozessen in der Gasphase: Spektroskopische Untersuchungen der Geschwindigkeit, Reaktionsprodukte und Mechanismen von Elementarreaktionen und die Modellierung der Oxidation von Kohlenwasserstoffen mit detaillierten Reaktionsmechanismen.* PhD thesis, University of Goettingen, 2003.

34. I. GY. Zsely, J. Zador, and T. Turanyi. On the similarity of the sensitivity functions of methane combustion models. *Combustion Theory and Modelling*, 9(4):721–738, 2005.

Numerical Simulation of Catalytic Reactors by Molecular-Based Models

Olaf Deutschmann and Steffen Tischer

Abstract Investigations in the field of high-temperature catalysis often reveal complex interactions of heterogeneous, homogeneous, and radical chemistry coupled with mass and heat transfer. The fundamental aspects as well as several applications of high-temperature catalysis are covered in the light of these interactions. Benefits of molecular-based numerical simulations are discussed. Furthermore, this chapter looks at challenges associated with parameter estimation.

1 Background

Understanding and optimization of heterogeneously catalyzed reactive systems require the knowledge of the physical and chemical processes on a molecular level. In particular, at short contact times and high temperatures, at which reactions occur on the catalyst and in the gas-phase, the interactions of transport and chemistry become important.

High-temperature catalysis is not a new concept; the Oswald process for the NO production by oxidation of ammonia over noble metal gauzes at temperatures above 1,000°C and residence times of less than a micro second has been technically applied for decades; total oxidation of hydrogen and methane (catalytic combustion) over platinum catalysts were even used before Berzelius proposed the term "catalysis." Recently, however, high-temperature catalysis has been extensively discussed again, in particular in the light of the synthesis of basic chemicals and hydrogen, and high-temperature fuel cells.

O. Deutschmann · S. Tischer (✉)
Institute for Chemical Technology and Polymer Chemistry, Karlsruhe Institute of Technology, Engesserstr. 20, 76128 Karlsruhe, Germany
e-mail: deutschmann@kit.edu; Steffen.Tischer@kit.edu

H.G. Bock et al. (eds.), *Model Based Parameter Estimation*, Contributions in Mathematical and Computational Sciences 4, DOI 10.1007/978-3-642-30367-8_11, © Springer-Verlag Berlin Heidelberg 2013

Catalytic partial oxidation (CPOX) of natural gas over noble metal catalysts at short contact times offers a promising route for the production of synthesis gas [1,2], olefins [3,4], and hydrogen. For instance, synthesis gas, also catalytically produced by steam and autothermal reforming, is needed in (gas-to-liquids) plants for synthetic fuels, which are currently under development. CPOX of gasoline, diesel, or alcohols to synthesis gas or hydrogen may soon play a significant role in mobile applications for reduction of pollutant emissions and auxiliary power units.

For any fuel other than hydrogen, catalytic reactions are likely to occur in the anode of a solid oxide fuel cell (SOFC) leading to a complex chemical composition at the anode–electrolyte interface [5]. Primarily the products of the electrochemical reactions, H_2O and CO_2, drive the catalytic chemistry in the anode. For the application of hydrocarbon and alcohol containing fuels, the understanding of the catalytic kinetics is vital for the precise prediction of fuel utilization and performance [6]. Coupling of the thermo catalytic reactions with the electrochemical processes and mass and heat transport in the cell will exemplarily be discussed for an anode-supported SOFC operated with methane containing fuels and a Ni/YSZ anode structure.

2 Fundamentals

Catalytic reactors are generally characterized by the complex interaction of various physical and chemical processes. Monolithic reactors can serve as example, in which partial oxidation and reforming of hydrocarbons, combustion of natural gas, and the reduction of pollutant emissions from automobiles are frequently carried out. Figure 1 illustrates the physics and chemistry in a catalytic combustion monolith that glows at a temperature of about 1,300 K due to the exothermic oxidation reactions. In each channel of the monolith, the transport of momentum, energy, and chemical species occurs not only in flow (axial) direction, but also in radial direction. The reactants diffuse to the inner channel wall, which is coated with the catalytic material, where the gaseous species adsorb and react on the surface. The products and intermediates desorb and diffuse back into the bulk flow. Due to the high temperatures, the chemical species may also react homogeneously in the gas phase. In catalytic reactors, the catalyst material is often dispersed in porous structures like washcoats or pellets. Mass transport in the fluid phase and chemical reactions are then superimposed by diffusion of the species to the active catalytic centers in the pores.

The temperature distribution depends on the interaction of heat convection and conduction in the fluid, heat release due to chemical reactions, heat transport in the solid material, and thermal radiation. If the feed conditions vary in time and space and/or heat transfer occurs between the reactor and the ambience, a non-uniform temperature distribution over the entire monolith will result, and the behavior will differ from channel to channel [7].

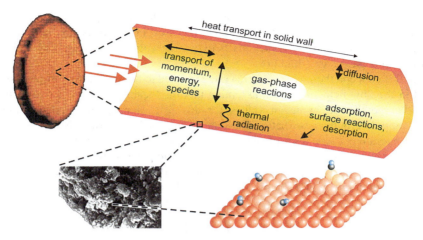

Fig. 1 Catalytic combustion monolith and physical and chemical process occurring in the single monolith channel

Today, the challenge in catalysis is not only the development of new catalysts to synthesize a desired product, but also the understanding of the interaction of the catalyst with the surrounding reactive flow field. Sometimes, the exploitation of these interactions can lead to the desired product selectivity and yield. A detailed introductions into fluid dynamics and transport phenomena can be found in [8–12], and into the coupling with heterogeneous reactions in [12, 13].

2.1 Heterogeneous Reaction Mechanisms

The development of a reliable surface reaction mechanism is a complex process. A tentative reaction mechanism can be proposed based on experimental surface science studies, on analogy to gas-phase kinetics and organo-metallic compounds, and on theoretical studies, increasingly including DFT calculations. This mechanism should include all possible paths for the formation of the chemical species under consideration in order to be "elementary-like" and thus applicable over a wide range of conditions. The mechanism idea then needs to be evaluated by numerous experimentally derived data, which are compared with theoretical predictions based on the mechanism. Here, the simulations of the laboratory reactors require appropriate models for all significant processes in order to evaluate the intrinsic kinetics. Sensitivity analysis leads to the crucial steps in the mechanism, for which refined kinetic experiments and data may be needed.

Since the early 1990, many groups have developed surface reaction mechanisms for high-temperature catalysis, following this concept, which has been adapted from modeling homogeneous gas-phase reactions in particular in the fields of combustion [10] and pyrolysis [14] of hydrocarbons. Consequently, this concept becomes handy when high-temperature processes in catalysis are considered, in particular the radical interactions between the solid phase (catalyst) and the surrounding gas-phase (fluid flow).

In this concept, the surface reaction rate is related to the size of the computational cell in the flow field simulation assuming that the local state of the active surface can be represented by mean values for this cell. Hence, this model assumes randomly distributed adsorbates. The state of the catalytic surface is described by the temperature T and a set of surface coverages Θ_i. The surface temperature and the coverages depend on time and the macroscopic position in the reactor, but are averaged over microscopic local fluctuations. Under those assumptions the molar net production rate $\dot{s}_i$ of a chemical species on the catalyst is given as

$$\dot{s}_i = \sum_{k=1}^{K_s} v_{ik} k_{f_k} \prod_{j=1}^{N_g+N_s+N_b} c_j^{v'_{jk}}. \tag{1}$$

Here, K_s is the number of surface reactions, c_i are the species concentrations, which are given, e.g., in $\mathrm{mol\,m^{-2}}$ for the N_s adsorbed species and in, e.g., $\mathrm{mol\,m^{-3}}$ for the N_g and N_b gaseous and bulk species. According to the relation

$$\Theta_i = c_i \sigma_i \Gamma^{-1}. \tag{2}$$

where Γ is surface site density with a coordination number σ_i describing the number of surface sites which are covered by the adsorbed species, the variations of surface coverages follow

$$\frac{\partial \Theta_i}{\partial t} = \frac{\dot{s}_i \sigma_i}{\Gamma}. \tag{3}$$

Since the reactor temperature and concentrations of gaseous species depend on the local position in the reactor, the set of surface coverages also varies with position. However, no lateral interaction of the surface species between different locations on the catalytic surface is modeled. This assumption is justified by the fact that the computational cells in reactor simulations are usually much larger than the range of lateral interactions of the surface processes.

Since the binding states of adsorption of all species vary with the surface coverage, the expression for the rate coefficient k_{f_k} is commonly extended by two additional parameters, μ_{ik} and ε_{ik} [12, 15]:

$$k_{f_k} = A_k T^{\beta_k} \exp\left[\frac{-E_{ak}}{RT}\right] \prod_{i=1}^{N_s} \Theta_i^{\mu_{ik}} \exp\left[\frac{\varepsilon_{ik}\Theta_i}{RT}\right]. \tag{4}$$

Numerical Simulation of Catalytic Reactors by Molecular-Based Models 231

A crucial issue with many of the surface mechanisms published is thermodynamic consistency [13]. Lately, optimization procedures enforcing overall thermodynamic consistency have been applied to overcome this problem [16].

In particular oxidation reactions, in which radical interactions play a very significant role, have been modeled extensively using this approach such as oxidation of hydrogen [17–23], CO [24–26], and methane [27–32] and ethane [4, 33–35] over Pt, formation of synthesis gas over Rh [32, 36]. Lately, mechanisms have been established for more complex reaction systems, for instance, three-way catalysts [37] or Chemical Vapor Deposition (CVD) reactors for the formation of diamond [38, 39], silica [40], and nanotubes [41]. In most of these reactions, adsorption and desorption of radicals is included and these steps are significant not only for the heterogeneous reaction but also for homogeneous conversion in the surrounding fluid. In most cases, the catalyst acts as sink for radicals produced in the gas-phase, and hence radical adsorption inhibits gas-phase reactions as exemplarily discussed below for oxy-dehydrogenation of alkanes by high-temperature catalysis.

2.2 Homogeneous Reactions

In many catalytic reactors, the reactions do not exclusively occur on the catalyst surface but also in the fluid flow. In some reactors even the desired products are mainly produced in the gas phase, for instance in the oxidative dehydrogenation of paraffins to olefins over noble metals at short contact times and high temperature as discussed below [4,35,42–47]. Such cases are dominated by the interaction between gas-phase and surface kinetics and transport. Therefore, any reactor simulation needs to include an appropriate model for the homogeneous kinetics along with the flow models. With v'_{ik}, v''_{ik} being the stoichiometric coefficients and an Arrhenius-like rate expression, the chemical source term of homogeneous reactions can be expressed by

$$R_i^{\mathrm{hom}} = M_i \sum_{k=1}^{K_g} (v''_{ik} - v'_{ik}) A_k T^{\beta_k} \exp\left[\frac{-E_{ak}}{RT}\right] \prod_{j=1}^{N_g} \left(\frac{Y_j \rho}{M_j}\right)^{a_{jk}}. \tag{5}$$

Here, A_k is the pre-exponential factor, β_k is the temperature exponent, E_{ak} is the activation energy, and a_{jk} is the order of reaction k related to the concentration of species j, which itself is expressed in terms of mass fractions Y_j, density ρ and molar mass M_j. Various reliable sets of elementary reactions are available for modeling homogeneous gas phase reactions, for instance for total [10] and partial oxidation, and pyrolysis of hydrocarbons. The advantage of the application of elementary reactions is that the reaction orders a_{jk} in (5) equal the stoichiometric coefficients v'_{ik}.

2.3 Coupling of Chemistry and Mass and Heat Transport

The chemical processes at the surface can be coupled with the surrounding flow field by boundary conditions for the species-continuity equations at the gas–surface interface [12, 15]:

$$n(j_i + \rho v_{\text{Stef}} Y_i) = R_i^{\text{het}}. \tag{6}$$

Here n is the outward-pointing unit vector normal to the surface, j_i is the diffusion mass flux of species i and R_i^{het} is the heterogeneous surface reaction rate, which is given per unit geometric surface area, corresponding to the reactor geometry, in $\text{kg m}^{-2} \text{s}^{-1}$. The Stefan velocity v_{Stef} occurs at the surface if there is a net mass flux between the surface and the gas phase [27, 48]. The calculation of R_i^{het} requires the knowledge of the amount of catalytically active surface area in relation to the geometric surface area, here denoted by $F_{\text{cat/geo}}$, at the gas–surface interface:

$$R_i^{\text{het}} = \eta F_{\text{cat/geo}} M_i \dot{s}_i. \tag{7}$$

Here, $\dot{s}_i$ is the molar net production rate of gas phase species i, given in $\text{mol m}^{-2} \text{s}^{-1}$; the area now refers to the actual catalytically active surface area. $F_{\text{cat/geo}}$ can be determined experimentally, e.g. by chemisorption measurements. The effect of internal mass transfer resistance for catalyst dispersed in a porous media is included by the effectiveness factor η [11, 49]. For more detailed models for transport in porous media it is referred to literature [50, 51].

Modeling the flow field in laminar and turbulent flows is discussed in many textbooks [8, 10] and review articles we refer to. Even though the implementation of (5)–(7) in those fluid flow models is straight forward, an additional highly nonlinear coupling is introduced into the governing equations describing the flow field (leading to considerable computational efforts. The nonlinearity, the large number of chemical species, and the fact that chemical reactions exhibit a large range of time scales, in particular when radicals are involved, make the solution of those equation systems challenging. In particular for turbulent flows, but sometimes even for laminar flows, the solution of the system is too CPU time-consuming with current numerical algorithms and computer capacities. This calls for the application of reduction algorithms for large reaction mechanisms, for instance by the extraction of the intrinsic low dimensional manifolds of trajectories in chemical space [52], which can be applied for heterogeneous reactions [53]. Another approach is to use "as little chemistry as necessary." In these so-called adaptive chemistry methods, the construction of the reaction mechanism includes only steps relevant for the application studied [54].

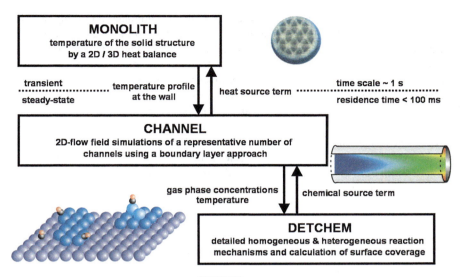

Fig. 2 Structure of the code DETCHEM^{MONOLITH} [57]

2.4 Monolithic Catalysts

As an example of a modeling a high-temperature catalyst, catalytically coated monolithic structures as given in Fig. 1 are discussed. An efficient approach, which still includes all fundamental aspects, is often used for modeling catalytic monoliths, which is based on the combination of simulations of a representative number of channels with the simulation of the temperature profiles of the solid structure treating the latter one as continuum [55, 56]. This approach is the basis for the computer code DETCHEM^{MONOLITH} [57], which has been applied to model the transient behavior of catalytic monoliths. The code combines a transient three-dimensional simulation of a catalytic monolith with a 2D model of the single-channel flow field based on the boundary layer approximation. It uses detailed models for homogeneous gas-phase chemistry, heterogeneous surface chemistry, and contains a model for the description of pore diffusion in washcoats.

The numerical procedure as sketched in Fig. 2 is based on the following ideas: The residence time of the reactive gas in the monolith channels is much smaller than the unsteadiness of the inlet conditions and the thermal response of the solid monolith structure. Under these assumptions, the time scales of the channel flow are decoupled from the temporal temperature variations of the solid, and the following procedure can be applied: A transient multi-dimensional heat balance is solved for the monolithic structure including the thermal insulation and reactor walls, which are treated as porous continuum. This simulation of the heat balance provides the temperature profiles along the channel walls. At each time step the reactive flow through a representative number of single channels is simulated including detailed transport and chemistry models. These single-channel simulations also calculate the heat flux from the fluid flow to the channel wall due to convective and conductive heat transport in the gaseous flow and heat released by chemical reactions. Thus, at each time step, the single-channel simulations provide the source terms for the heat balance of the monolith structure while the simulation of the heat balance provides the boundary condition (wall temperature) for the single-channel simulations. At each time step, the inlet conditions may vary. This very efficient iterative procedure enables a transient simulation of the entire monolith without sacrificing the details of the transport and chemistry models, as long as the prerequisites for the time scales remain valid. Furthermore, reactors with alternating channel properties such as flow directions, catalyst materials, and loadings can be treated.

2.5 Experimental Evaluation of Models Describing Radical Interactions

The coupling of several complex models introduces a large number of parameters into the simulations. Hence, agreement between predicted and experimentally observed overall conversion and selectivity alone is not sufficient to evaluate individual sub models. Time and locally resolved profiles provide a more stringent test for model evaluation. Useful data arise from the experimental resolution of local velocity profiles by laser Doppler anemometry/velocimetry (LDA, LDV) [58, 59] and of spatial and temporal species profiles by in situ, non-invasive methods such as Raman and laser induced fluorescence (LIF) spectroscopy. For instance, an optically accessible catalytic channel reactor can be used to evaluate models for heterogeneous and homogeneous chemistry as well as transport by the simultaneous detection of stable species by Raman measurements and OH radicals by Planar laser-induced fluorescence (PLIF) [60, 61].

2.6 Mathematical Optimization of Reactor Conditions and Catalyst Loading

In a chemical reactor, the initial and boundary conditions can be used to optimize the performance of the reactor, i.e., maximize the conversion, the selectivity or the yield of certain product species. In particular, at the inlet of the catalytic monolith, the mass or molar fractions of the species, the initial velocity, or the initial temperature can be controlled to optimize one product composition. Furthermore, it may be possible to control the temperature profile $T_{wall}(z)$ at the channel wall, and vary the loading with catalyst along the channel, i.e., $F_{cat/geo}(z)$. Moreover, the length of the catalytic monolith z_{max} can be optimized. Recently, algorithms have been established to not only optimize those control parameters but also to be applied to achieve a better understanding of the interactions between heterogeneous and homogeneous chemical reactions in catalytic reactors [62–64]. Radical interaction may play a decisive role as shown in the example given below.

3 Applications

3.1 Synthesis Gas from Natural Gas by High-Temperature Catalysis

A class of tube-like reactors is the monolith or honeycomb structure, which consists of numerous passageways with diameters reaching from a tenth of a millimeter to few millimeters. The flow field in the thin channels of this reactor type is usually laminar. The catalytic material is mostly dispersed in a washcoat on the inner channel wall. Monolith channels are manufactured with various cross-sectional shapes, e.g., circular, hexagonal, square or sinusoidal. In the next sections, the configuration of catalytic monolithic reactors will be used to discuss the interaction of gas-phase and surface chemistry in high-temperature catalysis.

The high-temperature catalytic partial oxidation (CPO) of methane over Rh based catalysts in short contact time (milliseconds) reactors has been intensively studied, because it offers a promising route to convert natural gas into synthesis gas (syngas, H_2 and CO), which can subsequently be converted to higher alkanes or methanol or be used in fuel cells [1,65,66]. The indirect route for syngas formation has meanwhile been accepted; at the catalyst entrance total oxidation occurs to form steam as long as oxygen is available at the surface, then methane is steam-reformed to hydrogen. Basically no dry reforming occurs and the surface acts as sink for radicals inhibiting significant gas-phase reactions at pressures below 10 bar [67]. Also the transient behavior during light-off of the reaction has been revealed. Exemplarily, Fig. 3 shows the time-resolved temperature and species profiles in a single channel of a catalytic monolith for partial oxidation of methane for the production of synthesis gas and the temperature distribution of the solid structure during light-off [36].

Since natural gas contains higher alkanes and other minor components besides methane, conversion and selectivity can be influenced by those other components. Consequently, conversion of methane in steam reforming of pure methane and in steam reforming of natural gas (North Sea H) differ as shown in Fig. 4. Here, fuel/steam mixtures, molar steam to carbon ratio of 2.5 and 4, diluted by 75% Ar, 10,000 h^{-1} space velocity, were fed into a furnace containing a Rh coated honeycomb catalyst [68].

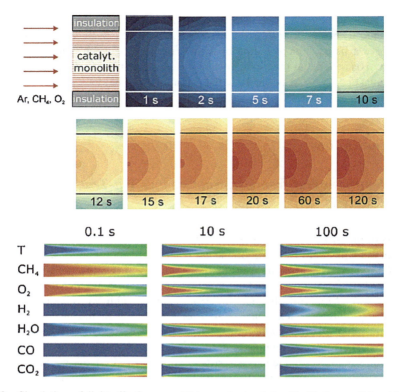

Fig. 3 Simulation of light-off of a monolithic reactor coated with Rh for partial oxidation of methane to synthesis gas [36]. Temperature of the solid structure (*top*) and gas-phase temperature and species mole fractions in a single channel in the center of the monolith (*below*), *red* = maximum, *blue* = minimum

3.2 Olefin Production by High-Temperature Oxidative Dehydrogenation of Alkanes

While gas-phase reactions are not significant in CPOX of methane at atmospheric pressure, CPOX of ethane to ethylene over platinum coated catalysts at short contact times [69, 70] is characterized by complex interactions of homogeneous gas-phase and heterogeneous surface reactions [4, 44, 71]. The principal picture of the reaction process is shown in Fig. 5. At the catalyst entrance oxygen is completely consumed at the surface within 1 mm primarily producing CO_2 and H_2O. This total combustion of ethane leads to a rapid temperature increase from RT to 1,000°C. The high temperature drives the pyrolysis of ethane in the gas-phase. After a decade of discussions on the reaction pathways, most studies today conclude that most of the ethylene (desired product) is actually homogeneously produced in the gas-phase. Further downstream, additionally, reforming and shift reactions occur.

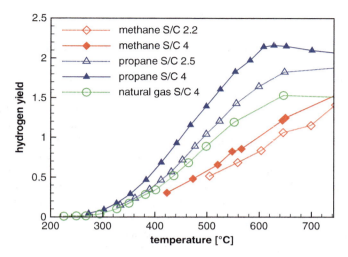

Fig. 4 Comparison of experimentally determined hydrogen yields in steam-reforming of methane with S/C 2.5 (*open diamond*) and S/C 4 (*filled diamond*), steam reforming of natural gas with S/C 4 (*open circle*), steam reforming of propane with S/C 2.5 (*open triangle*) and S/C 4 (*filled triangle*); Rh catalyst, taken from [68]

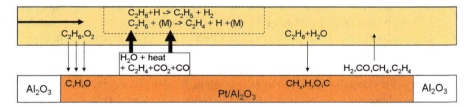

Fig. 5 General picture of oxy-dehydrogenation of ethane over Pt at millisecond contact times and temperatures of ∼900°C; inlet ethane/oxygen ratio ∼2. The *upper* and *lower panel* show gas-phase and surface processes, respectively, adapted from Zerkle et al. [4]

Based on the molecular understanding of interaction of gas-phase and surface chemistry in oxy-dehydrogenation of ethane over Pt [4], Minh et al. [62] used a recently developed optimization code to find the optimal Pt catalyst loading on along the flow direction in a Pt/Al$_2$O$_3$ coated honeycomb catalyst. The gas-phase mechanism used consists of 25 reactive species (mainly C1 and C2 species) involved in 131 reversible reactions and one irreversible reaction, and the surface reaction mechanism consists of another 82 elementary-step like reactions involving another 19 surface species [4]. This mechanism was later also used to study on-line catalyst addition effects [71]. The Pt coated monolith had a diameter of 18 mm and a length of 10 mm. Each channel has a diameter of 0.5 mm. The monolith is fed with ethane/oxygen/nitrogen mixture of varying C/O ratio at 5 standard liters per minute leading to a residence time of few milliseconds.

3.2.1 Formulation of an Optimal Control Problem

The initial and boundary conditions can be used to optimize the performance of the reactor, i.e., maximize the conversion, the selectivity or the yield of certain product species. In particular, at the inlet of the catalytic monolith, the mass or molar fractions of the species, the initial velocity, or the initial temperature can be controlled to optimize one product composition. Furthermore, it may be possible to control the temperature profile $T_{wall}(z)$ along the channel wall, and vary the loading with catalyst along the channel, i.e., $F_{cat/geo}(z)$. Moreover, the length of the catalytic monolith z_{max} can be optimized. Here, the objective function to be maximized is the mass fraction of ethylene at the outlet of the channel. The control considered here is the catalyst loading, expressed, which is a function of the axial coordinate z. For practical reasons, there are often equality and inequality constraints on the control and state variables, such as an upper and lower bounds for the catalyst loading and the (trivial) fact that the sum of all mass fractions must be one.

In the case considered, the inlet gas temperature is $T_{gas} = 600\,K$, and the wall temperature $T_{wall}(z)$ is kept fixed at $1,000\,K$. As constraint the $F_{cat/geo}(z)$ is required to be between 0, i.e., no catalyst, and 100, i.e., highly loaded. The optimization was started with a constant $F_{cat/geo}(z)$ profile of 20.0 leading to an objective value of 0.06. In the optimal solution the objective value is 0.19. Figure 6 shows the standard and optimal profiles of $F_{cat/geo}(z)$ and average mass fraction profiles of ethylene. Figure 7 reveals the mass fraction profiles of ethane and ethylene with the optimal profile of $F_{cat/geo}(z)$. Platinum is a very efficient catalyst for the oxidation of ethane. In the first two mm of the catalyst, oxygen is almost completely consumed by surface reactions (catalytic oxidation of ethane) leading to the total oxidation products CO_2 and H_2O, ethylene, and some CO. Ethylene however is substantially produced in the gas-phase as well. However, the conversion in the gas-phase only occurs if a sufficiently large radical pool is build up, which takes a certain time/distance (so-called ignition delay time). Furthermore, some of the ethylene formed by surface reactions adsorb on the surface as well, where in the region around 2 mm mainly reforming reactions occur but the total reaction rate is much smaller than in the first mm of the catalyst where oxygen was still available. Since the production of ethylene by gas-phase reactions really takes-off further downstream (most of the ethylene is produced in the gas-phase) due to the radicals available there and due to the fact that the surface is relatively inactive in the region around 2 mm, a plateau appears in that region around 2 mm (Fig. 6) due to the competition between ethylene production in the gas-phase and (partial) oxidation on the surface, both at relatively low rates. The optimization of the catalyst loading proposes a very low loading in this region, because here the catalyst does not only oxidize ethylene but also adsorbs radicals from the gas-phase, which are needed to initiate ethylene formation in the gas-phase. The optimization proposes relatively low catalyst loading in the very active initial catalyst section, which can be understood as follows: Within the first mm of the catalyst, where oxygen is available, the process is limited by mass-transfer of ethane and even more of oxygen to the surface. Here, primarily, oxidation of ethane occurs at a very high

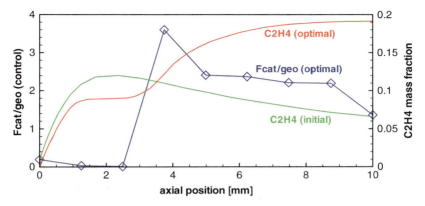

Fig. 6 Catalyst loading expressed by $F_{cat/geo}(z)$ and the radial averaged mass fraction of ethylene at the initial and at optimal solutions in oxy-dehydrogenation of ethane over Pt coated honeycomb monoliths, taken from Minh et al. [62]

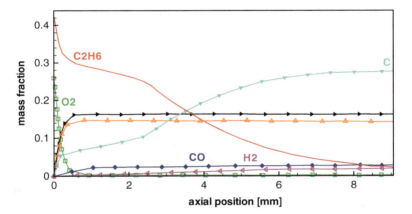

Fig. 7 Average mass fractions of major species at the optimal solution in optimization of catalyst loading in oxy-dehydrogenation of ethane over Pt coated honeycomb monoliths, taken from Minh et al. [62]

rate, the catalyst is very active. Consequently, catalyst is needed here but not at a high loading; this effect has also been observed in several experiments.

The hydrocarbon/oxygen (C/O) ratio can easily be used to tune the product selectivity in CPOX of alkanes. The group of L.D. Schmidt has studied a variety of fuels and basically found the same trend as shown in Fig. 8 for CPOX of n-decane at millisecond contact times over Rh catalysts [46]. At low C/O ratio, synthesis gas is the primary product; actually its maximum is reached close to the point where the reaction switches to total combustion, a flame is often formed in that transition region. With increasing C/O ratio, more and more olefins are formed, primarily

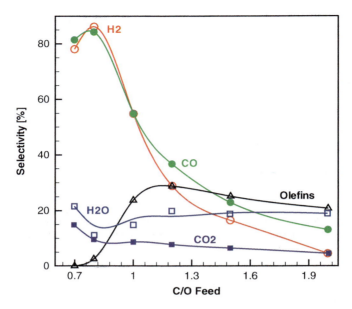

Fig. 8 Effect of the n-decane/oxygen feed ratio on the product selectivity in CPOX of n-decane at millisecond contact times over Rh catalysts [46]

α-olefins, but at higher C/O ratio a larger variety of other olefins as well, and eventually acetylene, benzene, and PAH are formed at very low oxygen content. Coking of the catalyst and formation of soot in the gas-phase quickly becomes a technical problem. Those studies over noble metal based catalysts and at millisecond contact times were recently extended to CPOX of biodiesel [72]. Also, autothermal reforming of ethanol was realized at short contact times and high temperatures in a two-stage reactor [73, 74]. A Rh—ceria catalyst on alumina foams/spheres served as first stage for reforming with some oxygen addition. The second reactor stage consists of Pt—ceria on alumina spheres to accelerate the water–gas shift reaction for maximizing the hydrogen yield.

3.3 Hydrogen Production from Logistic Fuels by High-Temperature Catalysis

The production of hydrogen and synthesis gas (syngas, H_2 and CO) from logistic fuels such as gasoline, diesel, and kerosene by catalytic partial oxidation (CPOX) and steam reforming (SR) is currently in the focus of both academic and industrial research. In contrast to the complex and costly supply of compressed and stored hydrogen for mobile fuel cell application, CPOX of liquid fuels allows production

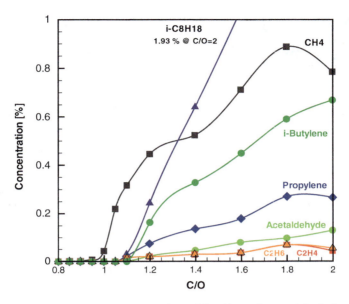

Fig. 9 Experimentally determined concentrations of the side products and the fuel remaining in the outlet stream of CPOX of i-octane over a Rh/alumina coated monolith as a function of C/O ratio, taken from [75]

and utilization of hydrogen through existing routes, which, in particular, is of interest for on-board applications. At operating temperatures around 1,000 K and higher, conversion of the fuel may not only occur on the solid catalyst but also in the gas-phase. Heterogeneous and homogeneous reactions in CPOX of hydrocarbons are coupled not only by adsorption and desorption of fuel and oxygen molecules and the products, respectively, but also by adsorption and desorption of intermediates and radicals. Therefore, mass transport of radicals and intermediates from/to the gaseous bulk phase and the catalytically active channel wall, mainly by radial diffusion in the small channels of the monolith being on the order of a quarter to one millimeter, is crucial for the interaction of heterogeneous and homogeneous reactions in CPOX reactors. Hartmann et al., for instance, studied the catalytic partial oxidation of iso-octane over rhodium/alumina coated honeycomb monolith serving as gasoline surrogate [75]. Very high hydrogen and carbon monoxide selectivity were found at stoichiometric conditions (C/O = 1), while at lean conditions more total oxidation occurs. At rich conditions (C/O > 1), homogeneous chemical conversion in the gas-phase is responsible for the formation of by-products such as olefins shown in Fig. 9 that also have the potential for coke formation, which was observed experimentally and numerically. This study also revealed that the chemical models applied—even though the most detailed ones available (857 gas-phase and 17 adsorbed species in over 7,000 elementary reactions) were used—need further improvement.

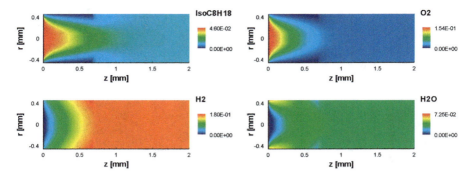

Fig. 10 Numerically predicted profiles of molar fractions of reactants and major products in the entrance region of the catalyst at C/O = 1.2 in CPOX of iso-octane over a Rh/alumina coated monolith, taken from Hartmann et al. [75]. Flow direction is from *left* to *right*

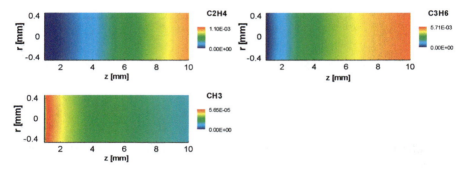

Fig. 11 Numerically predicted profiles of molar fractions of minor products and radicals along the entire catalyst at C/O = 1.2 in CPOX of iso-octane over a Rh/alumina coated monolith, taken from Hartmann et al. [75]. Flow direction is from *left* to *right*

Nevertheless, this combined modeling and experimental study revealed the role of surface, gas-phase, and radical chemistry in high-temperature oxidative catalytic conversion of larger hydrocarbons. From Fig. 10, it can clearly be concluded that the major products (syngas) are produced in the entrance region of the catalyst on the catalytic surface; radial concentration profiles are caused by a mass-transfer limited process. As soon as the oxygen is consumed on the catalytic surface—similar to CPOX of natural gas—hydrogen formation increases due to steam reforming, the major products are formed within few millimeters. At rich conditions (C/O > 1.0) a second process, now in the gas-phase, begins in the downstream part as shown in Fig. 11. The number of radicals in the gas-phase is sufficiently large to initiate gas-phase pyrolysis of the remaining fuel and formation of coke-precursors such as ethylene and propylene. In the experiment, the downstream part of the catalyst is coked-up, here the Rh surface cannot act as sink for radical.

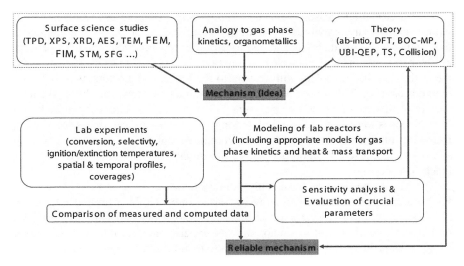

Fig. 12 Survey of the methodology of the development of a surface reaction mechanism

4 Model Parameters

One of the major questions for detailed reaction mechanisms is the source and the reliability of the kinetic parameters. Typically for molecular based models there are several 10s–100s of gas-phase and surface species involved. The number of reactions among them is even one order of magnitude higher. Thus it is virtually impossible to determine the kinetic parameters of each elementary step by direct experimental observation. The development of a reliable surface reaction mechanism follows an iterative scheme given Fig. 12.

A tentative reaction mechanism can be proposed based on experimental surface science studies, on analogy to gas-phase kinetics and organo-metallic compounds, and on theoretical studies, such as first principle calculations using density functional theory [76, 77]. This mechanism should include all possible paths for the formation of the chemical species under consideration in order to be elementary-like and thus be applicable over a wide range of conditions. The mechanism idea then needs to be evaluated by numerous experimentally derived data, which are compared with theoretical predictions based on the mechanism. Here, numerical simulations of the laboratory reactors require appropriate models for all significant processes in order to evaluate the intrinsic kinetics. Sensitivity analysis leads to information about crucial steps in the mechanism, for which refined kinetic experiment and data may be necessary. At the end of the iteration, an elementary-step reaction mechanism must be applicable to simulations of different kinds of experimental set-ups.

All experimental measurements and first principle calculations result in kinetic parameters with a comparatively small but significant error. Therefore, parameter fine tuning will nonetheless be necessary in order to improve agreement between experimental results and numerical simulation. The best way to do so is to solve the optimal control problem with respect to the kinetic parameters, with the same methods as described in Sect. 3.2 for reactor parameters. However, due to the large variety of different reactor set-ups, computer program development for each type of problem is very time consuming as long as there is no easy-to-use toolbox to formulate optimal control problems in computational fluid dynamics. Thus, the authors have chosen an approach where the numerical simulation is treated as a black-box process.

The kinetic parameters form a high-dimensional vector x in parameter space. Each component of this vector may be varied only within a limited range. The simulations of one or more reactor set-ups map vector x onto a result vector y^{sim}. We only assume continuity of this mapping. The experimental results shall be denoted by y^{exp}. Thus the objective function subject to be minimized can be written as

$$F(x) = \sum_i \left(\frac{y_i^{sim} - y_i^{exp}}{d_i} \right)^2 \tag{8}$$

where d_i denote user-defined values for normalization of the different results y_i, e.g. reference values or tolerances.

The steps taken into account during parameter fine-tuning are illustrated in Fig. 13. The first generation of parameters is generated randomly from the whole given parameter space (Fig. 13a). The point with lowest value of the objective function F will be memorized. With increasing number of generations, the search becomes more and more a random walk by searching preferably the vicinity of the optimal solution. The deviation of each parameter from the best solution follows a normal distribution with cut-off limits, where the variance decreases with increasing number of generations (Fig. 13b). Finally, the optimization procedure searches for improvements along the numerically computed gradient of F (Fig. 13c).

An example of achievements by this parameter fine-tuning is shown in Fig. 14. Here, an elementary-step like reaction mechanism over platinum, that describes the removal of pollutants from gasoline engines by three-way catalysis, was applied to test-bench experiments with exhaust gas recycling. The original mechanism was proposed by Chatterjee et al. [37]. The reaction mechanism consists of 30 reversible and one irreversible surface reactions. Fifty-two kinetic parameters were varied for fine-tuning. The remaining kinetic parameters have to be adjusted accordingly in order to ensure thermodynamic consistency of the resulting mechanism. A total of 153 target data points (3 gas mixtures at 17 temperatures, 3 components analyzed) have been selected for comparison of experiment and simulation. Thus, 51 different experiments had to be simulated numerically in each iteration step. After more than 10,000 iterations, agreement between experiment and simulation was improved significantly (Fig. 14).

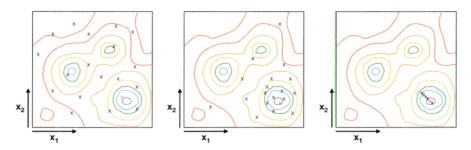

Fig. 13 Two-dimensional illustration of the parameter fine-tuning procedure to look for minimum of objective function; from *left* to *right*: (**a**) random seeds, (**b**) search preferably in the vicinity of best solution, (**c**) gradient search

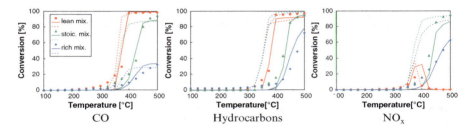

Fig. 14 Comparison of test-bench experiments using a three-way catalyst with exhaust gas recycling (*symbols*) with simulation results using original [37] (*dashed lines*) and improved mechanism (*solid lines*)

The random-walk approach seems to be feasible only for relatively small systems. The number iterations required to explore significant parts of the parameter space increases with number of parameters. Of course, one could limit the search to those parameters that exceed a given threshold in sensitivity analysis. Therefore, even in a 52-dimensional parameter space, a sufficient number of parameter variations in the most sensitive 3 or 4 parameters should occur during a few thousands of iterations. The crucial aspect of this kind of simulations is the computational time needed for a simulation of a single experiment. Molecular-based models in computational fluid dynamics usually result in time consuming simulations. Thus it would be desirable to have algorithms to solve the simulation problem and the optimal control problem at the same time without drastically increasing the simulation time. The aim of detailed reaction mechanisms is to describe a large variety of reactors with a large variety of operating conditions. Thus, the open question remains how a numerical tool box should look like that allows solving the optimal control problem to find the kinetic parameters given a large variety of experimental data (Fig. 15).

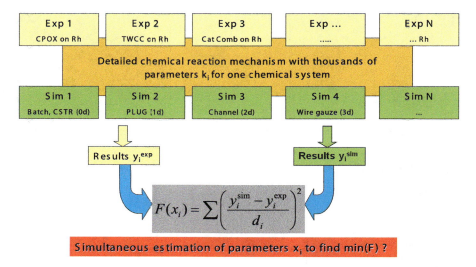

Fig. 15 A desired scheme for estimation of large sets of chemical parameters applicable to many experiments.

Acknowledgements The authors would like to thank J. Warnatz in memoriam (University of Heidelberg), L.D. Schmidt (University of Minnesota), and R.J. Kee (Colorado School of Mines) for fruitful collaborations and stimulating discussions.

References

1. D. A. Hickman and Lanny D. Schmidt. Synthesis gas formation by direct oxidation of methane over monoliths. *Journal of Catalysis*, 138(1):267–282, 1992.
2. O. Deutschmann and L. D. Schmidt. Modeling the partial oxidation of methane in a short-contact-time reactor. *American Institute of Chemical Engineering Journal*, 44(11):2465–2477, 1998.
3. Marylin C. Huff and Lanny D. Schmidt. Production of olefins by oxidative dehydrogenation of propane and butane over monoliths at short contact times. *Journal of Catalysis*, 149:127–141, 1993.
4. D. K. Zerkle, M. D. Allendorf, M. Wolf, and O. Deutschmann. Understanding homogeneous and heterogeneous contributions to the platinum-catalyzed partial oxidation of ethane in a short-contact-time reactor. *Journal of Catalysis*, 196(1):18–39, 2000.
5. H. Y. Zhu, R. J. Kee, V. M. Janardhanan, O. Deutschmann, and D. G. Goodwin. Modeling elementary heterogeneous chemistry and electrochemistry in solid-oxide fuel cells. *Journal of the Electrochemical Society*, 152(12):A2427–A2440, 2005.
6. E. S. Hecht, G. K. Gupta, H. Y. Zhu, A. M. Dean, R. J. Kee, L. Maier, and O. Deutschmann. Methane reforming kinetics within a Ni-YSZ SOFC anode support. *Applied Catalysis a-General*, 295(1):40–51, 2005.
7. J. Windmann, J. Braun, P. Zacke, S. Tischer, O. Deutschmann, and J. Warnatz. Impact of the inlet flow distribution on the light-off behavior of a 3-way catalytic converter. *SAE Technical Paper*, 2003-01-0937, 2003.

Numerical Simulation of Catalytic Reactors by Molecular-Based Models 247

8. R. B. Bird, W. E. Stewart, and E. N. Lightfoot. *Transport Phenomena*. John Wiley & Sons, Inc., New York, 2nd edition, 2001.

9. S.V. Patankar. *Numerical Heat Transfer and Fluid Flow*. McGraw-Hill, New York, 1980.

10. Jürgen Warnatz, R. W. Dibble, and Ulrich Maas. *Combustion, Physical and Chemical Fundamentals, Modeling and Simulation, Experiments, Pollutant Formation*. Springer-Verlag, New York, 1996.

11. R. E. Hayes and Stan T. Kolaczkowski. *Introduction to Catalytic Combustion*. Gordon and Breach Science Publ., Amsterdam, 1997.

12. R.J. Kee, M.E. Coltrin, and P. Glarborg. *Chemically Reacting Flow*. Wiley-Interscience, 2003.

13. O. Deutschmann. Computational fluid dynamics simulation of catalytic reactors: Chapter 6.6. In G. Ertl, H. Knözinger, F. Schüth, and J. Weitkamp, editors, *Handbook of Heterogeneous Catalysis, 2nd Ed.* Wiley-VCH, 2007.

14. Eliseo Ranzi, A. Sogaro, P. Gaffuri, G. Pennati, C. K. Westbrook, and W. J. Pitz. A new comprehensive reaction mechanism for combustion of hydrocarbon fuels. *Combustion and Flame*, 99:201–211, 1994.

15. Michael E. Coltrin, Robert J. Kee, and F. M. Rupley. *SURFACE CHEMKIN (Version 4.0): A Fortran Package for Analyzing Heterogeneous Chemical Kinetics at a Solid-Surface - Gas-Phase Interface, SAND91-8003B*. Sandia National Laboratories, 1991.

16. A. B. Mhadeshwar, H. Wang, and D. G. Vlachos. Thermodynamic consistency in microkinetic development of surface reaction mechanisms. *Journal of Physical Chemistry B*, 107(46):12721–12733, 2003.

17. W. R. Williams, C. M. Marks, and L. D. Schmidt. Steps in the reaction $H_2 + O_2 \rightleftharpoons H_2O$ on Pt - OH desorption at high-temperatures. *Journal of Physical Chemistry*, 96(14):5922–5931, 1992.

18. B. Hellsing, B. Kasemo, and V. P. Zhdanov. Kinetics of the hydrogen-oxygen reaction on platinum. *Journal of Catalysis*, 132:210–228, 1991.

19. Jürgen Warnatz. Resolution of gas phase and surface combustion chemistry into elementary reactions (invited lecture). *Proceedings of the Combustion Institute*, 24:553–579, 1992.

20. M. Rinnemo, O. Deutschmann, F. Behrendt, and B. Kasemo. Experimental and numerical investigation of the catalytic ignition of mixtures of hydrogen and oxygen on platinum. *Combustion and Flame*, 111(4):312–326, 1997.

21. Götz Veser. Experimental and theoretical investigation of h2 oxidation in a high-temperature catalytic microreactor. *Chemical Engineering Science*, 56:1265–1273, 2001.

22. P.-A. Bui, D. G. Vlachos, and P. R. Westmoreland. Modeling ignition of catalytic reactors with detailed surface kinetics and transport: Oxidation of h2/air mixtures over platinum surfaces. *Industrial & Engineering Chemistry Research*, 36(7):2558–2567, 1997.

23. Johan C. G. Andrae and Pehr H. Björnbom. Wall effects of laminar hydrogen flames over platinum and inert surfaces. *American Institute of Chemical Engineering Journal*, 46(7):1454–1460, 2000.

24. J. Mai, W. von Niessen, and A. Blumen. The CO+O2 reaction on metal surfaces. simulation and mean-field theory: The influence of diffusion. *Journal of Chemical Physics*, 93:3685–3692, 1990.

25. V. P. Zhdanov and B. Kasemo. Steady-state kinetics of co oxidation on pt: extrapolation from 10-10 to 1 bar. *Applied Surface Science*, 74(2):147–164, 1994.

26. P. Aghalayam, Y. K. Park, and D. G. Vlachos. A detailed surface reaction mechanism for CO oxidation on Pt. *Proceedings of the Combustion Institute*, 28:1331–1339, 2000.

27. Olaf Deutschmann, R. Schmidt, Frank Behrendt, and Jürgen Warnatz. Numerical modeling of catalytic ignition. *Proceedings of the Combustion Institute*, 26:1747–1754, 1996.

28. G. Veser, J. Frauhammer, Lanny D. Schmidt, and G. Eigenberger. Catalytic ignition during methane oxidation on platinum: Experiments and modeling. In *Studies in Surface Science and Catalysis 109*, pages 273–284. 1997.

29. P.-A. Bui, D. G. Vlachos, and P. R. Westmoreland. Catalytic ignition of methane/oxygen mixtures over platinum surfaces: comparison of detailed simulations and experiments. *Surface Science*, 386(2–3):L1029–L1034, 1997.

30. U. Dogwiler, P. Benz, and J. Mantzaras. Two-dimensional modelling for catalytically stabilized combustion of a lean methane-air mixture with elementary homogeneous and heterogeneous chemical reactions. *Combustion and Flame*, 116(1):243, 1999.
31. Preeti Aghalayam, Young K. Park, N. Fernandes, V. Papavassiliou, A. B. Mhadeshwar, and Dionisios G. Vlachos. A C1 mechanism for methane oxidation on platinum. *Journal of Catalysis*, 213(1):23–38, 2003.
32. D. A. Hickman and Lanny D. Schmidt. Steps in CH_4 oxidation on Pt and Rh surfaces: High-temperature reactor simulations. *American Institute of Chemical Engineering Journal*, 39:1164–1176, 1993.
33. M. Huff and L. D. Schmidt. Ethylene formation by oxidative dehydrogenation of ethane over monoliths at very short-contact times. *Journal of Physical Chemistry*, 97(45):11815–11822, 1993.
34. Marylin C. Huff, I. P. Androulakis, J. H. Sinfelt, and S. C. Reyes. The contribution of gas-phase reactions in the Pt-catalyzed conversion of ethane-oxygen mixtures. *Journal of Catalysis*, 191:46–45, 2000.
35. F. Donsi, K. A. Williams, and L. D. Schmidt. A multistep surface mechanism for ethane oxidative dehydrogenation on Pt- and Pt/Sn-coated monoliths. *Industrial & Engineering Chemistry Research*, 44(10):3453–3470, 2005.
36. R. Schwiedernoch, S. Tischer, C. Correa, and O. Deutschmann. Experimental and numerical study on the transient behavior of partial oxidation of methane in a catalytic, monolith. *Chemical Engineering Science*, 58(3–6):633–642, 2003.
37. D. Chatterjee, O. Deutschmann, and J. Warnatz. Detailed surface reaction mechanism in a three-way catalyst. *Faraday Discussions*, 119:371–384, 2001.
38. B. Ruf, F. Behrendt, O. Deutschmann, and J. Warnatz. Simulation of homoepitaxial growth on the diamond (100) surface using detailed reaction mechanisms. *Surface Science*, 352:602–606, 1996.
39. S. J. Harris and D. G. Goodwin. Growth on the reconstructed diamond (100) surface. *Journal of Physical Chemistry*, 97(1):23–28, 1993.
40. S. Romet, M. F. Couturier, and T. K. Whidden. Modeling of silicon dioxide chemical vapor deposition from tetraethoxysilane and ozone. *Journal of the Electrochemical Society*, 148(2):G82–G90, 2001.
41. C. D. Scott, A. Povitsky, C. Dateo, T. Gokcen, P. A. Willis, and R. E. Smalley. Iron catalyst chemistry in modeling a high-pressure carbon monoxide nanotube reactor. *Journal of Nanoscience and Nanotechnology*, 3(1–2):63–73, 2003.
42. Alessandra Beretta, Pio Forzatti, and Eliseo Ranzi. Production of olefins via oxidative dehydrogenation of propane in autothermal conditions. *Journal of Catalysis*, 184(2):469–478, 1999.
43. A. Beretta and P. Forzatti. High-temperature and short-contact-time oxidative dehydrogenation of ethane in the presence of Pt/Al_2O_3 and $BaMnAl_{11}O_{19}$ catalysts. *Journal of Catalysis*, 200(1):45–58, 2001.
44. A. Beretta, E. Ranzi, and P. Forzatti. Oxidative dehydrogenation of light paraffins in novel short contact time reactors. experimental and theoretical investigation. *Chemical Engineering Science*, 56(3):779–787, 2001.
45. R. Subramanian and L. D. Schmidt. Renewable olefins from biodiesel by autothermal reforming. *Angewandte Chemie-International Edition*, 44(2):302–305, 2005.
46. J. J. Krummenacher and L. D. Schmidt. High yields of olefins and hydrogen from decane in short contact time reactors: rhodium versus platinum. *Journal of Catalysis*, 222(2):429–438, 2004.
47. Lanny D. Schmidt, J. Siddall, and M. Bearden. New ways to make old chemicals. *American Institute of Chemical Engineering Journal*, 46:1492–1495, 2000.
48. L. L. Raja, Robert J. Kee, and L. R. Petzold. Simulation of the transient, compressible, gas-dynamic, behavior of catalytic-combustion ignition in stagnation flows. *Proceedings of the Combustion Institute*, 27:2249–2257, 1998.

Numerical Simulation of Catalytic Reactors by Molecular-Based Models 249

49. D. Papadias, L. Edsberg, and Pehr H. Björnbom. Simplified method of effectiveness factor calculations for irregular geometries of washcoats. a general case in a 3d concentration field. *Catalysis Today*, 60(1–2):11–20, 2000.
50. F. Keil. *Diffusion und Chemische Reaktionen in der Gas-Feststoff-Katalyse*. Springer-Verlag, Berlin, 1999.
51. F. J. Keil. Diffusion and reaction in porous networks. *Catalysis Today*, 53:245–258, 2000.
52. Ulrich Maas and S. Pope. Simplifying chemical kinetics: Intrinsic low-dimensional manifolds in composition space. *Combustion and Flame*, 88:239–264, 1992.
53. X. Yan and U. Maas. Intrinsic low-dimensional manifolds of heterogeneous combustion processes. *Proceedings of the Combustion Institute*, 28:1615–1621, 2000.
54. R. G. Susnow, A. M. Dean, W. H. Green, P. Peczak, and L. Broadbelt. Rate - based construction of kinetic models for complex systems. *Journal of Physical Chemistry A*, 101:3731–3740, 1997.
55. S. Tischer, C. Correa, and O. Deutschmann. Transient three-dimensional simulations of a catalytic combustion monolith using detailed models for heterogeneous and homogeneous reactions and transport phenomena. *Catalysis Today*, 69(1–4):57–62, 2001.
56. S. Tischer and O. Deutschmann. Recent advances in numerical modeling of catalytic monolith reactors. *Catalysis Today*, 105(3–4):407–413, 2005.
57. O. Deutschmann, S. Tischer, S. Kleditzsch, V.M. Janardhanan, C. Correa, D. Chatterjee, N. Mladenov, and H. D. Minh. *DETCHEM software package*. www.detchem.com, Karlsruhe, 2.2 edition, 2009.
58. C. Appel, J. Mantzaras, R. Schaeren, R. Bombach, and A. Inauen. Turbulent catalytically stabilized combustion of hydrogen/air mixtures in entry channel flows. *Combustion and Flame*, 140(1-2):70–92, 2005.
59. H. P. A. Calis, J. Nijenhuis, B. C. Paikert, F. M. Dautzenberg, and C. M. van den Bleek. CFD modelling and experimental validation of pressure drop and flow profile in a novel structured catalytic reactor packing. *Chemical Engineering Science*, 56(4):1713–1720, 2001.
60. C. Appel, J. Mantzaras, R. Schaeren, R. Bombach, B. Kaeppeli, and A. Inauen. An experimental and numerical investigation of turbulent catalytically stabilized channel flow combustion of hydrogen/air mixtures over platinum. *Proceedings of the Combustion Institute*, 29:1031–1038, 2002.
61. C. Appel, J. Mantzaras, R. Schaeren, R. Bombach, A. Inauen, B Kaeppeli, B. Hemmerling, and A. Stampanoni. An experimental and numerical investigation of homogeneous ignition in catalytically stabilized combustion of hydrogen/air mixtures over platinum. *Combustion and Flame*, 128(4):340–368, 2002.
62. H. D. Minh, H. G. Bock, S. Tischer, and O. Deutschmann. Optimization of two-dimensional flows with homogeneous and heterogeneously catalyzed gas-phase reactions. *Aiche Journal*, 54(9):2432–2440, 2008.
63. H. D. Minh, H. G. Bock, S. Tischer, and O. Deutschmann. Fast solution for large-scale 2-d convection-diffusion, reacting flows. *Computational Science and Its Applications - Iccsa 2008, Pt 1, Proceedings*, 5072:1121–1130 1266, 2008.
64. M. von Schwerin, O. Deutschmann, and V. Schulz. Process optimization of reactives systems by partially reduced sqp methods. *Computers & Chemical Engineering*, 24(1):89–97, 2000.
65. L. D. Schmidt, O. Deutschmann, and C. T. Goralski. Modeling the partial oxidation of methane to syngas at millisecond contact times. In *Natural Gas Conversion V*, volume 119 of *Studies in Surface Science and Catalysis*, pages 685–692. 1998.
66. R. Horn, K. A. Williams, N. J. Degenstein, and L. D. Schmidt. Syngas by catalytic partial oxidation of methane on rhodium: Mechanistic conclusions from spatially resolved measurements and numerical simulations. *Journal of Catalysis*, 242(1):92–102, 2006.
67. R. Quiceno, J. Perez-Ramirez, J. Warnatz, and O. Deutschmann. Modeling the high-temperature catalytic partial oxidation of methane over platinum gauze: Detailed gas-phase and surface chemistries coupled with 3d flow field simulations. *Applied Catalysis a-General*, 303(2):166–176, 2006.

68. B. T. Schädel, M. Duisberg, and O. Deutschmann. Steam reforming of methane, ethane, propane, butane, and natural gas over a rhodium-based catalyst. *Catalysis Today*, 142(1–2):42–51, 2009.
69. Marylin C. Huff and Lanny D. Schmidt. Ethylene formation by oxidative dehydrogenation of ethane over monoliths at very short contact times. *Journal of Physical Chemistry*, 97:11815–11822, 1993.
70. A. S. Bodke, D. A. Olschki, L. D. Schmidt, and E. Ranzi. High selectivities to ethylene by partial oxidation of ethane. *Science*, 285(5428):712–715, 1999.
71. D. K. Zerkle, M. D. Allendorf, M. Wolf, and O. Deutschmann. Modeling of on-line catalyst addition effects in a short contact time reactor. *Proceedings of the Combustion Institute*, 28:1365–1372, 2000.
72. J. J. Krummenacher, K. N. West, and L. D. Schmidt. Catalytic partial oxidation of higher hydrocarbons at millisecond contact times: decane, hexadecane, and diesel fuel. *Journal of Catalysis*, 215(2):332–343, 2003.
73. E. C. Wanat, K. Venkataraman, and L. D. Schmidt. Steam reforming and water–gas shift of ethanol on Rh and Rh-Ce catalysts in a catalytic wall reactor. *Applied Catalysis a-General*, 276(1–2):155–162, 2004.
74. G. A. Deluga, J. R. Salge, L. D. Schmidt, and X. E. Verykios. Renewable hydrogen from ethanol by autothermal reforming. *Science*, 303(5660):993–997, 2004.
75. M. Hartmann, L. Maier, and O. Deutschmann. Catalytic partial oxidation of iso-octane over rhodium catalysts: An experimental, modeling, and simulation study. *Combustion and Flame*, 157:1771–1782, 2010.
76. O. R. Inderwildi, D. Lebiedz, O. Deutschmann, and J. Warnatz. Coverage dependence of oxygen decomposition and surface diffusion on rhodium (111): A DFT study. *Journal of Chemical Physics*, 122(3), 2005.
77. A. Heyden, B. Peters, A. T. Bell, and F. J. Keil. Comprehensive DFT study of nitrous oxide decomposition over Fe-ZSM-5(vol 109, pg 1866, 2005). *Journal of Physical Chemistry B*, 109(10):4801–4804, 2005.

Model-Based Design of Experiments for Estimating Heat-Transport Parameters in Tubular Reactors

Alexander Badinski and Daniel Corbett

Abstract Heat-transport parameters in a two-dimensional heat-transport model are estimated from temperature data of a tubular reactor with a fixed catalyst bed. The reactor design is taken from Adler (Chem. Ing. Tech. 72:555–564, 2000). Using model-based design of experiment (DoE), two experimental control variables, the reactor wall temperature and the gas flow density, are optimized to yield minimal parameter uncertainties. Previously in the literature (Bauer, Theoretische und experimentelle Untersuchung zum Wärmetransport in gasdurchströmten Festbettreaktoren, Dissertation, Martin-Luther-Universität, Halle-Wittenberg, 2001; Grah, Entwicklung und Anwendung modularer Software zur Simulation und Parameterschätzung in gaskatalytischen Festbettreaktoren, Dissertation, Martin-Luther-Universität, Halle-Wittenberg, 2004), it was suggested that transient heating of the reactor wall (from ~ 30 to $\sim 300°C$) yields characteristics in the temperature data that are relevant for estimating heat-transport parameters. It is shown in this work that temperature data from stationary heating at maximum temperature gives much lower parameter uncertainties as when compared to transient heating. This insight allows a significant reduction in the experimental effort. Three to four experiments were previously performed to gather information used to estimate the set of heat-transport parameters for a specific catalyst bed. The number can be reduced down to one experiment when the gas flow density is allowed to change over time. Also, the time for a single experiment can further be reduced when the transient heating period is omitted.

A. Badinski (✉)
BASF SE, GVM/S Scientific Computing, 67056 Ludwigshafen, Germany
e-mail: alexander.badinski@basf.com

D. Corbett
Liquid Crystal Technology Group, Department of Engineering, Parks Road, Oxford OX1 3PJ, UK

H.G. Bock et al. (eds.), *Model Based Parameter Estimation*, Contributions
in Mathematical and Computational Sciences 4, DOI 10.1007/978-3-642-30367-8_12,
© Springer-Verlag Berlin Heidelberg 2013

1 Introduction

Heat transport plays an important role in the development process for the chemical industry. Tubes that are randomly filled with catalyst particles (e.g. pellets, rings), forming a fixed catalyst bed inside the tube, are often used as chemical reactors. In these reactors, the activity of catalytic reactions is strongly temperature dependent and hot-spots arising from exothermic reactions may lead to catalyst degradation and loss of the catalyst performance. Therefore, a quantitative understanding of the heat transport mechanisms is the basis for a rational design of such tubular reactors.

Different mathematical models may be used to model such tubular reactors varying in computational effort. The simplest is the "quasi-homogeneous" model where the gas and solid catalyst phases are considered in one continuously mixed phase [12]. The "heterogeneous" model describes the gas and catalyst as separate continuous phases [2]. The coupling between the two phases is done via some volume averaged interface where properties of the catalyst particles, like surface and shape, may effectively be considered. More elaborate models may be constructed and solved in the context of fluid dynamics simulations. There, a fixed bed of catalyst particles may generated virtually and interaction equations between the different phases are calculated locally.

Both the quasi-homogeneous and heterogeneous models use the symmetry of a tubular reactor which reduces the complexity to a two-dimensional problem. The attraction of the quasi-homogeneous model is the simplicity stemming from the fewest possible number of balance equations. The improvement of the heterogeneous model over the quasi-homogeneous model is evident when the dynamic behavior of the two phases is different, characterized by the different heat capacity of the phases. For example, this can be observed when the temperature of the gas and particle phase differ substantially. In the heterogeneous model, the simulated temperature profiles of the gas and catalyst phases correspond to the measured temperature profiles. This correspondence does not hold in the quasi-homogeneous model, where the simulated temperature profile is an effective temperature of the mixed gas and catalyst phase. Therefore, the localization and quantification of hot spots is limited with the quasi-homogeneous model. Another advantage of the heterogeneous model is that the heat and material transport properties are fundamentally founded. For the homogeneous model, these transport properties are effective and can only be determined from fitting the model to experimental data. Both of these models lack accuracy when the inhomogeneity of the fixed bed dominates the processes. In this case, a three dimensional complex fluid dynamic model may be used to directly describe the inhomogeneity of the fixed bed. However, the combination of complex fluid dynamic simulations and state-of-the-art optimization techniques is still held back by tremendous amount of required computer resources.

Since the scope of this work is to perform model-based DoE calculations where the numerical effort scales quadratically with the number of time-dependent state variables, we use the simplest quasi-homogeneous model and then carefully investigate the validity of the model.

The structure of this paper is as follows. The mathematical model of the tubular reactor and its approximations are outlined in Sect. 2. The experimental setup is briefly stated in Sect. 3. Computational details are presented in Sect. 4 and a discussion of the results is found in Sect. 5. This paper closes with a summary and future possible directions in Sect. 6.

2 Mathematical Model

2.1 Continuity Equations

The tubular reactor is fully described by the four state variables temperature T, gas density ρ^g, velocity v and pressure p. These four state variables can be specified by the energy, mass and momentum balance equation as well as an equation of state. Since the quasi-homogeneous model is used in this work, the rotational symmetry of the reactor model allows for a reduction of the three-dimensional problem to a two-dimensional one. All states are therefore functions of the two-dimensional reactor space spanned with the reactor radius r and the reactor axis z. v_r and v_z are then the gas velocities in the radial and axial directions, respectively, with the velocity vector $\mathbf{v} = (v_r, v_z)$. It should be noted that throughout this work, T will correspond to an effective temperature of the mixed phase as discussed in Sect. 1.

At the center of the quasi-homogeneous model is the definition of an effective heat capacity,

$$\tau = \rho^g c^g \varepsilon + \rho^p c^p (1 - \varepsilon), \tag{1}$$

where ε is the void volume fraction which is a measure of the porosity inside the tubular reactor and varies between 0 and 1 (0 corresponds to 0% and 1 to 100% gas volume in the reactor). ε is assumed to be constant over the reactor space. ρ^p is the catalyst particle density, and c^g and c^p are the specific heat capacities of the gas and the catalyst particles, respectively, at constant pressure.

For the mixed phase, the heat-balance equation is [12]

$$\tau \frac{\partial T}{\partial t} = \frac{1}{r} \frac{\partial}{\partial r} \left(r \lambda_r \frac{\partial T}{\partial r} \right) + \frac{\partial}{\partial z} \left(\lambda_z \frac{\partial T}{\partial z} \right) - \varepsilon \rho^g c^g \left(v_z \frac{\partial T}{\partial z} + v_r \frac{\partial T}{\partial r} \right), \tag{2}$$

where λ_r and λ_z (W m^{-1} K^{-1}) are the effective radial and axial heat conduction coefficients, respectively. The radial heat conduction term in (2) includes (a) heat conduction by molecular interaction of the gas molecules, (b) heat conduction through the catalyst particles, and (c) heat conduction via gas-flow that is driven by the cross-mixing of the gas flow perpendicular to the propagation direction along the z axis. For the axial heat conduction term in (2), a similar interpretation holds. The last term in (2) is the convective heat transport. Radiation can be neglected in the heat balance since the temperature is below 350°C.

The mass-balance equation is [11]

$$\frac{\partial \rho^g}{\partial t} = \frac{\partial}{\partial z}\left(D_z \frac{\partial \rho^g}{\partial z}\right) + \frac{1}{r}\frac{\partial}{\partial r}\left(r D_r \frac{\partial \rho^g}{\partial r}\right) - \left(\frac{\partial \rho^g v_r}{\partial r} + \frac{\partial \rho^g v_z}{\partial z}\right), \tag{3}$$

where D_r and D_z $(\mathrm{m}^2\,\mathrm{s}^{-1})$ are the radial and axial mass diffusion coefficients, respectively.

The momentum-balance equation is [11]

$$\rho^g \frac{\partial \mathbf{v}}{\partial t} = -\rho^g \mathbf{v}\nabla\cdot\mathbf{v} - \nabla p + \rho^g g_z - f^F, \tag{4}$$

where f^F comprises all dissipative forces in the reactor and g_z is the gravitational constant. It is convenient to introduce the mass current density defined as $\mathbf{G} = \rho^g \mathbf{v}$ $(\mathrm{kg}\,\mathrm{m}^{-1}\,\mathrm{s}^{-1})$.

2.2 Approximations to the Continuity Equations

In tubular reactors, the velocity of the gas is often considered only along the main axis (i.e. $v_r = 0$), where v_z remains a function of r and z. This approximation is justified as the axial velocity is generally much larger then the radial velocity of the gas. This approximation can be used to simplify both convective terms in (2) and (3).

The effective radial heat conduction λ_r in (2) can be parametrized in different forms. In general, the heat conduction may change along the radius of the reactor. Close to the reactor wall, the porosity increases as a result of the more spaced packing of the catalyst particles. This reduces the gas-flow driven cross-mixing perpendicular to the gas-flow propagation close to the wall and thereby reduces the radial heat conduction of the gas. The simplest model is to assume that λ_r is constant over the radius. More detailed models have been suggested for λ_r by specifying an analytic expression for λ_r that changes over the radius in some empirical form. Using literature studies as the basis [6], it is not straight forward to decide which of the empirical forms is most suited for the considered tubular reactors. The simplest model was therefore chosen in this work which assumes that λ_r is constant over r. This choice can also be justified by an analysis of the heat transport parameters, as further discussed in Sect. 5.3.

The mass-balance (3) can in principle be computed without approximations. In tubular reactors, however, the propagation velocity of the density v_z (or momentum density) is generally much faster than the propagation velocity of the temperature v_T. This can be assessed by estimating v_T for the considered reactor. To obtain an expression for v_T, the heat-balance (2) is divided by τ and v_T is defined as

Model-Based Design of Experiments for Estimating Heat-Transport Parameters 255

$$v_T = \frac{\varepsilon \rho^g c^g v_z}{\rho^g c^g \varepsilon + \rho^p c^p (1 - \varepsilon)}. \tag{5}$$

With some representative values $\varepsilon = 0.4$, $\rho^g = 1\,\mathrm{kg\,m^{-2}}$, $c^g = 1\,\mathrm{kJ\,kg^{-1}\,K^-}$, $\rho^p = 2{,}000\,\mathrm{kg\,m^{-3}}$ and $c^p = 1\,\mathrm{kJ\,kg^{-1}K^{-1}}$, we find that $v_T \approx 10^{-3}v_z$. This justifies well the assumption that the density (obtained from the mass balance) is in quasi steady-state when compared to the temperature (obtained from the energy balance). A further approximation to the mass balance is possible when the mass convection is much larger than the mass diffusion term. In this case, the mass diffusion can be neglected. Since the ratio of the convection to the diffusion term (the Peclet number $\frac{v_z L}{D_z}$) is between 10^2 and 10^4 for the considered reactor, neglecting the mass diffusion term is well justified. With these approximations, the mass-balance (3) simplifies to

$$\frac{\partial}{\partial z} \rho^g v_z = 0, \tag{6}$$

which implies that the mass current density $G_z = \rho^g v_z$ is constant in the axial direction. When G_z is specified, e.g. at the entrance of the reactor, then the product of ρ^g and v_z is known throughout the reactor. It is worth noting that considering (6) goes beyond the standard plug-flow approximation where the velocity is assumed to be constant over r. In (6), however, the velocity v_z may still vary over r.

The momentum-balance (4) can be solved when the dissipative forces are specified for the reactor. This leads to the extended Brinkman equation [6] where semi-empirical functions for the dissipative forces are specified. However, when the pressure drop is neglected across the reactor, one can avoid solving the momentum-balance equation altogether as the three remaining state variables can be determined by the energy and mass balance equations as well as an equation of state.

2.3 Boundary Conditions

For the solution of the energy balance (2), the following initial and boundary conditions are specified,

$$T(r, z, t = 0) = T_0(r, z), \tag{7}$$

$$T(r, z = 0, t) = c_1(t) + c_2(t)r^2 + c_3(t)r^3, \tag{8}$$

$$\left. \frac{\partial T}{\partial z} \right|_{z=L} = 0, \tag{9}$$

$$\left. \frac{\partial T}{\partial r} \right|_{r=0} = 0. \tag{10}$$

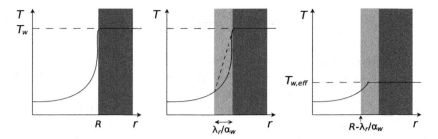

Fig. 1 Schematic of the temperature profile over the radius inside the reactor tube. *Left figure*: the temperature profile is plotted from the center of the reactor to the reactor wall. *Center figure*: when T_w, λ_r and α_w are specified, (11) defines an "auxiliary wall" at the radial position $R - \lambda_r/\alpha_w$ as discussed in the text. *Right figure*: the temperature at the auxiliary wall $T_{w,aux}$ is used as the new boundary value temperature in the simulation

T_0 in (7) is the initial temperature specified in Sect. 4.3. At the reactor entrance ($z = 0$), T in (8) is approximated as a cubic function with zero slope at $r = 0$ and time-dependent coefficients c_i, specified in Sect. 4.3. At the exit of the reactor ($z = L$), no heat conduction along the z axis is assumed due to the ending reactor bed, resulting in (9). At the reactor axis ($r = 0$), T is specified by the reactor symmetry, which prohibits the heat transfer across the axis, leading to (10). Following [12], at the reactor wall ($r = R$) T is described by the heat-transport equation across the wall,

$$\lambda_r \frac{\partial T}{\partial r}\bigg|_{r=R} = -\alpha_w(T - T_w)|_{r=R}. \quad (11)$$

α_w (W m^{-2}K^{-1}) is the wall heat-transfer coefficient, and T_w (°C) is the wall temperature. The left hand side of (11) is the energy flux into/out of the reactor volume, and the right hand side of that equation is the energy flux through the wall which is proportional to the difference of the temperature of the gas in contact with the reactor wall and T_w. This equation has a geometric interpretation illustrated in Fig. 1. When T_w, λ_r and α_w are specified, this equation defines a new wall temperature $T_{w,aux}$ at a distance λ_r/α_w away from the reactor wall, as indicated in Fig. 1. Following up an idea from [7], at this distance an "auxiliary wall" is considered that reduces the reactor volume in the simulation and thereby cuts out the volume of the reactor close to the wall. This leaves out the region where λ_r is likely not to be constant. When the auxiliary wall is small as compared to the remaining reactor volume ($\lambda_r/\alpha_w \ll R$), the use of (11) is justified and the heat-transfer coefficient has a well founded interpretation. In this work, (11) is used as a boundary condition for the energy-balance equation at the reactor wall, where T_w is specified in Sect. 4.3.

2.4 Parameter Representation

Following [2], page 87–92, the transport parameters are written as linear functions of the mass current density,

$$\lambda_r = L_1 + L_2 G_z, \tag{12}$$

$$\lambda_z = L_3 + L_4 G_z, \tag{13}$$

$$\alpha_w = A_1 + A_2 G_z, \tag{14}$$

where L_i and A_i are parameters. Bauer [2,3] further suggested empirical functions for L_i and A_i with 11 parameters and varying porosity ε and ratio R/d, where R is radius of the reactor and d some characteristic length of the catalyst particle. We have not followed this route in this work as neither R/d nor ε are varied in the experiment prohibiting a parameter estimation of all parameters. It is also known that these empirical functions are inaccurate for small R/d ratios [2,6]. The parameters L_1 and L_3 can be interpreted as the radial and axial heat conduction at no gas flow dominated by particle-particle interaction in the gas and catalyst phase. L_2 and L_4 characterize the intensities of the gas-flow driven cross mixing perpendicular to the propagation of the gas. A similar interpretation holds for parameters A_1 and A_2. All parameters have an associated uncertainty, which will later on be estimated, as well as minimized using DoE.

3 Experimental Setup

The reactor design used in this work was proposed by Adler [1]. A tube is randomly filled with catalyst particles forming a fixed catalyst bed. A mass current density enters this tubular reactor from above, travels through the catalyst bed of the reactor, and leaves the reactor at the bottom. The temperature of the reactor wall can be controlled with heated oil. The mass current density and the wall temperature are both controlled by the experimentalist. Experiments have been performed for three different designs of the catalyst particle shape. The different particle shapes are later on referred to as A, B and C.

In total 24 thermoelements are used to measure the temperatures in the reactor: at the reactor entrance five thermoelements are distributed along the radius, at the reactor wall five thermoelements are distributed along the z axis, and inside the reactor 14 thermoelements are distributed at three different levels on the z axis at varying radial positions. At installation of these thermoelements, all axial and radial positions are carefully measured. The novel idea of this reactor design is that the temperature is measured not only inside the fixed reactor bed, but also at the reactor entrance and at the reactor wall. This allows one to specify Dirichlet boundary conditions when solving the mathematical model for the temperature, further discussed in Sect. 2.3.

4 Computational Details

4.1 Numerical Methods

All calculations are done with the program package VPLAN [8]. The parameter estimation is based on the weighted least-square method using the generalized Gauss–Newton type method as implemented in PARFIT [4]. The weighted sum of squares

$$\text{WSS} = \min_{p_j} \sum_i r_i^2 \tag{15}$$

with

$$r_i = \frac{h_i - m_i}{\sigma_i} \tag{16}$$

is minimized over all parameters p_j such that the balance equations of Sect. 2 are satisfied. The sum in (15) is over all measured and simulated temperatures m_i and h_i, respectively, and σ_i is the one-sigma standard error of the measured temperature.

The linearized parameter covariance matrix from the Gauß–Newton method is also used for the sensitivity analysis of this work,

$$C_p = \text{cov}(p_i, p_j) = (J^T J)^{-1}, \tag{17}$$

where J is the Jacobian of the least squares term with elements $J_{ij} = \partial r_i / \partial p_j$.

The model-based DoE calculations minimize a measure of the parameter covariance matrix with respect to all experimental control variables such that the balance equations of Sect. 2 are satisfied. In this work, four different measures of C_p are considered [10],

$$\phi_A = \frac{1}{n} \text{trace} C_p \tag{18}$$

$$\phi_D = \det C_p \tag{19}$$

$$\phi_E = \max\{\lambda_i, \lambda_i \text{ is eigenvalue of } C_p\} \tag{20}$$

$$\phi_M = \max\{\sqrt{C_{p,ii}}, i = 1, \ldots, n\}. \tag{21}$$

To minimize the measure of C_p, the sequential programming method is used as implemented in the program package SNOPT [5].

The heat-transport equation is solved using the standard method of lines based on finite differences transforming the partial differential equation into ordinary differential equations. To do so, the integration domain is discretized in the radial ($i = 1, \ldots, N_R$) and axial ($j = 1, \ldots, N_L$) direction. The first derivatives are written as one sided differences (upwind scheme [9]), and the second derivatives are written as central differences. The resulting ordinary differential equations are then solved as initial value problems using a backward differentiation formula method [8].

Model-Based Design of Experiments for Estimating Heat-Transport Parameters   259

We find that using more than 18×18 grid points does not change the parameter values by more than 0.2%. Since this is much less than the statistical errors of the parameter values, we use this number of grid point for the results presented here.

To allow for a placement of the thermoelements inside the reactor independent of the discretization grid, the temperature observable h_i is calculated by interpolating the temperature at that position with a two dimensional plane passing through three of four adjacent temperature points.

4.2 Statistical Errors

Three main sources of statistical errors in the measured temperature profile are identified: an error from the temperature measurement device, an error from the uncertainty of the thermoelement position in the reactor, and an error from the spatial anisotropy of the fixed bed around the rotational axis of the reactor. The rotational anisotropy of the fixed bed causes the measured temperature profile to deviate from the rotational symmetry. This deviation is estimated in this work by measuring the temperature at eight different angles, all at a fixed axial and radial position in the reactor. We find experimentally that these eight measured temperatures fluctuate by about 5°C. This value holds approximately true for different temperature ranges as well as different mass flow densities. In our calculations, the fluctuation of the temperature around the rotational axis is assumed to be Gaussianly distributed with a two-sigma value of 5°C. As the anisotropy effect dominates the statistical errors of the measured temperature in the reactor, other statistical errors are neglected.

4.3 Initial and Boundary Values

The initial and boundary values for the temperature are stated in Sect. 2.3 and are determined as follows. The polynomial coefficients from (8) are obtained from a fit to five temperatures measured at the reactor entrance. Since the temperature changes over time, also the c_i change over time. When a polynomial of fourth order is used, no significant influence is seen on the estimated parameters.

The wall temperature T_w varies along the z axis and is obtained by a linear interpolation of five temperatures measured along the wall. As the five temperatures change over time, also the linear interpolation changes. In Fig. 2, these boundary values can be seen in a schematic plot of the simulated temperature profile at three different time snapshots.

The initial temperature T_0 at $t_0 = 0$ in (7) is obtained with a homotopy method. Before the simulation starts to model the experimental setup at $t_0 = 0$, an extra period of time is simulated where (2) is integrated using an initial temperature of $T_0 = 20$°C. During this extra period, the boundary conditions (8)–(10) are chosen from the experimental setup at $t = 0$. The length of that extra period is chosen such that (2) reaches a steady-state at $t_0 = 0$.

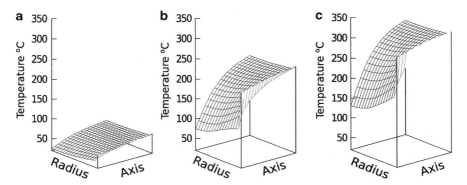

Fig. 2 The temperature profile plotted over the radius and z axis of the reactor for experimental setup C. *The left graph* is at $t = 1,000$ s, the *middle graph* at $t = 3,000$ s and the *right graph* at $t = 5,000$ s. The *bottom left* is the reactor entrance, the *bottom right* is the reactor wall, the *top right* is the reactor exit, and the *top left* is the reactor axis. In this plot, the gas flow enters from the *bottom left* and goes to the *top right*. The reactor wall is heated up over time

4.4 Influence of the Axial Heat Conduction

The heat-balance (2) describes the heat conduction in the axial and radial directions. When all available experimental data for an experimental setup are considered, it is found in each parameter estimation that the parameter covariance matrix is linearly dependent. It is therefore not possible to simultaneously estimate the axial and radial heat conduction parameters λ_r and λ_z from the given data. This problem was observed previously [2]. One remedy is to only consider the radial heat transport at a fixed axial position [2, 6]. This approach is problematic because the influence of the temperature profile at the entrance of the reactor onto the temperature profile at any axial position inside the reactor is neglected. In this work, the solution proposed in [2] is used where λ_z is eliminated by $\lambda_z = 0.1\lambda_r$. In our simulations, other relationships like $\lambda_z = 0$ or $\lambda_z = 0.5\lambda_r$ have been investigated and do not change the estimated parameters by more than 0.5% which is less than their statistical errors. The advantage of including a non-zero axial heat conduction is the well known observation that convection dominated solutions are stabilized by an additional diffusion term [9].

5 Results

5.1 Intuitive Versus Optimal Design of Experiments

For the three experimental setups, Table 1 compares the intuitive with the optimal experimental design. The two experimental control variables are considered: the mass current density G_i and the wall temperature T_w. The intuitive design (ID) was

Table 1 Experimental control variables used previously (intuitive design, ID) and obtained from the optimal design of experiment calculations (calculated optimal design, COD)

	Design	G_1 (kgs^{-1} m^{-2})	G_2 (kgs^{-1} 1m^{-2})	G_3 (kgs^{-1} m^{-2})	G_4 (kgs^{-1} m^{-2})	$T_w(t_0)$ °C	$T_w(t_{\text{end}})$ °C
A	ID	0.000	0.917	1.833	3.667	23–47	334–346
	COD	0.394	0.954	4	4	350	350
B	ID	0	0.917	1.833	3.667	22–69	344–347
	COD	0.442	1.012	4	4	350	350
C	ID	0.917	1.833	3.667	–	20–36	324–342
	COD	1.325	4	4	–	350	350

G_i is the mass current density for different experiments. T_w is the wall temperature at the beginning of the experiment (at t_0) and at the end of the experiment (at t_{end}). In the experimental setup, T_w changes approximately linearly in time between $T_w(t_0)$ and $T_w(t_{\text{end}})$. In the simulation, the measured values for T_w are used, which may deviate slightly from being linear in time

Table 2 Measures of the parameter covariance matrix evaluated for three catalysts

	Design	ϕ_A	ϕ_D	ϕ_E	ϕ_M
A	ID	0.00207	0.00034	0.00649	0.07059
	COD	0.00072	0.00009	0.00240	0.04253
B	ID	0.00149	0.00024	0.005406	0.06163
	COD	0.00048	0.00007	0.00180	0.03598
C	ID	0.00157	0.00028	0.00545	0.05362
	COD	0.00031	0.00005	0.00036	0.01620
	ROD	0.00114	0.00018	0.00405	0.04630

The second column specifies the type of experimental design: intuitive design (ID), calculated optimal design (COD), and realized optimal design (ROD). The measures are defined in (18)–(21)

previously chosen by the experimentalists in such a way that T_w changes transiently and approximately linear from $T_w(t_0) \approx 20°C$ to $T_w(t_{\text{end}}) \approx 350°C$. The calculated optimal design (COD) shows that T_w should be kept at the maximal possible wall temperature during the whole experimentation time. The rationale of this finding is that it is optimal for the parameter estimation to have huge variations in the temperature profile across the reactor. These huge variations are realized when T_w is maximal and when the temperature of the inflowing gas has room temperature.

From the optimal choice of the mass current densities G_i, as stated in Table 1, three observations are made. First, it is optimal to choose G_i to be always larger than zero. Secondly, at least one experiment should be done at a maximum rate of G_i (in this work, the maximum rate is $G_{\text{max}} = 4 \, \text{kgm}^{-2}\text{s}^{-1}$ is chosen). Thirdly, the spacing between different G_i is not uniform.

Since the stationary heating is found optimal in this work, we propose that the experiments done at different values for G_i can be put into a single experiment where G varies over time. This allows to reduce the number of experiments from about three to four down to one. This significantly reduces the experimentation time.

Table 2 summarizes the four different evaluated measures of the parameter covariance matrix defined in (18)–(21). It can be seen that the computed optimal design (COD) reduces the different measures by a factor of 3–5 (ϕ_A), 3–6 (ϕ_D), 3–14 (ϕ_E) and 2–5 (ϕ_M) when compared to the intuitive design. All presented design of experiment results are done by minimizing the ϕ_A measure. When one of the other measures is minimized, both Tables 1 and 2 look very similar.

Due to time constraints in the project, only the third catalyst setup is experimentally repeated with the proposed optimized design (one experiment with different G_i rates). In Table 2, these results are denoted as realized optimal design (ROD). We see that the measures of the parameter covariance matrix are substantially reduced when comparing the realized with the intuitive design (ROD vs. ID in Table 2). However, when comparing the ROD and COD, we find that the realized measures are a bit higher then the previously computed measures. This deviation is due to the fact that the inflowing gas is (accidentally) preheated by the inlet tube attached to the reactor entrance. This results in a lowering of the temperature variation in the ROD of the reactor. This preheating is not considered in the COD where the inflowing gas is assumed to have room temperature resulting in larger temperature variations in the reactor. In future reactor designs, a heat blocker between the gas inlet tube and the reactor entrance should therefore be considered.

5.2 Heat-Transport Parameters

Table 3 gives the estimated heat-transport parameters and their statistical errors for the three different catalysts. The statistical errors (two-sigma confidence intervals) are less than 8% of the parameter value. For a better comparison between the different catalysts, the total heat-transport coefficient κ is considered which combines the heat diffusion inside the reactor and the heat transport across the reactor wall [7],

$$\frac{1}{\kappa} = \frac{1}{\alpha_w} + \frac{R}{\lambda_r} \tag{22}$$

$$= \frac{1}{A_1 + A_2 G_z} + \frac{R}{L_1 + L_2 G_z}, \tag{23}$$

where R is the radius of the tubular reactor. The results are stated in Table 4 for the different catalyst, at a fixed mass current density of $G = 1.83\,\mathrm{kg\,m^{-2}\,s^{-1}}$. We find that κ ranges by more than 20% between the different catalysts. As the statistical uncertainties of the different κ are less than 6%, we can conclude that catalyst A has a significantly improved total heat transport when compared to the catalyst B and C.

5.3 Validation of the Model

To check the validity of the mathematical model, χ^2 tests are performed to investigate whether the residuals ($h_i - m_i$ in (15)) are Gaussian. With a confidence

Table 3 Estimated parameters for the different catalysts. The parameter uncertainties are denoted as $56(2) = 56\pm2$, where ±2 is the two-sigma confidence interval

Shape	Design	A_1 (WK^{-1})	A_2 (Wm^{2}sK^{-1}kg^{-1})	L_1 (WK^{-1}m^{-1})	L_2 (WmsK^{-1}kg^{-1})	Av. WSS —	# of data points	χ^2 —
A	ID	56(2)	77(1)	0.49(4)	1.36(5)	0.782	2,268	1.047
B	ID	72(3)	70(1)	0.62(4)	0.93(3)	0.891	2,268	1.047
C	ID	119(5)	55(2)	0.32(5)	0.80(3)	0.887	1,666	1.055
	ROD	131(4)	55(2)	0.32(5)	0.79(3)	1.221	1,680	1.055

The first column specifies the different types of catalysts. The second column specifies the type of experimental design: intuitive design (ID) and realized optimal design (ROD). The seventh column gives the average WSS which is obtained from the total WSS divided by the number of data points stated in the eighth column. The last column gives the normalized χ^2 value for a confidence level of 5% and the respective degrees of freedom

Table 4 The heat-transport parameters α_w and λ_r evaluated for $G = 1.83\,\text{kg/m}^2\text{s}$ and the derived total heat-transfer coefficient κ

	α_w $\left[\frac{W}{m^2 K}\right]$	λ_r $\left[\frac{W}{mK}\right]$	κ $\left[\frac{W}{m^2 K}\right]$
A	198(3)	2.98(12)	108(4)
B	200(3)	2.32(9)	96(4)
C	220(6)	1.78(8)	86(5)

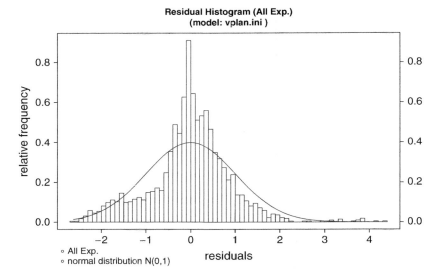

Fig. 3 Histogram plot of the normalized residuals for experiment A as defined in (16), using the parameter values stated in Table 3. The *blue line* is a Gaussian distribution with a normalized width of $\sigma = 1$

level of 0.05, the χ^2 tests are accepted for the different experimental setups, with the exception of the ROD of the experimental setup C. However, there the violation of the χ^2 test is not very strong. This can be seen in Table 3, when comparing the average WSS and the χ^2 value.

For the experimental setup C, as an example, Fig. 3 gives a histogram plot of the 1,666 residuals (each residual corresponds to one data point) which are in close agreement with a Gaussian shape. For the same setup, Fig. 4 compares the temperature profile at two different axial positions.

Furthermore, z tests are performed to assess that the mean value of the residuals is not significantly different from zero. Both, the χ^2 and the z test indicate that the quasi-homogeneous model (together with the model for the statistical error) give a good description of the experimental data.

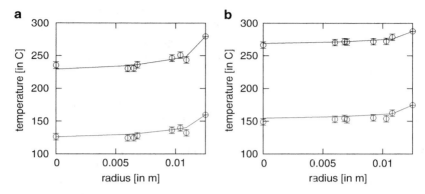

Fig. 4 Comparison between experimental and simulated temperature profiles at two different z levels in the reactor for setup C. The used parameter values are those from Table 3. The *left graph* is at $t = 2{,}000$ s and the *right graph* is at $t = 4{,}000$ s

5.4 Justification of α_w Model

It was shown in Sect. 2.3, that the heat transport parameters α_w and λ_r in the considered mathematical model are fundamentally founded when $\lambda_r/\alpha_w \ll R$. From Table 4, we see that this relationship is not strictly obeyed in our simulations as λ_r/α_w ranges between 0.008 and 0.015 m which is in the range of the reactor radius. As a result of that, we need to be careful in giving the physical interpretation of the heat transport parameters too much weight. However, in Sect. 5.3 it was shown that the mathematical model describes well the experimental data and that the parameters are well specified. Therefore, the parameters can still be used in comparing the different catalysts.

6 Summary and Outlook

Model-based optimal experimental design calculations were performed for the reactor proposed and designed by Adler to accurately determine heat-transport parameters in catalyst beds. The main outcome of the optimized design is the stationary heating of the reactor which allows to combine experiments at different mass current densities. This insight allowed a reduction in the experimentation time by a factor of about 3–4. In particular, the experimentation time which previously lasted about 2–3 days could therefore be reduced to about 1 day. At the same time, it was shown that the parameters can be determined even more accurately. In this work, no reactions were assumed inside the reactor. However, the results of this work can be transferred to systems where reactions takes place at the catalyst surface.

In this work, statistical uncertainties of the measured temperatures and hence of the heat-transport parameters were fully taken into account. This allowed a quantitative comparison of the heat-transport properties between different catalyst particles. Although the simplest possible mathematical model was chosen, it was shown that the model describes well the experimental data. At the same time, the parameters have a statistical error which is less than 8%.

In future studies it may be interesting to refine the mathematical model. The approximations made to the balance equations in Sect. 2.2 may be revisited, in particular when evaluating the mass balance, or Navier–Stokes equations. Although it was not found to be relevant for this work, it may still be appropriate to investigate whether different parametrizations of the radial heat conduction coefficient can be considered. Finally, the heat-transport parameters can be made dependent on material data (e.g. the heat capacity of the gas) which would allow for a better transferability of the obtained results, e.g. to different gases.

Acknowledgements Several people have contributed to that work. I like to thank Dr. H. Schultze for initializing that project at BASF and for fruitful discussions. Also, many thanks to Dr. A. Schimpf for providing the experimental data and the engineering background to this work. My gratitude also goes to Dr. S. Körkel for providing the software, and to Dr. A. Schreieck, and to Dr. C.-A. Winkler for valuable suggestions during writing of that work.

References

1. R. Adler, Stand der Simulation von heterogen-gaskatalytischen Reaktionsabläufen in Festbet-trohrreaktoren - Teil 1, Chem.-Ing.-Tech. Vol. 72 p. 555–564 (2000)
2. M. Bauer, Theoretische und experimentelle Untersuchung zum Wärmetransport in gasdurch-strömten Festbettreaktoren, Dissertation, Martin-Luther-Universität, Halle-Wittenberg (2001)
3. M. Bauer, R. Adler Novel Method for investigation and evaluation of heat transfer in fixed bed tubular reactors with gas flow, Heat and Mass Transfer 39, 421–427 (2003)
4. Bock, Kostina, Schlöder, GAMM-Mitt., Vol. 30, No. 2, 376–408 (2007)
5. P. E. Gill, W. Murray, and M. A. Saunders. SNOPT: An SQP algorithm for large-scale constrained optimization. Report NA 97-2, Dept. of Mathematics, University of California, San Diego (1997)
6. A. Grah Entwicklung und Anwendung modularer Software zur Simulation und Parame-terschätzung in gaskatalytischen Festbettreaktoren, Dissertation, Martin-Luther-Universität, Halle-Wittenberg (2004)
7. U. Grigull, H. Sandner, Wärmeleitung, Springer-Verlag (1979)
8. S. Körkel, Numerische Methoden für Optimale Versuchsplanungsprobleme bei nichtlinearen DAE-Modellen, Dissertation, Universität Heidelberg, Heidelberg (2002)
9. W. H. Press, S. A. Teukolsky, W. T. Vetterling, B. P. Flannery, Numerical Recipes, Cambridge Universtiy Press (2007)
10. F. Pukelsheim Optimal Design of Experiments, John Wiley & Sons, Inc. New York (1993)
11. R. B. Bird, W. E. Stewart, E. N. Lightfoot, Transport Phenomena, Wiley, 2 edition (2001)
12. VDI- Wärmeatlas, Herausgeber: Verein Deutscher Ingenieure, 10. Auflage, Springer-Verlag Berlin Heidelberg (2006)

Parameter Estimation for a Reconstructed SOFC Mixed-Conducting LSCF-Cathode

Thomas Carraro and Jochen Joos

Abstract The performance of a solid oxide fuel cell (SOFC) is strongly affected by electrode polarization losses, which are related to the composition and the microstructure of the porous materials. A model that can decouple the effects associated with the geometrical arrangement, shape, and size of the particles together with material distribution on one side and the material properties on the other can give a relevant improvement in the understanding of the underlying processes. A porous mixed ionic-electronic conducting (MIEC) cathode was reconstructed by focused ion beam tomography. The detailed geometry of the microstructure is used for 3D calculations of the electrochemical processes in the electrode and to calibrate a well-established reduced model obtained by averaging. We perform a model-based estimation of the parameters describing the main processes and estimate their confidence regions using the calibrated reduced model.

Keywords SOFC • Sensitivity • Parameter Estimation • 3D FEM model • 3D Reconstruction

1 Introduction

The electrochemical performance of solid oxide fuel cells (SOFCs) mostly depends on material composition and microstructure of the porous electrodes, which are the determinant causes of polarization losses. In this work a three-dimensional

---

T. Carraro (✉)
Institut für Angewandte Mathematik (IAM), Ruprecht-Karls-Universität Heidelberg,
62120 Heidelberg, Germany
e-mail: thomas.carraro@iwr.uni-heidelberg.de

J. Joos
Institut für Werkstoffe der Elektrotechnik (IWE), Karlsruher Institut für Technologie (KIT),
Adenauerring 20b, 76131 Karlsruhe, Germany
e-mail: jochen.joos@kit.edu

H.G. Bock et al. (eds.), *Model Based Parameter Estimation*, Contributions
in Mathematical and Computational Sciences 4, DOI 10.1007/978-3-642-30367-8_13,
© Springer-Verlag Berlin Heidelberg 2013

reconstruction of a mixed conducting $La_{0.58}Sr_{0.4}Co_{0.2}Fe_{0.8}O_{3-\delta}$ (LSCF)-cathode, that provides a detailed microstructure approximation, is considered. Mixed conductors have both electronic and ionic conducting properties, thus allowing the active zone of the electrodes to extend from the three phase boundary towards the inner volume of the electrode. The mechanisms that allow such an extension are related to oxygen ion diffusion within the MIEC and oxygen surface exchange at the interface between MIEC and gas phase. In this work the exchange of oxygen with the gas phase is described by the surface exchange coefficient k^δ and the transport of oxygen ions by the chemical diffusion coefficient D^δ as shown in [4, 18, 19].

In the last few years several groups have developed different methods for the reconstruction of a porous electrode [7, 9, 11, 22, 26, 27] in order to study the influence of the microstructure on the performance of the electrode. Two tomography techniques are considered for the reconstruction of SOFC electrodes: X-ray and focused ion beam (FIB). The latter is considered to be most suitable in our case (particle size between 300 and 600 nm), as it allows a finer resolution.

So far the reconstruction was mainly used to calculate different microstructural parameters: volume-specific surface area, volume/porosity fraction and tortuosity. These characteristic parameters are used to calculate effective parameters for homogenized models, which under some conditions [8] are adequate approximations of complex microstructures in case the quantity of interest has a macroscopic character, as e.g. the area specific resistance ASR_{cat}.

Simplified models are useful for homogeneous microstructures, but they are limited for microstructure optimization and are inadequate in case of inhomogeneous microstructures or in case the microstructure scale is too large in comparison with the scale of the electrode. In order to overcome these limitations, the development of detailed 3D models for the simulation of the electrochemical processes in an electrode is under investigation. Furthermore, the microstructure parameters used in homogenized models need to be determined by complex geometry models.

Some groups have started to develop micro-scale models to compute the performance of more general microstructures. For the anode side the lattice Boltzmann (LB) method has been used in [14, 23] and the finite volume method (FVM) in [21], while for the cathode side we have used the finite element method (FEM) for 3D calculations based on a real microstructure [11, 12]. In [4, 11] we have applied this model for electrochemical performance evaluation to a MIEC cathode reconstructed by means of focused ion beam coupled with scanning electronic microscope (FIB/SEM). In [5] we have done a sensitivity analysis with respect to the most important parameters.

A relevant prerogative of the sensitivity analysis is the characterization of the impact of different processes (gas diffusion, bulk diffusion, surface reaction) with respect to a performance index, e.g. ASR_{cat}. It is possible hence to distinguish based on a *quantitative* approach which process is most important and which region of the microstructure is relevant for the given process.

To perform a sensitivity analysis of a quantity of interest, the derivatives of this quantity with respect to the model parameters are calculated. The sensitivities are here used to estimate the confidence regions of the parameters as explained in Sect. 4.

Since in [5] we have shown that the sensitivities of the reduced model are a good approximation of the sensitivities calculated with the 3D model, we study here the parameter estimation problem using the calibrated reduced model.

The aim of this work is to study the possibility of estimating the parameters, which describe the diffusion and the reaction processes, simultaneously from a series of measurements of ASR_{cat} at different working conditions in the stationary case. An empirical functional dependence on temperature and oxygen partial pressure of the two parameters k^δ and D^δ leads to the estimation of five parameters as shown in Sect. 4. To study the feasibility of the parameter estimation problem, the essential approximation of the confidence region of the parameters is derived.

All the simulations are done by ParCell3D, a research finite element software based on the deal.II library [2], developed to simulate electrochemical processes in microstructures [4].

2 Model

We describe in this section the model and its parameters. The following processes are considered determining:

- Gas diffusion of oxygen in the porous structure (including a Knudsen term).
- Surface exchange (oxygen reduction) at the interface MIEC/pores.
- Bulk diffusion of oxygen ions in the lattice.
- Charge transfer at the interface cathode/electrolyte.
- Ionic conduction in the electrolyte.

Few approximations are made for the case under study. Due to high electronic conductivity of the cathode (e.g. at 800°C the conductivity of LSCF is 8, 899 S/m), a constant electrical potential is assumed. We also consider constant temperature condition. Three material parameters, which describe a mixed ionic-electronic conducting cathode as LSCF, are used in the model: (a) the surface exchange coefficient k^δ, (b) the chemical diffusion coefficient D^δ and (c) the oxygen ion equilibrium concentration in the perovskite lattice. The oxygen ion equilibrium concentration has been evaluated as a function of oxygen partial pressure for each temperature from measured data [24] as shown in [17].

The cells used in experiments have a base area of 1 cm^2 and a thickness of $30 \mu\text{m}$. For the computations we take a representative volume element (RVE) of the microstructure, as explained in [10, 11]. The RVE is divided in subregions corresponding to different materials: Ω_G is the corresponding region of the gas phase, Ω_M is the region of the bulk phase, while Ω_E is the electrolyte. Between these subregions we define the interfaces: Γ_{GM} is the surface between MIEC and gas, Γ_{ME} is the interface between MIEC and electrolyte and Γ_{GE} is the interface between electrolyte and gas. Furthermore we use the notation Γ_{CC} for the upper face of the current collector, Γ_{EL} for the bottom face of the electrolyte and Γ_0 for the side faces as shown in Fig. 1.

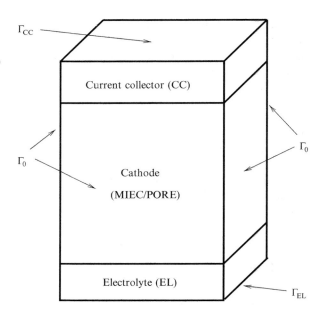

Fig. 1 Definition of the boundary surfaces for the application of the boundary conditions (10), (2d) and (7d)

2.1 Boundary Conditions

Four boundary conditions determine the working conditions of the cathode:

- Oxygen partial pressure on the cathode side p_{CG}.
- Oxygen partial pressure on counter electrode side p_{CE}.
- Potential of the cathode Φ_{MIEC}.
- Potential of the counter electrode Φ_{CE}.

Due to the high electronic conductivity of the porous LSCF-cathode analyzed in this study a constant electrical potential Φ_{MIEC} can be assumed. Its value is determined by the following expression:

$$\Phi_{MIEC} = U_{Nernst}(p_{CC}, p_{CE}) + \Phi_{CE} - \eta_{Model}, \tag{1}$$

where $U_{Nernst}(p_{CC}, p_{CE})$ is the difference of potentials between cathode and counter electrode at equilibrium and η_{Model} is the total voltage loss of the model, which is given, while the current I is calculated by the model. The area specific resistance of the model is calculated as $ASR_{Model} = \eta_{Model}/I$. Subtracting the specific resistance of the electrolyte $ASR_E = L_E/\sigma_E$, where L_E is the length of the electrolyte and σ_E is the ionic conductivity coefficient of the electrolyte, we get the area specific resistance of the cathode: $ASR_{cat} = ASR_{Model} - ASR_E$.

In the next sections we describe the processes considered in the model.

2.2 Gas Diffusion

In the pores (gas phase) we have diffusion and coupling between gas and MIEC (GM)

$$-\nabla \cdot (D_G \nabla u_G) = 0 \; in \; \Omega_{GAS}, \tag{2a}$$

$$D_G \frac{\partial u_G}{\partial n_G} = -\frac{k^\delta}{2} \left(f(u_G) - u_M \right) \; on \; \Gamma_{GM}, \tag{2b}$$

$$\frac{\partial u_G}{\partial n_G} = 0 \; on \; \Gamma_0 \cup \Gamma_{GE}, \tag{2c}$$

$$u_G = \bar{u}_G \; on \; \Gamma_{CC}, \tag{2d}$$

where u_G and u_M are respectively the oxygen and oxygen ions concentrations, D_G is the gas diffusion coefficient, k^δ is the surface reaction coefficient and $f(u_G)$ is an equilibrium value. Equation (2b) shows, hence, that the exchange between gas phase and MIEC phase depends on the difference between the actual value of oxygen ions in MIEC u_M and its equilibrium value $f(u_G)$, which depends on the concentration of oxygen molecules in the gas phase.

The gas phase is considered as a binary mixture of oxygen and nitrogen. The diffusion coefficient, which contains a binary diffusion term and a Knudsen term, is

$$D_G = \frac{p}{RT} \left[\frac{1 - (1 - \sqrt{M_{O_2}/M_{N_2}})u_G}{D_{O_2 N_2}} + \frac{1}{D^k_{O_2}(d_p)} \right]^{-1}, \tag{3}$$

where u_G is the oxygen concentration in the gas phase (pores), M_{O_2} and M_{N_2} are respectively the atomic masses of the oxygen and nitrogen molecules, $D_{O_2 N_2}$ is the binary diffusion coefficient and $D^k_{O_2}(d_p)$ the Knudsen diffusion coefficient, d_p is the average diameter of the pores, p and T are pressure and temperature and R is the ideal gas constant.

2.3 Bulk Diffusion

Since the electrical potential is assumed to be constant, we consider only diffusion of ions in the MIEC:

$$-\nabla \cdot (D_M \nabla u_M) = 0 \; in \; \Omega_M, \tag{4a}$$

$$D_M \frac{\partial u_M}{\partial n_M} = k^\delta \left(f(u_G) - u_M \right) \; on \; \Gamma_{GM}, \tag{4b}$$

$$D_M \frac{\partial u_M}{\partial n_M} = -\frac{1}{ASR_{CT}} \left(g(u_M) + u_E \right) \; on \; \Gamma_{ME}, \tag{4c}$$

$$\frac{\partial u_M}{\partial n_M} = 0 \; on \; \Gamma_0, \tag{4d}$$

where u_E is the electrical potential, D_M is the bulk diffusion coefficient in the MIEC, ASR_{CT} is the area specific resistance of the interface and $g(u_M)$ is the difference between the Nernst potential and the ionic potential of MIEC material.

2.4 Surface Exchange

Oxygen is exchanged through the boundary faces between gas phase and MIEC material. This process is described empirically by fluxes calculated from a variation of the concentration with respect to an equilibrium condition. This is described by the boundary conditions (2b) and (4b), where $f(u_M)$ is the reference value for the concentration of oxygen ions, which is detected experimentally [24]. The coefficient k^δ is the surface reaction coefficient, that limits the oxygen reduction.

2.5 Charge Transfer

Oxygen ions are incorporated at the interface MIEC/electrolyte (ME). The local charge transfer voltage depends on the local difference (η_{ct}) between the equilibrium reduction potential ($U_{Nernst,ME}$) and the interfacial discontinuity between the electronic potential of the electrolyte and ionic potential of the MIEC:

$$\eta_{ct} = U_{Nernst,ME} - \left(\Phi_M - u_{E,ME} \right), \tag{5}$$

where

$$U_{Nernst,ME} = \frac{RT}{2F} \log \sqrt{\frac{p_{M,ME}}{p_{CE}}},$$

Φ_M is the potential in the MIEC, $\Phi_{E,ME}$ is the potential in the electrolyte at the interface ME, $p_{M,ME}$ is the local equilibrium partial pressure of oxygen ions in MIEC at the interface ME and p_{CE} is the oxygen partial pressure in the counter electrode.

As both $U_{Nernst,ME}$ and Φ_M are functions of u_M, we write relation (5) as

$$\eta_{ct} = g(u_M) + u_E, \tag{6}$$

which is used in the boundary conditions (4c) and (7b) to describe the flux of oxygen ions at the interface ME. These fluxes depend on the interface parameter ASR_{CT}.

Parameter Estimation for a Reconstructed SOFC Mixed-Conducting LSCF-Cathode 273

As we consider a LSCF/CGO interface, the effect of ASR_{CT} is assumed negligible, but for numerical purposes we need to impose a small value. We use the value $ASR_{CT} = 10^{-4}\ \Omega\text{cm}^2$. In [5] we have shown, as expected, that the choice of a small ASR_{CT} value makes this parameter negligible for the model sensitivities.

2.6 Ionic Current

A constant chemical potential determined by p_{CE} (the oxygen partial pressure at the counter electrode) is assumed in the electrolyte, since an ideal reversible counter electrode is considered. The diffusion of oxygen ions in the electrolyte is induced by a gradient of the electrical potential (Ohm's law):

$$-\nabla \cdot (\sigma_E \nabla u_E) = 0 \; in \; \Omega_{EL}, \tag{7a}$$

$$\sigma_E \frac{\partial u_E}{\partial n_E} = -\frac{1}{2F} \frac{1}{ASR_{CT}} (g(u_M) + u_E) \; on \; \Gamma_{ME}, \tag{7b}$$

$$\frac{\partial u_E}{\partial n_E} = 0 \; on \; \Gamma_0 \cup \Gamma_{GE}, \tag{7c}$$

$$u_E = 0 \; on \; \Gamma_{EL}, \tag{7d}$$

where F is the Faraday constant.

2.7 Counter Electrode

An ideal reversible counter electrode (CE) with a constant potential is considered:

$$\Phi_{CE} = 0 \quad in \; \Omega_{CE}, \tag{8}$$

and the following boundary conditions at the interface CE are taken:

$$\Phi_{E,CE} = \Phi_{CE} \quad on \; \Gamma_{E,CE},$$

$$p_{E,CE} = p_{CE} \quad on \; \Gamma_{E,CE}. \tag{9}$$

2.8 Other Boundary Conditions

On the lateral boundaries of the RVE to close the problem we impose a no flux condition (insulation):

$$\frac{\partial u_G}{\partial n_G} = 0 \quad \text{on } \Gamma_0,$$

$$\frac{\partial u_M}{\partial n_M} = 0 \quad \text{on } \Gamma_0,$$

$$\frac{\partial u_E}{\partial n_E} = 0 \quad \text{on } \Gamma_0, \tag{10}$$

where Γ_0 are the four lateral boundaries as depicted in Fig. 1.

3 Numerical Method

3.1 Weak Formulation

To solve the model defined in the previous section we consider the discretization by the finite element method (FEM). The natural setting of the problem is thus the weak formulation of the systems of equations (2), (4) and (7). To this aim we introduce the following spaces

$$Y = H^1(\Omega_G) \times H^1(\Omega_M) \times H^1(\Omega_E),$$

$$Y_0 = H^1(\Omega_G) \times H^1(\Omega_M) \times H^1(\Omega_E) \cap \{(\varphi_G, \varphi_M, \varphi_E) : \varphi_{G|_{\Gamma_{CC}}} = \varphi_{E|_{\Gamma_{EL}}} = 0\},$$

$$Y_g = H^1(\Omega_G) \times H^1(\Omega_M) \times H^1(\Omega_E) \cap$$
$$\cap \{(u_G, u_M, u_E) : u_{G|_{\Gamma_{CC}}} = \bar{u}_G \text{ and } u_{E|_{\Gamma_{EL}}} = 0\},$$

and define $(u_G, u_M, u_E) \in Y_g$ as the solution of the weak formulation

$$\int_{\Omega_G} D_G(u_G)\nabla u_G \nabla \varphi_G + \frac{k^\delta}{2} \int_{\Gamma_{GM}} (f(u_G) - u_M) \varphi_G = 0 \tag{11a}$$

$$\int_{\Omega_M} D_M \nabla u_M \nabla \varphi_M - k^\delta \int_{\Gamma_{ME}} (f(u_G) - u_M) \varphi_M +$$
$$+ \frac{1}{ASR_{CT}} \int_{\Gamma_{ME}} (g(u_M) + u_E) \varphi_M = 0 \tag{11b}$$

$$\int_{\Omega_E} \sigma_E \nabla u_E \nabla \varphi_E + \frac{\frac{1}{ASR_{CT}}}{2F} \int_{\Gamma_{ME}} (g(u_M) + u_E) \varphi_E = 0 \tag{11c}$$

$\forall (\varphi_G, \varphi_M, \varphi_E) \in Y_0$. The weak formulation is discretized by the finite element method and the resulting discretized nonlinear problem is solved as explained in the next section.

3.2 Solver and Grid

The system of PDE that describes the complete coupled problem is discretized by the method of finite elements and solved numerically by a C++ program, called ParCell3D, which has been developed specifically to solve this problem [4]. The program relies for the finite element discretization on the library deal.II [2].

The model considered here is described by a stiff nonlinear system of equations, in which the stiffness intrinsically derives by the multiphysics nature of the problem.

The bad conditioning of the problems is due to the stiff coupling between the gas phase and the MIEC independently of the method used to discretize the problem. The high condition number of the matrix and the peculiar clustering of the eigenvalues demand hence for an iterative solver with an efficient preconditioner. At the present state of the development we use the GMRES solver [20] preconditioned by a domain decomposition method [15]. In the implemented domain decomposition method the RVE is divided in subregions. Each piece of RVE is solved in parallel using again a GMRES solver with a ILU [20] preconditioner.

The grid is created by segmentation of the 3D image obtained by the FIB/SEM technique described in [11]. The reconstructed volume of the microstructure used for the calculations is $5.25 \times 5.25 \times 30$ μm^3. This corresponds to 18 million voxels and 23 million degrees of freedom using trilinear finite elements. As can be seen in Figs. 3 and 4 the microstructure is active only in a region close to the three phase boundary. We have shown in [11] that the calculated quantity of interest (ASR_{cat}) does not change cutting the microstructure above the active region to obtain a smaller but still precise model. To save computational time we use almost half of the total height (14 μm). The model used here has almost 12 million degrees of freedom (Fig. 2).

3.3 Approximation of the RVE

At the lateral sides of the RVE (on Γ_0) we impose a no flux condition. If the dimension of the RVE is large enough, the ASR_{cat} value (our quantity of interest) is not affected by the boundary condition on the sides. A preliminary study of the dimension of the RVE has been done [17] for an artificial structure. A more accurate study with a larger reconstruction is planned, which is needed for such a study.

To save computational time we consider a model reduction by cutting the height of the cathode. In [10] (using three different real microstructures) the influence on ASR_{cat} of the variation of the model height from 0.35 to 30 μm was studied. The ASR_{cat} value for the studied microstructures does not vary significantly for a height larger than 3.5 μm. We have decided nevertheless to calculate a model of height 14 μm to be sure that we can neglect the influence of the cut.

Fig. 2 Schematic representation of the five slices shown in Figs. 3 and 4. The RVE is a volume of dimensions $5.25 \times 5.25 \times 14\,\mu m^3$

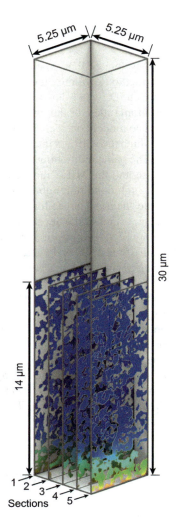

4 Parameter Estimation

4.1 Calibration of the Reduced Model

We have shown in [4] a comparison between our 3D model and a well established simplified model from Adler, Lane and Steele [1], known as ALS model. For the comparison we have calibrated the ALS model with the microstructure parameters (porosity, volume-specific surface area and tortuosity) calculated with our 3D model. For the calibration we need the values of the volume-specific surface area and porosity, which are calculated as postprocessing of the reconstruction. The

Parameter Estimation for a Reconstructed SOFC Mixed-Conducting LSCF-Cathode

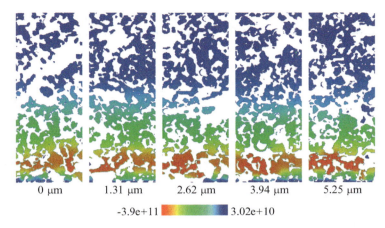

Fig. 3 Sensitivity of the solution (mol s/m^5) with respect to the parameter D^δ at $T = 800°$C

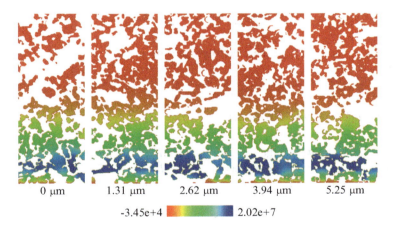

Fig. 4 Sensitivity of the solution (mol s/m^4) with respect to the parameter D^δ at $T = 800°$C

tortuosity on the contrary is not obtained by a simple postprocessing, it is calculated in our work using the same 3D microstructure model through a FEM simulation as explained in [11]. The results show that in case of homogeneous microstructures, as the one considered in this work, the ALS model can be used for predictive calculations of ASR_{cat}. Furthermore, in Part 2 of that work [5] we have shown that the derivatives of ASR_{cat} with respect to the parameters D^δ and k^δ differs 3–5% compared to our 3D model.

Thus we consider in this work the calibrated reduced ALS model as accurate enough to estimate the confidence regions of the fitted parameters. The use of this model will accelerate the uncertainty analysis and can be used as a preliminary study for a future optimal experimental design.

From the work of Adler and coauthors we consider the results of their asymptotic analysis, which gives a formula for the approximation of the total cell resistance R_{chem}

$$R_{\text{chem}} = \frac{RT}{2F^2} \sqrt{\frac{\tau \gamma^2}{(1 - \varepsilon)a c_{\text{mc}}^2 D^\delta k^\delta}}, \tag{12}$$

where R is the ideal gas constant, T the temperature, F the Faraday constant, γ the thermodynamic factor, c_{mc} the concentration of oxygen sites in the MIEC and the microstructure parameters are the tortuosity τ, the porosity ε and the volume-specific surface area a.

The ASR_{cat} value can than be derived by (12), subtracting the contribution of the electrolyte and dividing by two, since we consider only one electrode of the symmetric cell

$$ASR_{cat} = \frac{\left(R_{\text{chem}} - \dfrac{L_E}{\sigma_E} \right)}{2}, \tag{13}$$

where L_E is the length of the electrolyte and σ_E is the ionic conductivity of the electrolyte.

4.2 Least-Squares Problem

We have shown in [4] that the sensitivities of ASR_{cat} with respect to D^δ and k^δ have the same dependence on the temperature as ASR_{cat} itself. This means that it is not possible to estimate both D^δ and k^δ from measurements of ASR_{cat} taken by varying only the temperature.

It is known [3, 16, 24] nevertheless that k^δ and D^δ depend both on T, while only k^δ depends on p_{O_2} in the range of pressure considered in our experiment $(1 - 21 \text{ kPa})$. An empirical dependence of the parameters k^δ and D^δ on T and p_{O_2} can be found in [24]. We use the following parametrization

$$k^\delta = k_0 \exp\left(-\frac{E_{a,k}}{RT} \right) + p_{O_2}^\alpha, \tag{14}$$

$$D^\delta = D_0 \exp\left(-\frac{E_{a,D}}{RT} \right), \tag{15}$$

where $E_{a,k}$ and $E_{a,D}$ are the activation energies of k^δ and D^δ respectively, R is the gas constant, k_0 and D_0 are prefactors of the Arrhenius type of dependence on the temperature and α is an exponent for the dependence on p_{O_2} of k^δ.

The five parameters to be estimated are thus $D_0, E_{a,D}, k_0, E_{a,k}, \alpha$. We consider hence the possibility to measure ASR_{cat} at different temperatures T and different

Parameter Estimation for a Reconstructed SOFC Mixed-Conducting LSCF-Cathode 279

oxygen partial pressure p_{O_2}. The cell is operated at constant voltage η_{model}. From current measurements we obtain values of ASR_{cat}, which are used to make the fit of the model.

We have planned ASR_{cat} measurements for the cell under study using a symmetric cell, but we use in this work preliminarily only a virtual experiment to test the method. The virtual experiment consists of virtual measurements done by perturbation of "true" ASR_{cat} values with a random error ε. The "true" values are obtained by the model using reference values for the model parameters, as explained in [4]. We consider an error which is proportional to the measured value of the resistance:

$$ASR_{cat,meas} = ASR_{cat,true}(1 + \varepsilon), \tag{16}$$

where ε is a random error with zero mean and a given variance σ_{meas}.

ASR_{cat} is a function of k^δ, D^δ and several other material and microstructure parameters, see (13) and (12). The conductivity of the electrolyte σ_E is assumed to be known, as it can be measured with precision. The parameters ε, τ and a are calculated, specifically for the microstructure here considered, using our 3D FEM model. This is the most costly part of the procedure, but it is essential to permit the use of a reduced model. The material parameters C_{mc} and γ are calculated as in [4].

A least squares functional to be minimized by the Gauss–Newton method is defined

$$J(q, \xi) = \sum_{ij} \left(\widetilde{C_D}^{-1/2} \left(ASR_{cat}(q, \xi_{i,j}) - ASR_{cat,meas}(\xi_{i,j}) \right) \right)^2, \tag{17}$$

where $q := \{k_0, E_{a,k}, \alpha, D_0, E_{a,D}\}$ are the model parameters to be estimated and $\xi_{i,j} := \{T_i, p_{O_2,j}\}$ are design parameters to define the setup of the experiment. We have used $\widetilde{C_D} = C_D Z^2$ to approximate the covariance of the data, which is a proper approximation in case of relative errors as shown in [28]. C_D is the covariance of the multiplicative error ε and Z is the diagonal matrix with the model values $ASR_{cat}(\hat{q}, \xi_{i,j})$ on the diagonal. The asymptotic properties of this estimator can be found in [29].

The goal of this work is to study the feasibility of the parameter estimation defined above. For this reason we estimate the confidence regions of the parameters.

4.3 Confidence Regions

The estimation of the confidence regions of the parameters reveals quantitatively their correlation and the goodness of the fit.

For this purpose we estimate the linearized confidence region. This is an ellipsoid centered at the estimated parameter $\hat{q}$ and defined by the symmetric positive definite linearized covariance matrix [6]:

$$COV = \left(G^T \widetilde{C_D}^{-1} G \right)^{-1}, \tag{18}$$

where G is the Jacobian of ASR_{cat} with respect to q. In the estimation of the covariance matrix we need in G and Z the estimated value of the parameter $\hat{q}$ because the problem is nonlinear. A better approximation of the confidence regions can be obtained by a sequential optimal experimental design as shown in [6]. We do not address this point here, since it goes beyond the scope of this work. Considering uncorrelated measurements, C_D is a diagonal matrix with diagonal elements $C_D(i, i) = \sigma_{meas}^2$. To calculate the Jacobian G we need the sensitivities of ASR_{cat} with respect to the parameters.

The linearized confidence region is defined as

$$\mathcal{R}_L := \left\{ \hat{q} + \delta q : \delta q^T COV^{-1} \delta q \leq \chi_{(\alpha,\nu)}^2 \right\}, \tag{19}$$

where the confidence level is given by the χ^2 value with ν degrees of freedom and α significance level. The definition of the confidence level is motivated by the following observation. In the least squares case, with a linear observation operator and Gaussian random data, the fitted value is distributed as a χ^2 function with $\nu = N_Q$ degrees of freedom [25], where N_Q is the number of parameters.

4.4 Experiments

We consider virtual experiments with 20 measurements, each defined by a pair (T, p_{O_2}) combining the temperature values 600, 650, 700, 750, 800°C and the oxygen partial pressure values $1, 2, 10, 21$ kPa.

We assume that the measurements have a relative error of 0.1% with zero mean and Gaussian distribution, so we use formula (16) with $\sigma_{meas} = 0.001$.

We compare our results with a Monte Carlo simulation with 10,000 sampling points. Each point of the Monte Carlo simulation is the result of a parameter estimation using the functional (17) and fitting 20 measurements of ASR_{cat}. The measurement error is taken from the Gaussian distribution defined above. Since the errors are considered independent and identically distributed, we take 20 Gaussian distributions, one for each measurement, sampled with 10,000 points each.

5 Numerical Results

Figures 3 and 4 show the sensitivity of u_M with respect to the two parameters D^δ and k^δ. It can be noted that the microstructure is active only partially. The penetration depth (i.e. the length of the active region) depends on the critical ratio D^δ/k^δ, which controls the performance of the cell. It is thus crucial to estimate the coefficients which define these two important processes.

We use in this work a reduced model to calculate the confidence regions of the parameters. We have thus calibrated the ALS model [1] with the microstructure

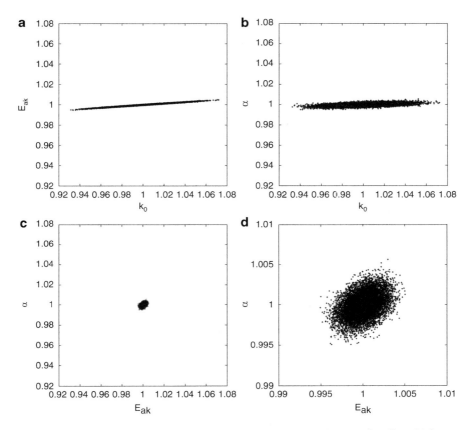

Fig. 5 Confidence region in the parameter space projected onto the planes (**a**) $k_0 - E_{a,k}$, (**b**) $k_0 - \alpha$, (**c**) $E_{a,k} - \alpha$. The plot (**d**) shows a closeup of (**c**)

parameters (volume-specific surface area, porosity and tortuosity) calculated with our 3D detailed FEM model. Since the sensitivities of the ALS model and the 3D model are comparable [4], the use of the ALS model for a study of the parameters' confidence regions is legitimate.

As can be seen in formula (12), we expect a strong correlation between k^δ and D^δ in the fitting of measured values of ASR_{cat}, thus a bad conditioning of the fitting problem. We first show that, in the case the fitting process is well behaving, the linearized covariance matrix gives an appropriate approximation of the confidence region. For this reason we consider a virtual experiment in which we estimate only the three parameters that define k^δ, i.e. k_0, $E_{a,k}$ and α from 20 measurements at different T and p_{O_2} as explained in Sect. 4.2. The value for the parameter D^δ has to be taken from literature or determined from an another experiment.

In Fig. 5 we report the 95% confidence regions of all pairs of parameters. These are sections of the ellipsoidal three dimensional confidence region. As the parameters have been normalized, the reference value for all the parameters is one.

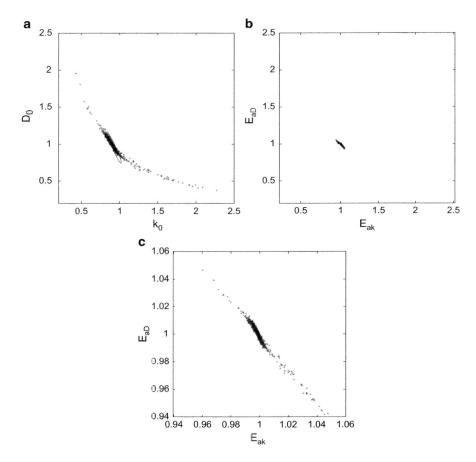

Fig. 6 Confidence region in the parameter space projected onto the planes: (**a**) $k_0 - D_0$, (**b**) $E_{a,k} - E_{a,D}$. The plot (**c**) shows a closeup of the plot (**b**)

The linearized confidence regions are compared with values obtained by a Monte Carlo simulation using 10,000 experiments as explained in Sect. 4.4. Furthermore, we observe that the variance in the estimation of k_0 is much larger than the variance of the other two parameters $E_{a,k}$ and α. An error in the data space is than amplified 50 times in the parameter space. We can also note that the linearized confidence region gives a good approximation of the confidence region sampled by the Monte Carlo method, which takes into account the nonlinearities.

If we consider the estimation of all five parameters, which define k^δ and D^δ, the covariance matrix has a condition number which is two orders of magnitude larger than the case of fitting only k^δ. The fitting is bad conditioned, so we can not expect to be able to approximate the confidence region using the linearized covariance matrix.

Figure 6 shows the correlations between k_0 and D_0 and between $E_{a,k}$ and $E_{a,D}$. It can be observed from the Monte Carlo sampling that the error of the parameters

k_0 and D_0 is in the range between -50% and $+200\%$, while $E_{a,k}$ and $E_{a,D}$ are within 5%. The linearized confidence regions do not approximate the tail of the distribution, which is clearly not Gaussian. This result was expected as the two parameters k^δ and D^δ appear in the formula (12) as a product. We want to show nevertheless that the estimation of the covariance matrix (18) gives a quantitative information on the well posedness of the parameter estimation problem and in case of well behaving of the system it can give a very good approximation of the uncertainty of the estimation, as shown in Fig. 5.

6 Conclusion

In previous works [4, 11] we have used a precise 3D reconstruction of a MIEC cathode to calculate its performance. It has been shown in that work that for homogeneous microstructures a simplified model as the ALS model can be used instead of the 3D model to estimate a macroscopic quantity as the ASR_{cat}. This is possible if the simplified model is calibrated using data obtained by the detailed 3D model. For this calibration we need the values of the volume-specific surface area, porosity and tortuosity as explained in Sect. 4. Furthermore we have shown in another work [5] that the sensitivities of ASR_{cat} calculated with the ALS model are very close to the results of the 3D model. It is hence possible to use the calibrated ALS model for the estimation of the parameters' confidence regions.

In this work we have studied the possibility to estimate the parameters that define the two most important processes, which are the bulk diffusion and the surface reaction. These two processes have been parametrized with five parameters. We have calculated an approximation of the confidence regions of the parameters, which have shown that the strong correlation of the two processes leads to a bad conditioning of the fitting procedure. The use of measurements of ASR_{cat} at different temperature and oxygen partial pressure in a range of $T = 600$–$800°C$ and $p_{O_2} = 1$–21 kPa is not enough to estimate both parameters k^δ and D^δ, while if D^δ is given, it is possible to estimate k^δ within an error of 5% if the measurements have an error of 0.1%.

The estimation of both parameters is possible using a technique revealing the time constants of the system, e.g. impedance measurements, as shown in [13].

The work done here will be extended with a study of optimal experimental design following [6], both in the stationary and the time dependent case.

Acknowledgements We thank Prof. Francesco Ciucci for the continuous and invaluable discussions and Dr. Bernd Rüger for helpful comments and discussions.

We are grateful to Dr. Holger Obermaier for his contribution on software optimization; Hartmut Häfner for the parallel computer support; the Steinbuch Centre for Computing (SCC) at the Karlsruhe Institute of Technology for providing the computing facility.

This work was funded by the Deutsche Forschungsgemeinschaft (DFG) through project "Modellierung, Simulation und Optimierung der Mikrostruktur mischleitender SOFC-Kathoden" (IV 14/16-1, RA 306/17-1) and by the Friedrich-und-Elisabeth-Boysen-Stiftung.

References

1. Adler, S.B., Lane, J.A., Steele, B.C.H.: Electrode kinetics of porous mixed-conducting oxygen electrodes. Journal of The Electrochemical Society **143**(11), 3554–3564 (1996)
2. Bangerth, W., Hartmann, R., Kanschat, G.: deal.II – a general purpose object oriented finite element library. ACM Trans. Math. Softw. **33**(4), 24/1–24/27 (2007)
3. Bouwmeester, H.J.M., den, M.W.O., Boukamp, B.A.: Oxygen transport in La 0.6Sr 0.4Co 1- yFe yO 3-d. Journal of Solid State Electrochemistry **8**(9), 599–605 (2004)
4. Carraro, T., Joos, J., Rüger, B., Weber, A., Ivers-Tiffée, E.: 3D FEM model for the reconstruction of a porous SOFC cathode: I. Performance quantification. Electrochimica Acta **77**(0), 315 – 323 (2012)
5. Carraro, T., Joos, J., Rüger, B., Weber, A., Ivers-Tiffée, E.: 3D FEM model for the reconstruction of a porous SOFC cathode: II. Parameter sensitivity analysis. Electrochimica Acta **77**(0), 309 – 314 (2012)
6. Ciucci, F., Carraro, T., Chueh, W.C., Lai, W.: Reducing error and measurement time in impedance spectroscopy using model based optimal experimental design. Electrochimica Acta **56**(15), 5416 – 5434 (2011)
7. Gostovic, D., Smith, J.R., Kundinger, D., Jones, K., Wachsman, E.: Three-dimensional reconstruction of porous LSCF cathodes. Electrochemical and Solid-State Letters **10**(12), B214–B217 (2007)
8. Hornung, U. (ed.): Homogenization and porous media. Springer-Verlag New York, Inc., New York, NY, USA (1997)
9. Iwai, H., Shikazono, N., Matsui, T., Teshima, H., Kishimoto, M., Kishida, R., Hayashi, D., Matsuzaki, K., Kanno, D., Saito M., M.H., K., E., N., K., Yoshida, H.: Quantification of Ni-YSZ anode microstructure based on dual beam FIB-SEM technique. ECS Transactions **25**(2 PART 3), 1819–1828 (2009)
10. Joos, J., Carraro, T., Ender, M., Rüger, B., Weber, A., Ivers-Tiffée, E.: Detailed Microstructure Analysis and 3D Simulations of Porous Electrodes. ECS Transactions **35**, 2357–2368 (2011)
11. Joos, J., Carraro, T., Weber, A., Ivers-Tiffée, E.: Reconstruction of porous electrodes by FIB/SEM for detailed microstructure modeling. Journal of Power Sources **196**(17), 7302–7307 (2011)
12. Joos, J., Rüger, B., Carraro, T., Weber, A., Ivers-Tiffée, E.: Electrode Reconstruction by FIB/SEM and Microstructure Modeling. ECS Transactions **28**(11), 81–91 (2010)
13. Leonide, A.: Modelling and parameter identification by means of impedance spectroscopy. Ph.D. thesis, Karlsruher Institut für Technologie (KIT) (2010)
14. Matsuzaki, K., Kanno, D., Teshima, H., Shikazono, N., Kasagi, N.: Three-dimensional numerical simulation of Ni-YSZ anode polarization using reconstructed microstructure from FIB-SEM images. ECS Transactions **25**(2), 1829–1836 (2009)
15. Quarteroni, A., Valli, A.: Domain Decomposition Methods for Partial Differential Equations. Oxford University Press, Oxford, UK (1999)
16. Ried, P., Bucher, E., Preis, W., Sitte, W., Holtappels, P.: Characterisation of La0.6Sr0.4Co0.2Fe0.8O3-d and Ba0.5Sr0.5Co0.8Fe0.2O3-d as Cathode Materials for the Application in Intermediate Temperature Fuel Cells. ECS Transactions **7**(1), 1217–1224 (2007)
17. Rüger, B.: Mikrostrukturmodellierung von elektroden für die festelektrolytbrennstoffzelle. Ph.D. thesis, Karlsruher Institut für Technologie (KIT) (2009)
18. Rüger, B., Joos, J., Weber, A., Carraro, T., Ivers-Tiffée, E.: 3D electrode microstructure reconstruction and modelling. ECS Transactions **25**(2), 1211–1220 (2009)
19. Rüger, B., Weber, A., Ivers-Tiffée, E.: 3D-modelling and performance evaluation of mixed conducting (MIEC) cathodes. ECS Transactions **7**(1), 2065–2074 (2007)
20. Saad, Y.: Iterative Methods for Sparse Linear Systems, 2nd edn. Society for Industrial and Applied Mathematics, Philadelphia, PA, USA (2003)

21. Shearing, P.R., Cai, Q., Golbert, J.I., Yufit, V., Adjiman, C.S., Brandon, N.P.: Microstructural analysis of a solid oxide fuel cell anode using focused ion beam techniques coupled with electrochemical simulation. Journal of Power Sources **195**(15), 4804 – 4810 (2010)
22. Shearing, P.R., Golbert, J., Chater, R.J., Brandon, N.P.: 3D reconstruction of SOFC anodes using a focused ion beam lift-out technique. Chemical Engineering Science **64**(17), 3928–3933 (2009)
23. Shikazono, N., Kanno, D., Matsuzaki, K., Teshima, H., Sumino, S., Kasagi, N.: Numerical assessment of SOFC anode polarization based on three-dimensional model microstructure reconstructed from FIB-SEM images. Journal of The Electrochemical Society **157**(5), B665–B672 (2010)
24. Søgaard, M., Hendriksen, P., Jacobsen, T., Mogensen, M.: Modelling of the polarization resistance from surface exchange and diffusion coefficient data. In: Proc. 7th European SOFC Forum, pp. 1–17. European Fuel Cell Forum, Lucerne, Switzerland (2006)
25. Tarantola, A.: Inverse problem theory and methods for model parameter estimation. SIAM, Philadelphia (2005)
26. Wilson, J.R., Gameiro, M., Mischaikow, K., Kalies, W., Voorhees, P.W., Barnett, S.A.: Three-dimensional analysis of solid oxide fuel cell Ni-YSZ anode interconnectivity. Microscopy and Microanalysis **15**(1), 71–77 (2009)
27. Wilson, J.R., Kobsiriphat, W., Mendoza, R., Chen, H.Y., Hiller, J.M., Miller, D.J., Thornton, K., Voorhees, P.W., Adler, S.B., Barnett, S.A.: Three-dimensional reconstruction of a solid-oxide fuel-cell anode. Nature Material **5**(7), 541–544 (2006)
28. Banks, H. T., Davidian, M., Samuels Jr, J.R.: An inverse problem statistical methodology summary. CRSC-TR07-14 (2007)
29. Davidian, M., Giltinan D. M.: Nonlinear Models for Repeated Measurement Data. Monographs on Statistics & Applied Probability, Chapman & Hall/CRC (1995)

An Application of Robust Parameter Estimation in Environmental Physics

Alexandra G. Herzog and Felix R. Vogel

Abstract This article presents a current research application of robust parameter estimation for inverse problems in air–sea gas exchange. The first part illustrates the interaction of measurements, analysis, and parameter estimation in the concept of the modeling cycle: Typical measurement techniques are classified and discussed with respect to application areas and reproducibility. Statistical analysis of the assumed error distributions is presented as the link to parameter estimation via choosing a suitable parameter estimator. In the case that the error distribution cannot be determined definitely and appears to be highly non-Gaussian the use of robust estimators is advised instead of the standard least-squares approach. Huber's M-estimator is used here as robust method. Reliable parameter reconstruction by the chosen estimator closes the modeling cycle by validating model against experiment.

In the second part all previously presented theoretical aspects are applied to the spectral reconstruction approach (SPERA), a special measurement technique for determining water-sided gas concentration fields via Laser-induced fluorescence. In this application the use of a robust estimator allows the reconstruction of local gas concentration as the model parameters of an inverse problem, for which least-squares methods failed. The article finishes with a brief outlook of applying optimal experimental design techniques to measurements in environmental physics.

1 Introduction

Environmental physics aims at improving our understanding of the strongly coupled non-linear system earth regarding intra- and inter-transport of the individual subsystems, in recent time with special emphasis on influence of human-induced

---

A.G. Herzog (✉) · F.R. Vogel
Institute for Environmental Physics, Im Neuenheimer Feld 229, 69120 Heidelberg, Germany
e-mail: aherzog@iup.uni-heidelberg.de; fvogel@iup.uni-heidelberg.de

H.G. Bock et al. (eds.), *Model Based Parameter Estimation*, Contributions
in Mathematical and Computational Sciences 4, DOI 10.1007/978-3-642-30367-8_14,
© Springer-Verlag Berlin Heidelberg 2013

changes like increasing concentration of greenhouse gases in the atmosphere. The key challenges can be subdivided roughly in two categories:

(a) On the one hand, new measurement techniques based on (mainly) well understood processes are developed, that rely on measurements of easily accessible quantities. Together with a reliable model description they allow the reconstruction of quantities of interest from estimated model parameters, for example greenhouse gas concentration from DOAS measurements [44], soil water content from radar backscatter [7], or paleotemperatures from concentrations of dissolved noble gases in groundwater [1].

(b) On the other hand, typical research topics in environmental physics are acquiring representative data sets that cover the range of the encountered multiscale processes, e.g. in turbulent transport phenomena, therefore demanding profound screening of processes from smallest, intermediate up to global scale. The immense amount of data necessary for a complete covering of encountered phenomena emphasizes the need for model descriptions of processes at least on small to intermediate scales to reduce the required data density to an acceptable level. These tasks typically combine a wide range of both direct measurement methods (e.g. temperature, humidity) and indirect methods described in (a) with modeling approaches. Moreover, several subdisciplines of environmental physics may be involved in the same study. An example may be the determination of global sea-air carbon dioxide flux [52]. Both categories often lead to inverse problems, which form a significant part of experimental analysis in environmental physics. Inverse problems explicitly consist of a theoretical process model and a set of usually experimental data. On the basis of these elements the model parameters are to be determined that show the least deviation between model prediction and experimental realization under the given conditions.

In this article, the application of techniques from model based parameter estimation is discussed, especially with a view to inverse problems under non-Gaussian error distributions of experimental data. Special emphasis is given to an illustrative example from current research in environmental physics. The example was chosen to represent a typical and often encountered type of problem that may be easily adapted to other situations ranging from physics over biosciences to economics.

The article is structured as following: The so-called "modeling cycle" is introduced to illustrate the interaction of experimental data acquisition, modeling, and model validation in combination with parameter estimation. The experimental techniques used in environmental physics are presented, followed by a rough sketch of theoretical aspects of inverse problems as far as necessary for the example application from air–sea interaction, the Spectral Reconstruction Approach.

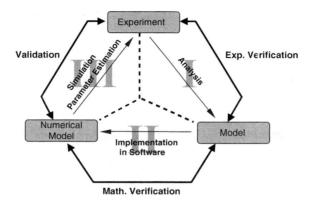

Fig. 1 The basic modeling cycle: experiment as initial step, after analysis of the process model candidates can be developed, these again are directly implemented in software or if necessary implemented as an approximated model in software, here called "numerical model," that allows simulation of the real experimental conditions. Iterative review and revision of these steps by physical and mathematical validation and verification finally leads to the desired model. Based on a graphical idea by Oberkampf et al. [43]

2 The Modeling Cycle

Hardly understood processes are common in young research areas such as environmental physics, chemical engineering, and biosciences.

Especially model development, comparison of model candidates, and evaluation of single models need to be repeatedly discussed, see for example the assessment of atmosphere-ocean general circulation models (AOGCM), by the IPCC expert meeting, see [51]. The search for a suitable model for a certain (experimental) problem is an iterative process of adapting, validating, and refining promising model candidates. The main goal is to reproduce the experimental data of the regarded (physical) processes satisfactorily and statistically significantly with respect to the process uncertainties and—if possible—with a model that can be theoretically deduced. This procedure can be illustrated in the so-called modeling cycle sketched here in a reduced form in Fig. 1. Models are developed by analyzing discrepancies of model predicted states and experimental measurements and by adapting the respective model candidate for improvement.

The basic modeling cycle presented here is a formalization of this process by implicitly including methods from parameter estimation for model validation. Essential for using these methods is that the model is implemented in software. Implemented models allow simulation of process states that can be compared to measured experimental data by estimating the corresponding model parameters and evaluating the residual in the solution. If a suitable model candidate is identified, parameter estimation techniques allow to determine the model parameters for a given set of data.

In the course of this article a specific problem from environmental physics is treated as an example problem in the framework of the modeling cycle in Sect. 5. As a start, the next sections present physical background as well as the used techniques as motivation and problem classification.

3 Experimental Challenges and Methods in Environmental Physics

Modeling of a given problem and later estimation of the characteristic model parameters relies on a sound background in the specific area of research. Experiments from environmental physics form the background of the studied application problem: Environmental physics deals with the planet Earth, one of the most complex directly explorable systems. Humankind as integral part of the system Earth affects the system and is affected by the system itself [32]. Despite these mutual interactions, knowledge of the system and the corresponding processes within is still very poor. The system Earth is subdivided into the compartments atmosphere, biosphere, lithosphere, cryosphere, and ocean. The arising dynamics are usually described by highly non-linear partial differential equation systems both for processes within the individual subsystems as well as for interactions of two or more subsystems. Typically, it is intended to describe subsystems by models as simple as possible. Herein, the identification of suitable model candidates and the dominating system parameters are the most challenging tasks.

Usually, the (sub)systems are too large for classical experimental studies in laboratories, for example the carbon cycle, especially the interaction and trends in the uptake of CO_2 in the terrestrial and marine biosphere [42] are extremely difficult to be designed in the necessary complexity. However, there are some exceptions for which the basic processes are still under investigation. An example here may be the transport of gaseous tracers over a gaseous–liquid interface, that forms the basis for understanding the transport of atmospheric gases across the air–sea interface under turbulent conditions, especially in combination with waves. Up to now, a reliable generalized transfer model has not been found [59].

Therefore, physical investigation in environmental physics has to adapt old and invent new methods to deal with these complex situations. The methods for investigation can be classified in the following four groups: (a) classic laboratory studies, (b) measurements that combine field studies for local information with lab analysis, and (c) remote sensing. The most recent group in this context is (d) virtual experiments, that are actually detailed simulations in cases where the elementary physics are already well understood and modeled, but for example the effects causing macroscopic intermittency are still unknown.

3.1 Classic Laboratory Measurements

The oldest classical investigation is the laboratory study. Classical laboratory experiments can be regarded as a physical simulation, with a simplified system and

An Application of Robust Parameter Estimation in Environmental Physics 291

simplified boundary conditions. Under controlled and reproducible (environmental) conditions (e.g. temperature, pressure, salinity) data of the process under investigation are systematically collected to gain information about parameter dependent quantities like temperature or pressure for model development. Examples from current research at Heidelberg are soil column investigation [45], or determination of vertical gas concentration profiles. The latter is used as an example for robust parameter estimation in Sect. 5.

3.2 Field Studies with Discrete Sampling Points

Field studies aim at investigating a certain quantity, e.g. concentration or velocity fields, in large areas, volumes or timescales. Usually measurements of the observed quantity are taken in discrete sample points. Sample point data is typically interpolated to generate continuous approximations of the quantities of interest, for example emission maps for anthropogenic CO_2 emissions. These emission estimates are part of the input for complex computer simulations. Yet, current computer models are not able to fully reconstruct the typical variations of the CO_2 concentration over Europe gained in point measurements, but the recent progress is evident, see for example [24] and [2]. Besides, sampling points are typically assumed to be representatives for specific conditions. Furthermore, these sampling points may serve as references for alignment of data gained by remote sensing.

The obtained data are analyzed both in-situ and in detailed lab studies, that may take several months up to years. Examples of large scale field studies are measurements and analysis of ice cores [47], and sea sediments [39], as well as oceanic observation cruises, see the cruises supported by SOPRAN, the German sub-project of the International SOLAS project (www.solas-int.org) for studies of e.g. local oceanic aerosol composition [4]. An additional category of field studies is the investigation of spatially strongly localized phenomena e.g. cave drip water [37], or measurements of trace gases in volcanic plumes [13].

Field measurement campaigns are usually very expensive due to the amount of scientific equipment that has to be transported to the desired locations. A severe problem in field studies are the given environmental conditions: Usually they cannot be controlled and must be monitored in all possible detail.

3.3 Remote Sensing Techniques

In recent years remote sensing systems, e.g. satellite observation, have gained more and more importance in the analysis of global processes, as they can usually cover large areas with fine local discretization, though in some cases, they do not achieve the precision of in-situ sample point measurements.

Remote sensing comprises a huge variety of completely different methods. Examples are the DOAS technique used for trace gas detection like HCHO with GOME [40], or remote sensing of oceanic wind fields [15], and wave slope via radar backscatter [58]. Another method is active microwave sensing that allows soil water measurements in the topmost soil layer, see e.g. [7]. Moreover, remote sensing is closely linked to image processing as data from remote sensing usually consists of images, see e.g. [50]. In the last 20 years especially digital image processing has evolved extremely rapidly, for a review on applications see for example the books [16] and [34], for classification and statistical methods in image processing see e.g. [20] and [12]. Also in laboratory studies such image-based techniques are used. Examples are the determination of bubble distributions in ice cores [55], in air–sea interaction, IR-imaging for measuring turbulent heat transfer [27, 28], as well as Laser-induced fluorescence (LIF) or particle image velocimetry (PIV) studies in fluid dynamics, see [29], and [10] in biomedical research.

Again it has to be noted, that remote sensing techniques like field studies are also subject to the environmental conditions during observation and therefore need to log these conditions as well. The most recent family of experiments is a product of theoretical model descriptions and computing power: the virtual experiments.

3.4 Virtual Experiments: Direct Numerical Simulation in Fluid Dynamics

Virtual experiments mark the current frontier of interaction of physical investigation and computer science. Here, nature is simulated completely, both the process to investigate and the measurement equipment are implemented in algorithms.

One representative of virtual experiments currently used in environmental physics and in fluid dynamics is the so-called "Direct Numerical Simulation" (DNS). The differential equation system for transport is used to determine the dynamical behavior of the system in question using an initial guess and boundary conditions. Such virtual experiments deliver full information on the flow fields and scalar transport quantities where direct measurements are extremely difficult if not impossible in this detail. In DNS, the transport equations are solved directly without any specific model assumptions on a computational mesh as fine resolved as necessary to cover all ranges of turbulence that are theoretically expected, see for example [54] for channel, or [22] for pipe flow. With full information on the turbulent scalar fields of transported quantities and vector flow fields, simplified parametric models can be tested or newly developed for certain conditions. Up to now, classical laboratory experiments have failed to deliver the necessary detailed information for reliable parameter estimation. This gap is closed by virtual experiments.

3.5 Statistical Analysis: A Link from Measurements to Parameter Estimation

Experimental studies and especially field studies have gained much from improved storing devices and processing of data in modern computers. Furthermore, implementation of mathematical analysis techniques for time-series, e.g. periodograms, FT- and wavelet analysis measurement [14], support research in environmental physics. The same is true for statistical analysis techniques: ANOVA, or significance tests like t- or F-test have facilitated determining estimates of the error distribution $\mathcal{E}$ of data sets. An identified error distribution sets the statistical framework for parameter estimation, especially the choice of a suitable norm for evaluating deviations of model prediction and measurements [51]. In more detail, this allows to identify the best-suited parameter estimator and allows stochastically-representative simulations. Even in the case of a linear model this is necessary, see for example the textbook [41]. A suitable parameter estimator can be linked to the maximum likelihood approach that determines the best matching norm for a given error distribution, for example the popular least squares approach with l_2-norm is the maximum likelihood estimator for a Gaussian distribution, the absolute norm l_1 for a Laplacian error distribution, for details on maximum likelihood see the articles [23] and [3].

A special problem are outliers in the data: If the error distribution differs from the Gaussian assumption, especially if it is a heavy-tailed distribution, the number of outliers from the Gaussian prediction increases. In the case of influential outliers, they can prevent the solver from finding an adequate estimate of the desired parameters. Possible solutions are presented in the following section. After this detailed presentation of data sources and a short introduction to the influence of error distributions the next section gives a short overview on parameter estimation techniques necessary for the application in Sect. 5.

The topic modeling of processes is not discussed here, as it comprises theoretical considerations and methods of vast range. Here, we refer the reader to discussions and publications in the respective communities, some examples in environmental physics are large scale modeling of global transport models in [51], and applications on intermediate to small scales in soil physics, see for example [56] and [48].

It is important to note again, that inverse modeling requires a suitable process model in advance, that allows a quantitative parametrization of the measured data sets. In the following we assume that such a model exists.

4 Parameter Estimation from Experimental Data

Inverse modeling is used to determine an optimal parameter set of a theoretical model with respect to given experimental observations. The model parameters usually represent single physical system characteristics or comprise several physical quantities.

Analysis of tomographic data is a typical field of application for inverse modeling. Current examples in environmental physics may be the tomographic DOAS method, e.g. [35], the reconstruction of paleotemperature from the isotopic composition of stalagmites [57], or the determination of soil parameters from measurements of its electromagnetic properties [60].

In the current section we give a very brief summary on inverse problems as far as necessary for Sect. 5. The term "inverse problem" is used here as a synonym for ill-posed parameter estimation problems. Generally, a function or a vector of a certain realization of the measurable quantities of interest, the so called states y, is parametrized by a functional relation G, with parameters p and control variables q,

$$y = G(p, q). \tag{1}$$

Herein, the number of parameters is assumed to be finite. The control variables q represent the environmental conditions such as temperature that influence the states y, but can be controlled by the experimenter either directly (e.g. temperature stabilization in lab studies) or indirectly (e.g. measuring only at certain conditions in field studies). G as the model of the measurement process may implicitly contain a system of ordinary (ODE) or partial differential equations (PDE), or a system of differential algebraic equations (DAE). This is the so-called forward problem, usually with a (locally) unique solution. The solution of the forward problem is called "simulation": G is determined for given p and q.

In an inverse problem, the parameters p of a given model are to be determined from a vector of indirect observations, the measured states y^{meas}, by minimizing the deviation of model and measurements by a suitable choice of parameters, e.g. see the textbook [36]. The measured values y^{meas} are assumed to be an additive composition of the model response y^{true} and a random error ϵ, that may depend on the y^{true},

$$y^{meas} = G(p_{true}, q) + \epsilon(y^{true}). \tag{2}$$

For a Gaussian distribution the relation is simplified to

$$y_G^{meas} = y^{true} + \epsilon, \tag{3}$$

with $\epsilon = \sqrt{\sigma}$ as the standard deviation of a distribution with zero mean and variance σ.

The deviation of model prediction and measurements, is called residual r,

$$r_i(p) := y_i^{meas} - G_i(p, q). \tag{4}$$

The parameters of a local solution should decrease the residuals. In the best case, the vector of residuals in a solution would correspond to the vector of measurement errors. A local solution is gained by minimizing the sum of a criterion function $\phi(r)$,

An Application of Robust Parameter Estimation in Environmental Physics 295

e.g. the best suited l_p-norm of the residuals r, with w as positive scale parameters or weights

$$\min_{p_i} \sum_{i=1}^{n} \phi \left(\frac{r_i(p)}{w_i} \right), \tag{5}$$

in the sense of a maximum likelihood-type estimation or M-estimation, a generalization of maximum likelihood estimators. In the least-squares approach the sum of the squared residuals is minimized. The criterion function for this case is

$$\phi_{LS}(r(p)) := \frac{r(p)^2}{2\,w}. \tag{6}$$

Here, the weights w correspond to the variance σ or the squared standard deviation of the Gaussian distribution. Typically the estimated measurement errors are used as weighting factors. In that way the measurements with lowest error receive the highest weight.

In the case that the error distribution shows deviations from the assumed one, so-called robust estimators can be used. Robust estimators can be classified by the definition of a breakdown point. The general idea of the breakdown point is the smallest fraction of model-inconsistent observations to consistent ones, for which the estimator can be expected to give correct results [6]. The fraction ranges from 0.0 to 0.5. The maximum possible breakdown point of 0.5 can be achieved by the median. For details on breakdown point definition and outlier detection see for example [17] and the textbook [46].

An example of a robust estimator is Huber's M-estimator introduced in the beginning of 1970s [31]. This estimator belongs to the class of iteratively reweighted least squares (IRLS) methods that can exclude large outliers from minimization by special criterion functions ϕ. The criterion function for Huber's M-estimator is

$$\phi_H(r(p)) := \begin{cases} r(p)^2/2, & \text{if } |r| \leq \gamma; \\ \gamma|r(p)| - \gamma^2/2, & \text{if } |r| > \gamma. \end{cases} \tag{7}$$

with γ a constant. If $\gamma \rightarrow \infty$ Huber's limit function is the least squares approach.

For an introduction to maximum likelihood estimators and an introduction to robust estimators see for example the corresponding chapters in the textbook [41] or in this collection the article of Kostina et al. [11].

In contrast to the forward problem of a simulation, in which the system response to a given input parameter set is calculated, the inverse problem tries to determine the parameters from a (measured) system response. Inverse problems are usually "ill-conditioned" in a Hadamard sense, meaning that the solution needs neither to be unique nor to exist [26]. Regularization strategies, like the Tikhonov regularization, may be used to deal with ill-posed inverse problems by introducing restrictions on the solution. The regularization may be achieved by adding a penalty term, e.g. for smoothness, or by multiplying with a non-singular regularization matrix W. For details on regularization strategies see for example the introductory chapters in the textbooks [36] and [25]. An introduction to inverse problems for natural scientists including regularization strategies is given for example in the textbook [53].

In the next section an example application is shown where the previously discussed strategies are used. In that problem, Huber's M-estimator allowed determining the parameters of an inverse problem, while the least squares estimator failed, implied by a strong non-Gaussian error distribution.

5 Application: The Spectral Reconstruction Approach

In this section the application of the modeling cycle framework is demonstrated in an example from small scale air–sea interaction: The reconstruction of water-sided depth-dependent gas concentrations from spectral measurements via inverse modeling. This novel technique called SPERA (Spectral Reconstruction Approach) has been developed at the Institute of Environmental Physics in the group of Prof. B. Jähne at the University of Heidelberg. The image processing software has been contributed by the Heidelberg Collaboratory for Imaging (HCI), while the parameter estimation part has been provided by the Numerical Optimization group of Prof. E. Kostina, University of Marburg.

In the combination of an experimental technique, image processing, and parameter estimation the resolution could be remarkably refined by a factor of five, giving concentrations in five individual depth layers instead of only one from a purely experimentally evaluated technique. First results have been presented in [30]. Small Scale Air–Sea Interaction as a sub-discipline of environmental physics tries to find simplified transport equations of heat, momentum, and mass, i.e. gas, especially near the air–water interface, in and next to the boundary layers, where the main resistance of transport is located. Important in this context is the determination of the transport velocities of gases like oxygen and carbon dioxide. Good estimates of these mass fluxes (commonly given as piston or transport velocities) under various conditions are of high interest for climate modeling with respect to greenhouse gases as they allow to predict quantitative gas transport from the atmosphere into the oceans and vice versa. In that way for example, the ocean acts as the largest sink of the greenhouse gas carbon dioxide, CO_2. Up to 30–40% of man-made CO_2 is estimated to be transferred into the oceans [19].

5.1 Setup and Basic Concept

The Spectral Reconstruction Approach (SPERA) is a technique that tries to capture 2D turbulent concentration fields in a fluid by analyzing the spectral composition of the integral fluorescence spectrum detected from above the investigated volume.

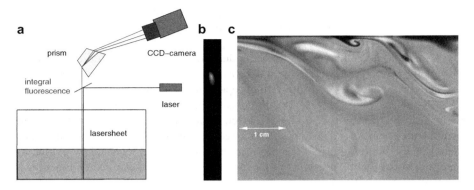

Fig. 2 (**a**) Sketch of the spectral setup in side view for spanwise spectra. Flow direction from *right* to *left*. (**b**) Integral spectrum taken in a setup sketched in (**a**). (**c**) Classical dynamic depth-dependent concentration 2D-profile parallel to main flow direction at a mean wind speed of $3\,\mathrm{ms}^{-1}$ gained by laser-induced fluorescence. Intensity marks local dissolved gas concentration. Image taken with a camera system in side-view not in spectral view. For details see text. Wind direction from *right* to *left*

The setup is shown in Fig. 2a. The water volume of interest is dyed with a combination of a fluorophore and an absorber dye. Fluorophore concentration is assumed to be proportional to local gas concentration. Fluorescence is induced by laser excitation that is coupled in via a dichroitic mirror. The detector system is located at the top of the water surface. In a classical setup no absorber dye would be present. In this standard side-setup concentration profiles are gained by a camera in side view, where the pixel position in the image is correlated to water depth. There intensity would correspond to local gas concentration. An image gained by this technique is shown in Fig. 2c. In the presence of an absorber dye this is no longer possible.

In SPERA concentration gradients of dissolved gases in the aqueous mass boundary layer are detected above the water surface by spectrally analyzing the emitted integral fluorescence. The depth information is lost in the recorded image as the spectrometer only sees a fluorescing line at the water surface, see Fig. 2b.

The lost depth information has to be extracted from the emitted spectrum by means of an inverse parameter estimation. The depth-dependent concentrations appear as the parameters of the inverse problem.

The technique does not give enough information in its basic setup to solve the PDE system of transport in three dimensions, but allows to measure two dimensional concentration fields. SPERA is based on a solely experimental technique developed in [9] that allows to capture the concentration in the topmost fluid layer the thickness of which is determined by the concentration of the added absorber dye.

In combination with robust parameter estimation techniques this experimental method can decompose the boundary layer in finer sublayers to achieve a concentration resolution with depth in one single measurement. The method used is actually a tomographic method with one significant alteration: the position of the detector is fixed and is not varied over a whole range of radial positions. Here, the integral fluorescent spectrum is detected only at one position above the water surface with a spectrometer. The integral spectrum is the overlay of spectra emitted at every depth layer in the water sample below.

From that integral spectrum the individual spectra coming from different depth layers can be reconstructed: In the structure of each spectrum the local concentration at the corresponding depth z and the depth information z itself is coded like a fingerprint, so each depth and each concentration has its own uniquely formed fluorescence spectrum. This depth and concentration coding is achieved by the combined usage of a fluorophore and a second dye that absorbs in the emission band of the fluorophore. So the depth information can be regained by extracting the individual depth spectra from the integral spectrum like a vector function is expanded in its different vector components.

5.2 Measuring

Turbulent processes are unpredictable and have to be captured stochastically. Therefore, one problem in environmental physics is to store the vast number of data sets capturing smallest to largest timescales or spatial scales to identify the dominating frequencies in the fluctuations and the dominating length scales so to identify the quantities of importance. Especially for quantities that are transported by a fluid like mass a suitable measurement technique has to be developed that does not affect the flow field. Typical measurement techniques that have been used are the so-called Laser-induced fluorescence (LIF) for measuring concentration and temperature fields, and the Particle Image Velocimetry or PIV to obtain in-situ velocity fields. These two techniques are commonly used in fluid mechanics and even allow 3D measurements of the desired quantities without disturbing the measured quantity. In SPERA, the integral fluorescence of a LIF measurement plane is recorded with a spectrometer setup sketched in Fig. 2. The resolution of the wavelength is about 1 nm, which is quite coarse, but the sensitivity of the camera setup with a CMOS camera allows detection of lowest signal intensities [33]. Usually the signal is so low that the Poisson statistic of the fluorescence is dominating the signal fluctuations without additional averaging. In Fig. 3 depth-dependent fluorescence spectra are plotted.

5.3 Modeling

In SPERA, the model function state represents a theoretical description of the integral fluorescence intensity L measurable above the water surface. This continuous spectrum is discretized according to the discrete wavelength resolution of the detecting spectrometer setup. The emission model of the dye-system gives the

An Application of Robust Parameter Estimation in Environmental Physics

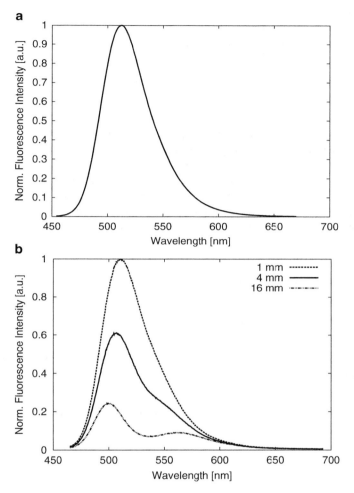

Fig. 3 Depth-dependent fluorescence spectra of a fluorophore combined with an absorber dye measured by illumination with a single laser beam. Spectra normalized by intensity at the water surface: (**a**) measured at the water surface, (**b**) spectra at water depth of 1 mm, 4 mm, and 16 mm. Measurement device Ocean Optics Spectrometer 2000

following approximation of the fluorescence spectrum $\frac{\partial L(z,\lambda)}{\partial z} \Delta z$ in a single depth layer of thickness Δz normalized to one degree of spatial angle:

$$\frac{\partial L_{em}(z,\lambda)}{\partial z} = \frac{1}{4\pi} \Phi_q f(\lambda) \varepsilon(\lambda_L) c(z) E_L + O\left(\left(\varepsilon(\lambda_L) \int_0^z c(z')\, dz' \right)^2 \right). \quad (8)$$

The factor $1/4\pi$ is the normalization constant of the spatial angle, Φ_q is the quantum yield of the fluorophore, $f(\lambda)$ represents the normalized fluorescence spectrum, $\varepsilon(\lambda_L)$ is the extinction coefficient of the fluorophore at laser wavelength λ_L, $c(z)$ is the concentration of the fluorophore, E_L the emittance of the laser. This approximation is valid if the rest term in the last approximation can be neglected

$$\varepsilon(\lambda_L) \int_0^z c(z')\, dz' \ll 1. \tag{9}$$

This can be achieved by reducing the concentration of the fluorophore accordingly. A dye concentration satisfying this inequality guarantees a linear relation between fluorescence intensity and absorption. In fluorescence spectroscopy this case can be expected for most dyes up to an absorption of 0.05.

An additionally added absorber dye with extinction coefficient $\varepsilon_A(\lambda)$ modifies this relation in every layer by an exponential factor $\exp(-\varepsilon_A(\lambda)c_A z')$. The fluorescence-absorption spectrum that can be measured at the water surface is calculated as the integral of the fluorescence-absorption spectra of the individual layers,

$$L^{int}(z; \lambda) = \int_0^{z=\infty} L_{absem}(z'; \lambda)\, dz' \tag{10}$$

$$= \int_0^{z=\infty} \frac{\partial L_{em}(z'; \lambda)}{\partial z'} \exp(-\varepsilon_A(\lambda)c_A z')\, dz'. \tag{11}$$

For the integral fluorescence spectrum this eventually leads to the discretized form of an infinite sum,

$$L^{int}(z, \lambda) \doteq C f(\lambda) \cdot \int_0^{z=\infty} c(z') \exp(-\varepsilon_A(\lambda)c_A z')\, dz' \tag{12}$$

$$\doteq C f(\lambda) \cdot \sum_{i=0}^{\infty} c(z_i) \exp(-\varepsilon_A(\lambda)c_A z_i)\, \Delta z_i, \tag{13}$$

with $C := \frac{1}{4\pi} \Phi_q \, \varepsilon(\lambda_L)\, E_L$ as summarizing constant; the maximum water depth is assumed to be infinity. This infinite sum can be split up into summands,

$$L^{int}(\lambda) = \int_0^{\hat{z}} \frac{\partial L}{\partial z}(\lambda; z)\, dz + \int_{\hat{z}}^{\infty} \frac{\partial L}{\partial z}(\lambda; z)\, dz, \tag{14}$$

$$\doteq \sum_{i=0}^{\hat{i}} \frac{\Delta L}{\Delta z}(\lambda; z_i)\, \Delta z + \sum_{i=\hat{i}}^{\infty} \frac{\Delta L}{\Delta z}(\lambda; z_i)\, \Delta z. \tag{15}$$

now the question remains from which discretized depth z_i with $i = \hat{i}$ the rest term can be neglected, or in other words, which depth no longer significantly contributes to the integral spectrum. This is equivalent to the fact that the corresponding depth

An Application of Robust Parameter Estimation in Environmental Physics

layer can no longer be resolved. The usual limit for influence on the result is a proportion of the term of approximately $p = 1\%$, with respect to the value of the total sum.

An estimation of the error term is shown here for an equal discretization of the depth layers with $\Delta z_i = \Delta z = $ constant and $z_i = i \cdot \Delta z$. Furthermore, a homogeneous distribution of the concentration c of the fluorescent dye is assumed for estimation: $c = $ constant. As concentration changes are assumed to be small, this approximation is acceptable for a rough estimation of the error term. All constant terms are collected in the expression $\mathcal{K} := c\, C f(\lambda) \Delta z$. For calculation of the error term the discretized infinite sum is rearranged. Herein, the expression $x(\lambda) := \exp(-\varepsilon_A(\lambda)c_A \Delta z)$ is used,

$$L^{int}(z;\lambda) \doteq \sum_{i=0}^{\infty} c\, C f(\lambda) \exp(-\varepsilon_A(\lambda)c_A z_i)\Delta z_i \tag{16}$$

$$\doteq \mathcal{K} \cdot \sum_{i=0}^{\infty} \exp(-\varepsilon_A(\lambda)c_A \Delta z)^i \tag{17}$$

$$\doteq \mathcal{K} \cdot \sum_{i=0}^{\infty} x(\lambda)^i. \tag{18}$$

This is the geometric series. For $|x(\lambda)| < 1 \,\forall\, \lambda$ this infinite series is convergent. Therefore, there exists a certain depth L after which the rest terms can be neglected,

$$L \leq \frac{\ln(p)}{\ln(x(\lambda))}. \tag{19}$$

Since the absolute value of both p and $x(\lambda)$ is smaller than one and both are positive values, the logarithm is negative and therefore L always positive. Together with the above made assumptions on the concentration of both fluorophore and absorber dye this sum can be further approximated to,

$$\underbrace{L_j^{int}(\lambda_j)}_{=:L_j} \doteq \mathcal{K} \cdot \sum_i \underbrace{c(z_i)}_{=:c_i} \underbrace{\left| \underbrace{f^{fluo}(\lambda_j)}_{=:f_j} \underbrace{exp(-\varepsilon_A(\lambda_j)c_A\, z_i)}_{=:\alpha_j} \right|}_{=:\mathcal{B}(z_i,\lambda_j)}, \tag{20}$$

This can be rewritten as a linear equation system,

$$\begin{pmatrix} L_0^{int}(\lambda_0) \\ \vdots \\ L_{L-1}^{int}(\lambda_{L-1}) \end{pmatrix} = K \cdot \begin{pmatrix} \mathcal{B}(z_0,\lambda_0) & \cdots & \mathcal{B}(z_{N-1},\lambda_0) \\ \vdots & & \vdots \\ \mathcal{B}(z_0,\lambda_{L-1}) & \cdots & \mathcal{B}(z_{N-1},\lambda_{L-1}) \end{pmatrix} \begin{pmatrix} c_0 \\ \vdots \\ c_{N-1} \end{pmatrix}. \tag{21}$$

This formulation forms the basic model for the simulation and the formulation of the inverse problem.

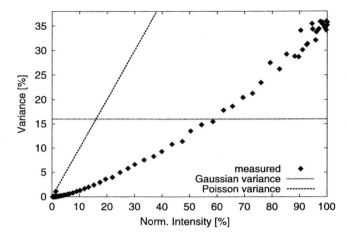

Fig. 4 Variance–mean plot for the measured data sets plotted as diamonds; theoretical distributions: *dashed*: Poisson distribution, *dotted*: Gaussian distribution; Measurement device: Varian Eclipse Carrier, AG Herten, BioQuant, Heidelberg

5.4 Statistical Analysis of Experimental Error

As previously discussed, the actual distribution of errors is unknown a priori. This is also the case in SPERA: The noise distribution of the measured data is expected to show a strong Poisson-like characteristic due to fluorescence being subject to Poisson distribution. Yet, due to detector influence this Poisson distribution is altered by the intrinsic detector statistics. For more details on statistics in LIF measurements see [18]. They showed that the actual distribution in LIF-signals is an overlay of the statistics of all involved processes. If the compound error distribution is unknown, a maximum likelihood estimator can not be determined. In Fig. 4, the variance–mean plots of one of the tested spectrometers is shown. There the measured values do support neither the Gaussian nor the Poisson assumptions, and do not seem to fit a known probability distribution. Given such a case, using a robust estimator seems appropriate, that can deal with a variety of long-tailed error distributions and guarantees sufficient robustness against outliers.

5.5 Formulating the Parameter Estimation Problem

After the estimator, here Huber's-M-estimator, has been chosen, the parameter estimation problem can be set up,

$$\min_{c_i} \sum_{i=0}^{L-1} [1/s_i]\, \phi_{Huber}\left(\left(L_{model}^{int}(\lambda_i, c_i) - L_{meas}^{int}(\lambda_i)\right)\right). \qquad (22)$$

An Application of Robust Parameter Estimation in Environmental Physics 303

Table 1 Comparison of estimators for simulated data with errors from gamma distribution

Profile	Estimator	c_0	c_1	c_2
(I)	True values	$1.0 \cdot 10^{-5}$	$5.0 \cdot 10^{-6}$	$1.0 \cdot 10^{-7}$
	Huber	$1.00001 \cdot 10^{-5}$	$4.99972 \cdot 10^{-6}$	$1.00105 \cdot 10^{-7}$
	LS	$-88.2453 \cdot 10^{-5}$	$2515.35 \cdot 10^{-6}$	$-17176.6 \cdot 10^{-7}$
(II)	True values	$1.0 \cdot 10^{-7}$	$5.0 \cdot 10^{-6}$	$1.0 \cdot 10^{-5}$
	Huber	$1.00046 \cdot 10^{-7}$	$5.00003 \cdot 10^{-6}$	$0.99998 \cdot 10^{-5}$
	LS	$7716.61 \cdot 10^{-7}$	$-1994.89 \cdot 10^{-6}$	$128.287 \cdot 10^{-5}$

Table 2 Comparison of estimators for simulated data with errors from log-normal distribution

Profile	Estimator	c_0	c_1	c_2
(I)	True values	$1.0 \cdot 10^{-5}$	$5.0 \cdot 10^{-6}$	$1.0 \cdot 10^{-7}$
	Huber	$0.999999 \cdot 10^{-5}$	$5.00036 \cdot 10^{-5}$	$0.990427 \cdot 10^{-7}$
	LS	$0.998000 \cdot 10^{-5}$	$6.12988 \cdot 10^{-6}$	$-0.185964 \cdot 10^{-7}$
(II)	True values	$1.0 \cdot 10^{-7}$	$5.0 \cdot 10^{-6}$	$1.0 \cdot 10^{-5}$
	Huber	$0.99402 \cdot 10^{-7}$	$5.00182 \cdot 10^{-6}$	$0.99987 \cdot 10^{-5}$
	LS	$1.01200 \cdot 10^{-7}$	$5.01480 \cdot 10^{-6}$	$0.99806 \cdot 10^{-5}$

The s_i represent estimates of the variance of the measurement errors and act as scaling or weighting factors. Additional constraints on the solution are not used. For implementation in software the linear parameter estimation problem is written in matrix form:

$$
\min \left[K \cdot \begin{pmatrix} \frac{1}{\sigma_0} & \cdots & 0 \\ \vdots & & \vdots \\ 0 & \cdots & \frac{1}{\sigma_{L-1}} \end{pmatrix} \cdot \phi \left(\begin{pmatrix} \mathcal{B}(z_0, \lambda_0) & \cdots & \mathcal{B}(z_{L-1}, \lambda_0) \\ \vdots & & \vdots \\ \mathcal{B}(z_0, \lambda_{L-1}) & \cdots & \mathcal{B}(z_{L-1}, \lambda_{L-1}) \end{pmatrix} \begin{pmatrix} c_0 \\ \vdots \\ c_{L-1} \end{pmatrix} - \begin{pmatrix} L_0^{meas}(\lambda_0) \\ \vdots \\ L_{L-1}^{meas}(\lambda_{L-1}) \end{pmatrix} \right) \right].
$$

5.6 Results of the Parameter Estimation

The previously formulated parameter estimation problem has been solved with both least squares estimator and Huber's M-estimator. The results are shown in Tables 1 and 2 for a three layer concentration reconstruction of simulated data sets with known solution. Profiles I and II represent two typical cases, invasion and evasion of a gaseous tracer. Gamma and log-normal distribution were used as error distributions in simulations.

Least squares estimation failed three out of four data sets, presented here, while Huber's M-estimator succeeded in parameter reconstruction for all test cases.

This result is not a surprising one having the previous sections in mind. Even for this relatively simple application the error distribution and therefore the appropriate choice of an estimator has a remarkable effect on reconstructing the solution successfully, in spite of correct model assumptions. Summarizing, robust estimators

304 Alexandra G. Herzog and Felix R. Vogel

that respect heavy-tailed noise distributions are far better suited than the least squares approach under such conditions.

For more details on results of the SPERA problem see [30].

5.7 Optimal Experimental Design as Extension of the Modeling Cycle

Besides statistical analysis and parameter estimation, also "optimal experimental design" or "optimal control" are techniques that may be used with SPERA. These techniques are described in detail for example in [38] and are just mentioned here briefly for completion of the modeling cycle.

Optimal experimental design focuses on maximizing the significance of parameters estimated from experimental data. This is accomplished by minimizing the confidence ellipsoid of a parameter of interest, hence the error estimate of this parameter. The confidence ellipsoid can be minimized by minimizing a criterion function on the covariance matrix of the model-constraint system, for example the largest eigenvalue of the covariance matrix. In that way optimal experimental design corresponds to a non-linear optimization problem. The parameters to be determined are the control parameters and functions q (measurement parameters like temperature, pressure, humidity, etc.), the weights w, as well as the number and spatial/temporal positioning of the measurements. The optimal solution of an experimental design problem determines the best suited run of the controllable parameters. This run is typically extremely difficult to find by experimental trial and error. The potential of optimal experimental design has been demonstrated in various applications in science and industry, see e.g. the articles [5, 8], or [49]. In that respect, optimal experimental design can be regarded as an extension of the modeling cycle as presented in Fig. 1, Sect. 2. It is a technique that uses simulated virtual experiments on the basis of an available process model to predict optimal experiments for determining the model parameters, e.g. with lowest errors. These predicted experiments can now be validated by conducting real experiments accordingly and analyzing deviations from prediction and actual realization. Therefore it can be included as a part IV in the modeling cycle, see Fig. 5.

In laboratory studies optimized experiments can be realized relatively easy as controllable environmental conditions are common. In field studies the situation is more difficult: environmental conditions are almost impossible to control or reproduce. Nonetheless, there are approaches to choose at least conditions (e.g. times when to take samples) under which measuring would be most preferable. Under such terms, the application of optimal experimental design seems possible to a certain degree. This has already been discussed for field studies, e.g. in [21] 20 years ago. Optimal experimental design for SPERA is also limited, yet possible: In wind-wave tunnel experiments with their dynamics usually far away from stationarity, controlling steering functions like temperature or pressure is difficult.

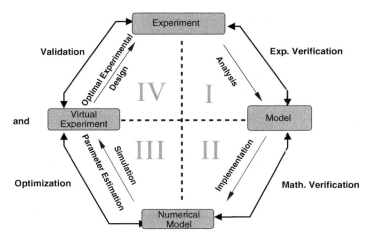

Fig. 5 The iteration cycle of model development as basic concept extended with optimal experimental design as part IV. For this extension a suitable model formulation is absolutely required

However, for measurements in convection tanks in which pressure and temperature can be controlled much more readily this problem is avoided. So for this special configuration SPERA may be optimized to gain maximum information out of the data sets.

Summarizing, whenever an experimental setup allows steering of control parameters or functions optimal experimental design can be applied.

6 Summary

The specific error distribution of a measurement technique can differ from the standard assumption of a Gaussian distribution. This will affect the choice of an appropriate estimator if these data sets are intended to determine model parameters in an inverse problem. In the case of heavily tailed non-Gaussian error distribution the use of robust estimators is indicated. Huber's M-estimator is one example of a robust iteratively reweighted least squares method, that may be applied to such a problem.

In an example from current research in environmental physics, the spectral reconstruction approach (SPERA), Huber's M-estimator improved parameter reconstruction significantly: Previous tests with a least squares estimator failed. Simulated data sets with known model parameters were used for a detailed illustration of this problem for SPERA, see Table 2. Here, the basic combination of an experimental method from air–sea interaction research and robust parameter estimation allowed the successful reconstruction of the desired vertical concentration profiles.

This article tried to present the basics concepts of robust parameter estimation for practical applications especially in environmental physics including the effects of error distributions on the choice of the estimator. Furthermore, an example application from current research was provided using a robust estimator that is widely available, e.g. in standard software packages like Matlab, so that the transfer to the reader's research problems is facilitated.

Acknowledgements We gratefully acknowledge financial support by the DFG Research Training Group 1114 'Optical Techniques for Measurement of Interfacial Transport Phenomena' and the Heidelberg Graduate School of Mathematical and Computational Methods for the Sciences, as well as Synmikro, Marburg. Moreover, we want to thank our colleagues for their support, especially Tanja Binder (Marburg), Pascal Bohleber, Evelyn Böhm, Samuel Hammer, Barbara May, Gabriele Schenk, and Martin Wieser (all Institute of Environmental Physics, Heidelberg). Furthermore, special thanks go to Felix Friedl for performing the measurements shown in Fig. 3 and the group of PD D.P. Herten, BioQuant, Heidelberg, for giving us the opportunity to use their fluorescence spectrometer.

References

1. W. Aeschbach-Hertig, F. Peeters, U. Beyerle, and R. Kipfer. Interpretation of dissolved atmospheric noble gases in natural waters. *Water Res. Research*, 35, 1999.
2. R. Ahmadov, C. Gerbig, R. Kretschmer, S. Körner, C. Rödenbeck, P. Bousquet, and M. Ramonet. Comparing high resolution WRF-VPRM simulations and two global CO_2 transport models with coastal tower measurements of CO_2. *Biogeosciences*, 6:807–817, May 2009.
3. J. Aldrich. R. A. Fisher and the making of maximum likelihood 1912-1922. *Stat. Sci.*, 12:162–176, 1997.
4. J. D. Allan, D. O. Topping, N. Good, M. Irwin, M. Flynn, P. I. Williams, H. Coe, A. R. Baker, M. Martino, N. Niedermeier, A. Wiedensohler, S. Lehmann, K. Mller, H. Herrmann, , and G. McFiggans. Composition and properties of atmospheric particles in the eastern atlantic and impacts on gas phase uptake rates. *Atmos. Chem. Phys.*, 9:9299–9314., 2009.
5. A. E. Altmann-Dieses, J. P. Schlöder, H. G. Bock, and O. Richter. Optimal experimental design for parameter estimation in column outflow experiments. *Water Resour. Res.*, 38:1186–1197, 2002.
6. R. Andersen. *Modern Methods for Robust Regression*. Sage Publications, 2008.
7. K. Anderson and H. Croft. Remote sensing of soil surface properties. *Prog. Phys. Geog.*, 33(4):457 – 473, 2009.
8. H. Arellano-Garcia, J. Schöneberger, and S. Körkel. Optimale Versuchsplanung in der chemischen Verfahrenstechnik. *Chem. Ing. Tech.*, 79:1625–1638, 2007.
9. W. E. Asher and J. F. Pankow. Direct observation of concentration fluctuations close to a gas-liquid interface. *Chem. Eng. Sci.*, 44:1451–1455, 1989.
10. A. Berthe, D. Kondermann, C. Christensen, I. Goubergrits, C. S. Garbe, K. Affeld, and U. Kertzscher. Three-dimensional, three-component wall-PIV. *Exp. Fluids*, 48:983–997, 2010.
11. T. Binder and E. Kostina. Gauss-Newton methods for robust parameter estimation. In H. G. Bock et al., editor, *Model Based Parameter Estimation*, Contributions in Mathematical and Computational Sciences 4, pages 53–85. Springer, 2012.
12. C. M. Bishop. *Pattern Recognition and Machine Learning*, volume 1. Springer, 2007.
13. N. Bobrowski, G. Hünninger, F. Lohberger, and U. Platt. Idoas: A new monitoring technique to study the 2D distribution of volcanic gas emissions. *J. Volcanol. Geotherm. Res.*, 4:329–338, 2006.

An Application of Robust Parameter Estimation in Environmental Physics 307

14. G. E. P. Box, G. M. Jenkins, and G. C. Reinsel. *Time Series Analysis: Forecasting and Control.* Wiley, 4 edition, 2008.
15. D. B. Chelton, M. G. Schlax, M. H. Freilich, and R. F. Milliff. Satellite measurements reveal persistent small-scale features in ocean winds. *Science*, 303:978–983, 2004.
16. E. R. Davies. *Maschine Vision. Theory, Algorithms, Practicalities*, volume 3. Academic Press, 2005.
17. P. L. Davies and U. Gather. Unmasking multivariate outliers and leverage points. *Stat. Journ.*, 5(1):1–17, 2007.
18. N. Dimarcq, V. Giordano, and P. Cerez. Statistical properties of laser-induced fluorescence signals. *Appl. Phys. B Lasers O.*, 59:135–145, 1994.
19. M. A. Donelan and R. Wanninkhof. Gas transfer at water surfaces - conepts and issues. In M. A. Donelan, W. M. Drennan, E. S. Saltzman, and R. Wanninkhof, editors, *Gas Transfer at Water Surfaces*. American Geophysical Union, 2002.
20. R. O. Duda, P. E. Hart, and D. G. Stork. *Pattern Classification*, volume 2. John Wiley and Sons, 2000.
21. L. L. Eberhardt and J. M. Thomas. Designing environmental field studies. *Ecol. Mono.*, 61:53–73, 1991.
22. B. Eckhardt, A. Schmiegel, H. Faisst, and T. Schneider. Dynamical systems and the transition to turbulence in shear flows. *Phil. Trans. R. Soc. A*, 366(1868):1297–1315, 2008.
23. R. A. Fisher. An absolute criterion for fitting frequency curves. *Mess. Math.*, 41:155–160, 1912.
24. C. Geels, M. Gloor, P. Ciais, P. Bousquet, P. Peylin, A. T. Vermeulen, R. Dargaville, T. Aalto, J. Brandt, J. H. Christensen, L. M. Frohn, L. Haszpra, U. Karstens, C. Rödenbeck, M. Ramonet, G. Carboni, and R. Santaguida. Comparing atmospheric transport models for future regional inversions over Europe. Part 1: Mapping the CO_2 atmospheric signals. *Atmos. Chem. Phys. Discuss.*, 6:3709–3756, May 2006.
25. P. E. Gill, W. Murray, and M.H. Wright. *Practical Optimization.* Academic Press, INC, 1981.
26. J. Hadamard. Sur les problèmes aux dérivées partielles et leur signification physique. *Princeton University Bulletin*, 13:49–52, 1902.
27. T. Hara, E. VanInwegen, J. Wendelbo, C. S. Garbe, U. Schimpf, B. Jähne, and N. Frew. Estimation of air-sea gas and heat fluxes from infrared imagery based on near surface turbulence models. In C. S. Garbe, R. A. Handler, and B. Jähne, editors, *Transport at the Air Sea Interface - Measurements, Models and Parametrizations*. Springer-Verlag, 2007.
28. H. Haußecker, U. Schimpf, C. S. Garbe, and B. Jähne. Physics from IR image sequences: Quantitative analysis of transport models and parameters of air-sea gas transfer. In E. Saltzman, M. Donelan, W. Drennan, and R. Wanninkhof, editors, *Gas Transfer at Water Surfaces*, volume 127 of *Geophysical Monograph*. American Geophysical Union, 2002.
29. I. Herlina and G. H. Jirka. Experiments on gas transfer at the air–water interface induced by oscillating grid turbulence. *J. Fluid. Mech.*, 594:183–208, 2008.
30. A. G. Herzog, T. Binder, B. Jähne, and E. A. Kostina. Estimating water-sided vertical gas concentration profiles by inverse modeling. In *2. International Conference on Engineering Optimization, Lisbon*, September 2010.
31. P. J. Huber. Robust statistics: A review. *Annals of Mathematical Statistics*, 43:1041–1067, 1972.
32. IPCC. Climate change 2007: The physical science basis. Contribution of working group I to the fourth assessment report of the intergovernmental panel on climate change, 2007.
33. B. Jähne. *Digitale Bildverarbeitung, 6. Auflage.* Springer, 2005.
34. B. Jähne, H. Haußecker, and P. Geißler, editors. *Handbook of Computer Vision and Applications.* Academic Press, 1999.
35. M. Johansson, B. Galle, C. Rivera, and Y. Zhang. Tomographic reconstruction of gas plumes using scanning DOAS. *B. Volcanol.*, 71:1169–1178, 2009.
36. J. Kaipio and E. Somersalo. *Statistical and Computational Inverse Problems.* Applied Mathematical Sciences. Springer, 1 edition, 2004.

37. T. Kluge, D. F. C. Riechelmann, M. Wieser, C. Spötl, J. Sültenfuß, A. Schröder-Ritzrau, S. Niggemann, and W. Aeschbach-Hertig. Dating cave drip water by tritium. *J. Hydrol.*, 394:396–406, 2010.

38. S. Körkel. *Numerische Methoden für Optimale Versuchsplanungsprobleme bei nichtlinearen DAE-Modellen*. PhD thesis, Univ. Heidelberg, 2002.

39. J. Lippold, J. Grützner, D. Winter, Y. Lahaye, A. Mangini, and M. Christl. Does sedimentary $^{231}Pa/^{230}Th$ from the bermuda rise monitor past atlantic meridional overturning circulation? *Geophys. Res. Lett.*, 36:L12601, 2009.

40. T. Marbach, S. Beirle, U. Platt, P. Hoorand F. Wittrock, A. Richter, M. Vrekoussis, M. Grzegorski, J. P. Burrowsand, and T. Wagner. Satellite measurements of formaldehyde from shipping emissions. *Atmos. Chem. Phys.*, 9, 2009.

41. D.C. Montgomery, E. A. Peck, and G. G. Vinning. *Introduction to Linear Regression Analysis*. Wiley InterScience, 4 edition, 2006.

42. R. R. Nemani, C. D. Keeling, H. Hashimoto, W. M. Jolly, S. C. Piper, C. J. Tucker, R. B. Myneni, and S. W. Running. Climate-driven increases in global terrestrial net primary production from 1982 to 1999. *Science*, 300:1560–1563, June 2003.

43. W. L. Oberkampf, T. G. Trucano, and C. Hirsch. Verification, validation, and predictive capability in computational engineering and physics. *Appl. Mech. Rev.*, 57(5):345–385, 2004.

44. U. Platt and J. Stutz. *Differential Optical Absorption Spectroscopy, Principles and Applications*. Physics of Earth and Space Environments. Springer-Verlag, Berlin, Heidelberg, New York, 2000.

45. F. Rezanezhad, H. J.Vogel, and K.Roth. Experimental study of fingered flow through initially dry sand. *Hydr. Earth Syst. Sci. Disc.*, 3:2595–2620, 2006.

46. P. J. Rousseeuw and A. M. Leroy. *Robust Regression and Outlier Detection*. John Wiley & Sons, 1987.

47. U. Ruth, D. Wagenbach, R. Mulvaney, H. Oerter, W. Graf, H. Pulz, and G. Littot. Comprehensive 1000 year climatic history from an intermediate depth ice core from the south dome berkner island, antarctica: methodics, dating, and first results. *Ann. Glaciol.*, 39:146–154, 2004.

48. K. Schneider-Zapp, O. Ippisch, and K. Roth. Numerical study of the evaporation process and parameter estimation analysis of an evaporation experiment. *Hydrol. Earth Syst. Sci.*, 14:765–781, 2010.

49. Schöneberger, J. C., H. Arellano-Garcia, G. Wozny, S. Körkel, and H.Thielert. Model-based experimental analysis of a fixed-bed reactor for catalytic SO_2 oxidation. *Ind. & Eng. Chem. Res.*, 48(11):5165–5176, 2009.

50. R. A. Schowengerdt. *Remote Sensing: Models and Methods for Image Processing*, volume 3. Academic Press, 2006.

51. T. Stocker, Q. Dahe, G.-K. Plattner, M. Tignor, and P. Midgley. IPCC expert meeting on assessing and combining multi model climate projections. Boulder, Colorado, USA, 25-27 January 2010.

52. T. Takahashi, S. C. Sutherland, C. Sweeney, A. Poisson, N. Metzl, B. Tilbrook, N. Bates, R. Wanninkhof, R. A. Feely, C. Sabine, J. Olafsson, and Y. Nojiri. Global sea-air CO_2 flux based on climatological surface ocean pCO_2, and seasonal biological and temperature effects. *Deep Sea Res. Pt II*, 49(9-10):1601–1622, 2002.

53. A. Tarantola. *Inverse Problem Theory and Methods for Model Parameter Estimation*. SIAM, 2005.

54. W. Tsai and L. P. Hung. Three-dimensional modeling of small-scale processes in the upper boundary layer bounded by a dynamic ocean surface. *J. Geophys. Res.*, 112:C02019, 2007.

55. K. J. Ueltzhöffer, V. Bendel, J. Freitag, S. Kipfstuhl, D. Wagenbach, S. H. Faria, and C. S. Garbe. Distribution of air bubbles in the EDML and EDC (antarctica) ice cores, using a new method of automatic image analysis. *J. Glaciol.*, 56(196):339–348, 2010.

56. J. Šimůnek and S.A. Bradford. Vadose zone modeling: Introduction and importance. *Vadose Zone J.*, 7:581–586, 2008.

57. A. Wackerbarth, D. Scholz, J. Fohlmeister, and A. Mangini. Modelling the $\delta^{18}o$ value of cave drip water and speleothem calcite. *Earth Planet. Sci. Lett.*, 299:387–397, 2010.

58. E. J. Walsh, C. W. Wright, M. L. Banner, D. C. Vandemark, B. Chapron, J. Jensen, and S. Lee. The southern ocean waves experiment.Part III: Sea surface slope statistics and near-nadir remote sensing. *J. Phys. Oceanogr.*, 38:670–685, 2008.
59. R. Wanninkhof, W. E. Asher, D. T. Ho, C. Sweeny, and W. R. McGillis. Advances in Quantifying Air-Sea Gas Exchange and Environmental Forcing. *Ann. Rev. Mar. Sci.*, 1:213–244, 2009.
60. U. Wollschläger, T. Pfaff, and K. Roth. Field-scale apparent hydraulic parameterisation obtained from tdr time series and inverse modeling. *Hydrol. Earth Syst. Sci.*, 13:1953–1966, 2009.

Parameter Estimation in Image Processing and Computer Vision

Christoph S. Garbe and Björn Ommer

Abstract Parameter estimation plays a dominant role in a wide number of image processing and computer vision tasks. In these settings, parameterizations can be as diverse as the application areas. Examples of such parameters are the entries of filter kernels optimized for a certain criterion, image features such as the velocity field, or part descriptors or compositions thereof. Subsequently, approaches for estimating these parameters encompass a wide range of techniques, often tuned to the application, the underlying data and viable assumptions. Here, an overview of parameter estimation in image processing and computer vision will be given. Due to the wide and diverse areas in which parameter estimation is applicable, this review does not claim completeness. Based on selected key topics in image processing and computer vision we will discuss parameter estimation, its relevance, and give an overview over the techniques involved.

1 Introduction

In most image processing and computer vision tasks, one starts off with visual data such as an image or a sequence thereof. Of course, more general modalities are also becoming increasingly more common. These include spectral image sequences for example in satellite remote sensing or volumetric or even spectral volumetric time series, for example in state-of-the-art medical imaging devices. Cheap consumer devices are also transitioning beyond simple 2D capturing apparatuses. The Microsoft Kinect device, which captures intensity images and the scene depth at the same time, is an example of this. Also, first consumer grade cameras that capture light fields are on the horizon. All of these devices require adapted image processing

---

C.S. Garbe (✉) · B. Ommer
Interdisciplinary Center for Scientific Computing (IWR), University of Heidelberg,
Speyerer Str. 6, 69115 Heidelberg, Germany
{Christoph.Garbe,Bjoern.Ommer}@iwr.uni-heidelberg.de

H.G. Bock et al. (eds.), *Model Based Parameter Estimation*, Contributions
in Mathematical and Computational Sciences 4, DOI 10.1007/978-3-642-30367-8_15,
© Springer-Verlag Berlin Heidelberg 2013

and computer vision algorithms. Common to all of them is the classical inverse problem they lead to: n-dimensional data is acquired and some model parameters need to be estimated to best describe the data. Of course, the model will strongly depend on the application, as will the metric in which "best" is described by some sort of optimality condition. In this contribution, some common problems in image processing and computer vision will be described. Common methodologies for solving the resulting parameter estimation problem are presented.

This article is organized into two main parts, the first focusing on parameter estimation in low-level image processing, the second on high-level computer vision. In Sect. 2.1, parameter optimization for filter kernels will be introduced. The estimation of image motion or optical flow is outlined in Sect. 2.2. The reconstruction of images and optical flow fields is discussed in Sects. 2.4 and 2.5 respectively. The estimation of confidence and situation measures for optical flows is presented in Sect. 2.3. The segmentation of images based on their underlying motion is touched upon in Sect. 2.6. A combination of such approaches in a single functional is presented in Sect. 2.7 which is concerned with joint estimation of optical flow, segmentation and denoising of image sequences.

Parameter estimation in high level computer vision is presented in Sect. 3. Central to the analysis are key modeling decisions which are explained in Sect. 3.1. In Sect. 3.2 a compositional approach to object categorization is presented. The problem of object detection in cluttered scenes is discussed in Sect. 3.3.

2 Low-Level Image Processing

2.1 Optimization of Filter Kernels

A fundamental operation in image processing represents the filtering of intensity images. Such filtering is used for computing derivative filters of first order for motion estimating and for edge detection, and second or higher order for feature extraction of curvature information. Filter design is a well-established area in time-series signal processing and subject of standard textbooks [103, 107, 113]. The extension from 1-D signal processing to image processing is not trivial, however. This is largely due to uncertainty of design criteria for higher-dimensional signals and much more involved mathematical problems.

For edge detection and motion estimation, the computation of precise gradients of image intensities is vital. It can be show that the highly accurate computation of both, orientation and magnitude of gradients, are not feasible. Therefore, design choices have to be made. Very often, subspace problems, which are orthogonal to the gradient directions have to solved. Hence the precise direction of the gradients is more important that their magnitude. Making such a design choice leads to the formulation of a optimization problem, yielding the appropriate filter kernels. These filters are discretized by finite differences using convolution kernels optimized with respect to the model assumptions, scales and/or noise present in the data.

2.2 Optical Flow Estimation

For the estimation of motion from digital image sequences, a number of different techniques has been proposed. Generally, they can be categorized into four groups:

1. The class of *gradient based* techniques relies on computing spatio-temporal derivatives of image intensity, which can either be of first order [46,65] or second order [93,115].
2. *Region-based matching* may be employed when under certain circumstances (aliasing, small number of frames, etc.) it is inappropriate to compute derivatives of grey-values. In this approach the velocity is defined as a shift giving the best fit between image regions at different times [7,55,80].
3. The *energy-based methods* rely on the output energy of velocity-tuned filters. These methods are often referred to as *frequency-based methods* owing to their design in the Fourier domain [1,50,62].
4. Another class of methods is called *phase-based*, because velocity is defined in terms of phase behavior of band-pass filter output and phase information is used to estimate the optical flow [51,122].

Overviews of these estimators including error analysis can be found in [11, 12, 60]. One widely used technique to estimate the local optical flow $v(t, x)$ corresponding to an image sequence $u : [0, 1] \times \Omega \rightarrow \mathbb{R}$ on an image domain $\Omega \subset \mathbb{R}^n$ $(n = 2, 3)$ is the *first order gradient based* approach [46,65]. Together with phase based techniques [51], this approach offers the best performance with respect to accuracy [11,53]. Here, constancy of gray values $u(t, x(t))$ along trajectories $x(t)$ is assumed, leading to the constraint equation

$$0 = \frac{\mathrm{d}}{\mathrm{d}t} u(t, x(t)) = \nabla u(t, x(t)) \cdot v(1, x(t)) + \partial_t u(t, x). \qquad (1)$$

This constraint equation is generally known as the brightness change constraint equation (BCCE). The BCCE gives us one constraint for two unknowns for image sequences or three unknowns for volume sequences. Thus it is an ill posed problem, which is also known as the aperture problem and only the component of the velocity orthogonal to gray-value structures can be computed from (1). Let us assume that v is at least locally constant. One approach to solve the aperture problem for locally constant v was presented by Guichard [59]. The aperture problem can also be solved *locally* with the Lucas–Kanade approach [82,83] or with the *structure tensor approach* [21] which minimizes the local energy functional

$$\int_0^1 \int_\Omega w(x - y, t - s) \left(\nabla u(s, y) \cdot v(s, y) + \partial_t u(s, y) \right)^2 \mathrm{d}y \, \mathrm{d}s. \qquad (2)$$

Here $w(\cdot, \cdot)$ is a window function, indicating the local spatio-temporal neighborhood. These local approaches offer relatively high robustness with respect to noise

Fig. 1 On a traffic scene plotted motion vectors indicate regions, where a local flow estimater achieves reliable results. For significantly large regions the velocity can not be computed

and allow for a computation of confidence and type measures, which characterize the quality of estimates. Generally, they do not lead to dense flow fields (cf. Fig. 1). However, *global* variational estimators as discussed below lead by design to fully dense flow fields but are known to be more sensitive to noise.

2.2.1 Global, Variational Methods for Optical Flow Estimation

The study of variational methods in optical flow estimation started with the classical work of Horn and Schunk in 1984 [65]. They considered minimizers of the energy functional

$$E[v] = \int_\Omega [(\partial_t u, \nabla u) \cdot (1, v)]^2 + \alpha |\nabla v|^2 \delta x \qquad (3)$$

acting on image intensities $u : [0, 1] \times \Omega \to \mathbb{R}$ at decoupled time steps $t \in [0, 1]$. Thereby they implicitly assume the optical flow field $v : [0, 1] \times \Omega \to \mathbb{R}^n$ to be spatially smooth. Here, the scalar α denotes the constant weighting parameter of the regularizer. Nagel and Enkelmann [94] replaced the second, regularizing term $\alpha |\nabla v|^2$ by a quadratic form

$$\int_\Omega \nabla v \cdot J^\sigma[u] \nabla v \delta x \qquad (4)$$

which involves the local structure tensor $J^\sigma[u]$ of the image (σ indicates the involved filter width) and allows for a significant change in v across image edges indicated by steep image gradients. Alternatively, one can replace the quadratic regularization by a BV type regularization $\int_\Omega |\nabla u| \delta x$ as presented by Cohen [36] or other convex regularizers as considered by Weickert and Schnörr [123]. Connections to shape optimization have been exploited by Schnörr [108]. Weickert et al. [26] proposed a

Parameter Estimation in Image Processing and Computer Vision 315

combination of local flow estimation and global variational techniques to combine the benefits of robustness and dense field representation, respectively. A broader comparison of different regularization techniques involving quasi-convex functionals has been given by Hinterberger et al. [63]. In particular they consider $W^{1,p}$-approximation or BV type functionals. Applying multi-grid methods in the solution of the Euler Lagrange equations for the above combined global-local method was presented by Bruhn et al. [25]. In all these approaches the choice of the regularization terms basically determines the class of admissible flow fields. A rigorous analysis of assumptions on the flow field under which global minimizers of the non regularized variational problem can be found was given by Lefébure and Cohen [76].

In general, numerical algorithms for the minimization tend to get stuck in local minima. Alvarez et al. [84] proposed to consider a scale space approach to solve this problem. Starting to correlate coarse representations of subsequent images in an image sequence via an optical flow field or a deformation, one proceeds on successively finer representation until the actual fine scale images are properly matched.

2.3 Confidence and Situation Measures for Optical Flows

In order to detect artifacts and erroneous flow vectors in optical flow fields, one can analyze confidence and situation measures. Confidence measures are used to estimate the correctness of flow fields, based on information derived from the image sequence and/or the displacement field. Based on proposed techniques, two kinds of confidence measures can be distinguish: situation and confidence measures. Situation measures are used to detect locations, where the optical flow cannot be estimated unambiguously. This is contrasted by confidence measures, which are suited for evaluating the degree of accuracy of the flow field based. Situation measures can be applied e.g., in image reconstruction [87], to derive dense reliable flow fields [110] or to choose the strength of the smoothness parameter in global methods (e.g., indirectly mentioned in [72]). Confidence measures are important for quantifying the accuracy of the estimated optical flow fields. A successful way to obtain robustness to noise in situation and confidence measures is also discussed in [70].

Confidence measures employed are generally chosen as innate to the flow estimation technique. By combining flow methods with non-inherent confidence measures [70] were able to show considerable improvements for confidence and situation measures. Altogether the results of the known measures are only partially satisfactory as many errors remain undetected and a large number of false positive error detections have been observed. Based on a derived optimal confidence map they obtain the results in Fig. 2 for Lynn Quam's Yosemite sequence [61], and the Street [90] test sequences.

For variational methods, the inverse of the energy after optimization has been proposed as a general confidence measure in [27]. For methods not relying on global smoothness assumptions, e.g., local methods, a new confidence measure based on linear subspace projections was proposed in [69]. The idea is to derive a

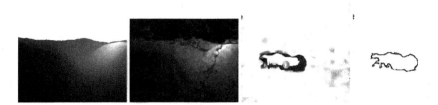

Fig. 2 Comparison of optimal confidence measure (*left*) to best known confidence measure (*right*) for Yosemite and Street sequences

spatio-temporal model of typical flow field patches using e.g., principal component analysis (PCA). Using temporal information the resulting eigenflows represent complex temporal phenomena such as a direction change, a moving motion discontinuity or a moving divergence. Then the reconstruction error of the flow vector is used to define a confidence measure.

2.4 Image Restoration

An active field of research is the restoration of damaged paintings or, in case of digital images, the reconstruction of blank image regions based on image information outside this area. It was first proposed by Masnou and Morel [87] and named "disocclusion." The term "inpainting" was introduced by Bertalmio et al. [17], Ballester et al. [9] proposed a variational approach based on the continuation of contours of equal luminance in the image, also called isophote lines. A variational approach based on level set perimeter and mean curvature was presented by Ambrosio and Masnou in [3]. Other approaches have been proposed for image inpainting, e.g., TV-inpainting and curvature-driven diffusion inpainting suggested by Chan and Shen [34, 35].

In general the problem of inpainting is stated as follows: Given an image $u_0 : \Omega \to \mathbb{R}$ and an inpainting domain $D \subset \Omega$, one asks for a restored image intensity $u : \Omega \to \mathbb{R}$, such that $u|_{\Omega \setminus D} = u_0$ and $u|_D$ is a suitable and regular extension of the image intensity u_0 outside D. The simplest inpainting model is based on the construction of a harmonic function u on D with boundary data $u = u_0$ on ∂D. This model is equivalent to the minimization of the Dirichlet functions $E_{\text{harmon}}[u] = \frac{1}{2} \int_D \|\nabla u\|^2 \delta x$ for given boundary data, as can be derived from Dirichlet's principle. Due to standard elliptic regularity the resulting intensity function u is smooth inside D. This means that edge information present on the boundary will not be restored in the inpainted area. An overview on first order variational functionals related to this problem has been given by Chan and Shen [31, 32]. To resolve this shortcoming, TV-type inpainting models have been proposed [29, 30]. They are based on the functional $E_{\text{TV}}[u] = \frac{1}{2} \int_D \|\nabla u\| \delta x$, which allows for steep transitions on some edge contour. The resulting image intensity is a Bounded Variation (BV)

Parameter Estimation in Image Processing and Computer Vision 317

function and thus characterized by jumps along rectifiable edge contours. In a weak sense, it solves the geometric PDE $h = 0$ with Neumann boundary conditions, where $h = \mathrm{div}\,(\|\nabla u\|^{-1}\nabla u)$ is the mean curvature on level sets or edge contours. Thus the resulting edges will be straight lines.

This issue of straight lines has been overcome in later work by [30, 33]. Here, the functional of the energy is based on the curvature of the intensity level curves. This enforces a smooth transition between the level curves. The equations obtained from such models are highly nonlinear and of higher (fourth) order. Recently, Bredies et al. [24] proposed an approach for higher order TV which significantly improves on stair-casing effects often found in TV regularization. Such an approach has been applied to denoising depth images of Time-of-Flight (ToF) imagers [78].

In many applications the assumption of a sharp boundary ∂D of the corrupted region turns out to be a significant restriction. In fact the reliability of the given image intensity gradually deteriorates from the outside to the inside of the inpainted region. This can be reflected by a relaxed formulation of the variational problem. One considers the functional

$$e^{\epsilon}[u] = \int_{\Omega} |u - u_0|\, H_{\epsilon} + \lambda(1 - H_{\epsilon})\|\nabla u\|^{p}\delta x\,, \qquad (5)$$

where $\lambda > 0$, $p = 1$ or 2, and H_{ϵ} is a convoluted characteristic function χ_D and $\epsilon > 0$ indicates the width of the convolution kernel.

Frequently, one aims for a better continuation of image structures from outside the destroyed region. Mumford [92] phrased this problem in terms of a minimization of elastic energy of curves or surfaces $\mathcal{M}$ treated as elastic rods or shells and C^1 continuity conditions on ∂D. Morel and Masnou [86, 88] have further exploited the relation to a minimization of the Willmore functional

$$E[\mathcal{M}] = \frac{1}{2} \int_{\mathcal{M}} h^2\, da, \qquad (6)$$

where h is the mean curvature of the manifold $\mathcal{M}$. They explicitly construct continuations of level lines in an occluded image region D as parameterized minimizers of the Willmore energy. Ambrosio and Masnou [4] revisited this problem in the context of geometric measure theory and derived minimizers of an implicit formulation of the Willmore energy

$$E[u] = \int_{D} \mathrm{div}\, \left(\frac{\nabla u}{|\nabla u|}\right)^2 |\nabla u|\delta x. \qquad (7)$$

General properties of such variational problems have been discussed by Bellettini et al. in [13]. Chen et al. [35] presented a finite difference relaxation algorithm for this Willmore functional and applied it to image inpainting. Esedoglu and Shen [44] proposed a phase field approximation of the Willmore energy and used a parabolic relaxation in their concrete minimization algorithm. Bertalmio et al. [16] and Ballester et al. [9] relaxed the Willmore functional and studied the simultaneous extension of the image's normal field $\theta = \frac{\nabla u}{|\nabla u|}$ and the intensity u. This resulted to the minimization of the energy

$$E[\theta, u] = \int_D (|\nabla_{(t,x)}u| - \theta \cdot \nabla u) + (\text{div } \theta)^p(a + b|\nabla G * u|)\delta x, \qquad (8)$$

where G is a Gaussian smoothing kernel. For θ they impose the constraint $|\theta| \le 1$. Obviously, the first energy integrant is zero if θ coincides with the image normal. Thus, for $p = 2$ the second term approximates the Willmore energy on the ensemble of level sets on D. Dirichlet boundary conditions for u and θ are assumed.

A different approach has been considered by Bertalmio, Bertozzi and Sapiro [18] connecting fluid dynamics and image inpainting. A vorticity formulation of the stationary Navier–Stokes equations with zero viscosity can be written in terms of the stream function Ψ

$$\nabla^\perp \Psi \cdot \nabla \Delta \Psi = 0, \qquad (9)$$

where $\nabla^\perp$ denotes the orthogonal gradient and $v = \nabla^\perp \Psi$ the velocity field of the flow. The connection to image processing is drawn by replacing Ψ by the image intensity u. One tries to find a solutions of the above equation under boundary conditions for the intensity u and the direction of level lines $\nabla^\perp u$. In terms of physics, this can be thought of as the solution to the transport of the outside image into the hole D which solves the stationary Navier–Stokes equations. The corresponding algorithm is based on a third order, parabolic relaxation

$$\partial_t u - \nabla^\perp u \cdot \nabla \Delta u = 0. \qquad (10)$$

Recently, texture inpainting has attracted attention. Bertalmio et al. [19] proposed a technique to first decomposed the image into texture and structure and then propagated into the inpainting domain both these classes in different ways. This idea to decompose texture and structure is also applied in [58]. Some statistical approaches have been presented in [40] to perform a texture synthesis and structure propagation.

2.5 Optical Flow Reconstruction

In Sect. 2.2 a number of different approaches for estimating optical flow have been presented. Even the most advanced techniques cannot accurately estimate correct flow fields under all conditions. A powerful tool for detecting and removing incorrect flow vectors from the flow field are confidence measures [27, 70]. Discarting erroneous flow vectors results in accurate but sparse motion fields. However, many applications require dense flow fields. This reconstruction of missing vectors in optical flow fields is based on information from the surrounding areas is addressed in this section. The tasks is similar to that addresses in the previous section for image reconstruction, where it was called "inpainting."

Parameter Estimation in Image Processing and Computer Vision 319

The reconstruction of motion fields has lately been proposed in the field of video completion. In case of large holes with complicated texture previously used methods are often not suitable to obtain good results. Instead of reconstructing the frame itself by means of inpainting, the reconstruction of the underlying motion field allows for the subsequent restoration of the corrupted region even in difficult cases.

Hence, the reconstruction of motion fields called "motion inpainting" was first introduced for video stabilization by Matsushita et al. in [89]. The idea is to continue the central motion field to the edges of the image sequence where the field is lost due to camera shaking. This is done by a basic interpolation scheme between four neighboring vectors and a fast marching method. An extension of inpainting to higher dimensional surfaces has also been presented [10].

The reconstruction of optical flow fields can be accomplished by a simple extension of these inpainting functionals for images, e.g., TV-inpainting on two dimensional vector fields. However, these methods sometimes fail in situations where the course of the motion boundary is unclear, e.g., if round motion boundaries or junctions occur. Since image edges often correspond to motion edges the information drawn from the image sequence can be important for the reconstruction, especially in such cases where the damaged vector field does not contain enough information to uniquely determine the optical flow of motion boundaries.

Hence, in the special case of optical flow, the image sequence provides a source of information in addition to the corrupted vector field, which can be used to guide the reconstruction process in ambiguous cases. Optical flow fields have been used for the reconstruction of images in [15]. The resulting functional is nonlinear and can be minimized by means of the finite elements method. This techniques compares favorably to diffusion based and TV inpainting methods, see Fig. 3.

2.6 Motion Segmentation

A common task in image processing is the segmentation of images. Image segmentation in its typical form is the process of assigning a label to every image pixel in a way that pixels with the same label share certain visual characteristics. Image segmentation is closely related to the task of classification. For classification, one tries to assign to each pixel in the image a label or object class, where the classes are agreed in advance. These labels can also be probabilities of the pixel to belonging to certain objects. Depending on the cues that distinguish the object of interest from the background, segmentation can be based on features such as edge information, intensity, color, texture or motion. Well known approaches are:

- Variational Approach (Mumford and Shah functional [91], Geodesic active contours [68], Segmentation from motion[37–39]).
- Multi-resolution techniques (Pyramid linking [104], Wavelet coefficient analysis [73, 79, 116]).

The segmentation from motion can be achieved by *iterative algorithms* based on interleafed motion estimation steps and segmentation steps [111]. Well known

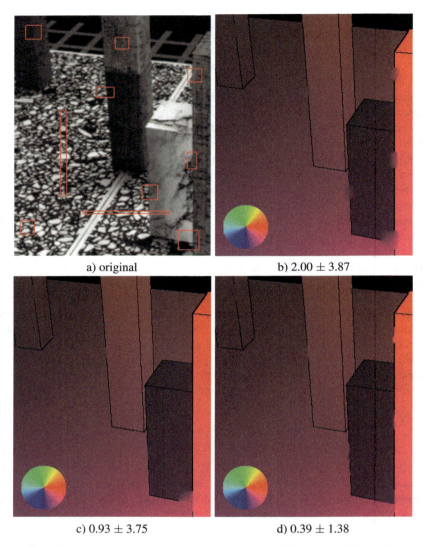

Fig. 3 Comparison of the proposed inpainting algorithm to diffusion and TV inpainting; the numbers indicate the average angular error within the corrupted regions after reconstruction; (**a**) Original corrupted Marble sequence, (**b**) Reconstruction result of diffusion based motion inpainting, (**c**) Reconstruction result of TV based motion inpainting, (**d**) Reconstruction result of image based motion inpainting. Taken from [15]

techniques are furthermore a *layered representation* of image sequences [119–121], *variational approaches*, for instance motion competition [38], and other techniques such as *tensor voting* [95,96] or *algebraic methods for multi-body motion models* [117]. The iterative algorithm introduced by [111] describes a probabilistic relaxation framework for robust multiple motion estimation and segmentation.

Parameter Estimation in Image Processing and Computer Vision 321

[119–121] present a set of techniques for segmenting images into coherently moving regions in a layered representation. Cremers and Soatto [38, 39] gave an extension of the Mumford and Shah functional from intensity segmentation to motion based segmentation in terms of a probabilistic framework implemented by level sets. The geometric prior favors motion boundaries of minimal length and the likelihood term takes into account the orthogonality between the spatio-temporal gradient and the velocity vector. The classical pyramid linking segmentation, as discussed in [28, 105], is based on the Gaussian pyramid of the image. Several improvements were introduced by [104]. Instead of using a Gaussian pyramid, they propose a continuous scale-space. The segmentation by clustering in the feature space has been described by [116]. The feature extraction is based on local variance estimates of the wavelet coefficients of the image.

2.7 Joint Estimation of Optical Flow, Segmentation and Denoising

Rather than denoising images, computing optical flow and performing a segmentation step, all separately, all these components can be combined and computed concurrently [106, 114]. This approach is based on an extension of the well known Mumford–Shah functional which originally was proposed for the joint denoising and segmentation of still images. Given a noisy initial image sequence $u_0 : D \to \mathbb{R}$ on the space-time domain $D = [0, T] \times \Omega$ the following energy is considered

$$E_{\text{MSopt}}[u, w, S] = \int_D \frac{\lambda_u}{2} (u - u_0)^2 \, d\mathcal{L} + \int_{D \backslash S} \frac{\lambda_w}{2} \left(w \cdot \nabla_{(t,x)} u \right)^2 \, d\mathcal{L}$$
$$+ \int_{D \backslash S} \frac{\mu_u}{2} \| \nabla_{(t,x)} u \|^2 \, d\mathcal{L} + \int_{D \backslash S} \frac{\mu_w}{2} \| P_\delta[\zeta] \nabla_{(t,x)} w \|^2 \, d\mathcal{L} + \nu \, \mathcal{H}^d(S)$$

for a piecewise smooth de-noised image sequence $u : D \to \mathbb{R}$, and a piecewise smooth motion field $w = (1, v)$ and a set $S \subset D$ of discontinuities of u and w. λ_u and λ_w are the weighting factors for the fidelity terms of u and w, while μ_u and μ_w are those for the smoothness terms respectively. The first term models the fidelity of the denoised image-sequence u, the second term represents the fidelity of the flow field w in terms of the optical flow equation (1). The smoothness of u and w is required on $D \backslash S$ and finally, the last term is the Hausdorff measure of the set S. A suitable choice of the projection $P_\delta[\zeta]$ leads to an anisotropic smoothing of the flow field along the edges indicated by ζ [106].

The model is implemented in [106, 114] using a phase-field approximation in the spirit of Ambrosio and Tortorelli's approach [5]. Thereby the edge set S is replaced by a phase-field function $\zeta : D \to \mathbb{R}$ such that $\zeta = 0$ on S and $\zeta \approx 1$ far from S. As in the original Ambrosio–Tortorelli model, a scale parameter ϵ controls the thickness of the region with small phase field values. The Euler–Lagrange equations of the corresponding parameters yield a system of three partial differential equations for the image-sequence u, the optical flow field v and the phase field ζ:

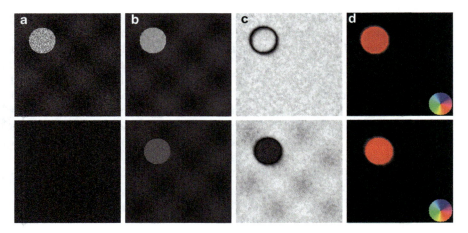

Fig. 4 Noisy test sequence: From top to bottom frames 9 and 10 are shown. (**a**) original image sequence, (**b**) smoothed images, (**c**) phase field, (**d**) estimated motion (*color coded*). Taken from [114]

Fig. 5 Pedestrian video: frames from original sequence (*left*); phase field (*middle*); optical flow, color coded (*right*) [114]

$$-\text{div}_{(t,x)}\left(\frac{\mu_u}{\lambda_u}(\zeta^2 + k_\epsilon)\nabla_{(t,x)}u + \frac{\lambda_w}{\lambda_u}w(\nabla_{(t,x)}u \cdot w)\right) + u = u_0$$

$$-\epsilon \Delta_{(t,x)}\zeta + \left(\frac{1}{4\epsilon} + \frac{\mu_u}{2\nu}\|\nabla_{(t,x)}u\|^2\right)\zeta = \frac{1}{4\epsilon} \quad (11)$$

$$-\frac{\mu_w}{\lambda_w}\text{div}_{(t,x)}\left(P_\delta[\zeta]\nabla_{(t,x)}v\right) + (\nabla_{(t,x)}u \cdot v)\nabla_{(x)}u = 0$$

Neumann boundary conditions are considered in this application. For details on this approximation and its discretization, we refer to [106].

In Fig. 4 we show results from this model on a noisy test-sequence where one frame is completely missing. But this does not hamper the restoration of the correct optical flow field shown in the fourth column, because of the anisotropic smoothing of information from the surrounding frames into the destroyed frame.

Furthermore, in Fig. 5 we consider a complex, higher resolution video sequence showing a group of walking pedestrians. The human silhouettes are well extracted

Parameter Estimation in Image Processing and Computer Vision

and captured by the phase field. The color–coded optical flow plot shows how the method is able to extract the moving limbs of the pedestrians.

3 High-Level Computer Vision

Object recognition in images and videos poses one of the long standing key challenges of computer vision. The problem itself is twofold since recognition involves localizing instances of an object class in novel, cluttered scenes (detection) and classifying these instances as belonging to one of several potential classes (categorization). Developing appropriate object models, which represent the appearance and geometry of each object class and thereby help to distinguish objects from another, constitutes the central problem of recognition. It is common practice to automatically learn these models from unsegmented training images [47, 56, 99], from bounding box segmentations [85, 100], from segmented training images [77], or from manually annotated training images [45]. Since the complexity of object models has to scale with the complexity of object classes, object detection and categorization become particularly challenging when object categories are featuring large intra-class variability. Additionally, the complexity of this problem depends on several other factors, such as the level of supervision during training, the between-class similarity, and the constraints that can be imposed on scenes (e.g., constraints on variation in scale or viewpoint).

3.1 Key Modeling Decisions

Visual recognition can be pursued on different levels of semantic granularity. One extreme strategy is exemplar detection (e.g., [81]), where exactly the same query object is sought in scenes with different environmental conditions such as background, lighting, occlusion, viewpoint etc. The other extreme is category-level object recognition where all instances of a category are to be recognized. Therefore, the granularity of the set of categories controls the complexity of the recognition task as it defines the within-class variability. Influential papers such as [6, 47] have focused research in the field of category-level object recognition on principled probabilistic object models with semi-local feature descriptors. The general goal is to represent objects by learning local appearance features and their spatial configuration and comprising both in a common model. Within this coarse fundamental modeling framework, the current approaches to object categorization can be characterized by the core modeling decisions they make.

1. *Local Descriptors:* A classical way to capture image region information are *appearance patches*, i.e., subsampled image patches that are vector quantized into a large codebook (e.g.,[2, 47, 77]). Complex edge histogram features such as *SIFT* features [81] are another popular choice. In [98] we have proposed a low

dimensional representation of image patches that is based on compact local edge and color histograms of subpatches. Another popular descriptor is *geometric blur* [14]. This feature weights edge orientations around a feature point using a spatially varying kernel. Moreover, edge contour based methods have been proposed in [48, 102]. Opelt et al. [102] extract curve fragments from training images and they apply Adaboost to learn strong object detectors.

2. *Spatial Model:* A second choice concerns the model that combines all local features with their spatial distribution to represent object shape. It should be emphasized that this notion of shape is not based on the object boundary but on the geometry of object parts that are distributed all over the object. Object models have to deal with two problems, simultaneously. On the one hand individual local appearance descriptors in a test image are to be matched against those from a learned model. On the other hand the co-occurrence and the spatial relations between individual features have to be taken into account to represent the global object geometry. The simplest approach is, therefore, to histogram over all local descriptors found in an image (e.g., [41]) and to categorize the image directly based on the overall feature frequencies. On the one hand such *bag of features* methods offer robustness with respect to alteration of individual parts of an object (e.g., due to occlusion) at low computational costs. On the other hand they fail to capture any spatial relations between local image patches and they often adapt to background features. By making the restricting assumption that the spatial structure of objects is limited in its variation with respect to the image, Lazebnik et al. [75] can improve the performance of the bag of features approach using a spatially fixed grid of feature bags. At the other end of the modeling spectrum we find *constellation models*: Originally, Fischler and Elschlager [49] have proposed a spring model for coupling local features. Inspired by the *Dynamic Link Architecture* for cognitive processes, Lades et al. [71] followed the same fundamental idea when proposing their face recognizer. Lately increasingly complex models for capturing part constellations have been proposed [47]. However the complexity of such a joint model of all parts causes only small numbers of parts to be feasible. To incorporate larger numbers of parts, [2, 77] use a simpler object model and a comparably large codebook of distinctive parts. Leibe and Schiele [77] use a probabilistic Hough voting strategy to distinguish one category from the background. In [99] we advance the idea of large numbers of parts by grouping parts prior to spatially coupling the resulting compositions in a graphical model. Conflicting categorization hypotheses proposed by compositions and the spatial model are then reconciled using probabilistic inference in the underlying Bayesian network. The processing pipeline for this automatic scene analysis is presented in Fig. 6. Finally, Berg et al. [14] describe and regularize the spatial distortion resulting from matching an image to a training sample using thin plate splines.

3. *Hierarchies:* For a long time, research on object recognition has aimed at building hierarchical models [52]. Despite this effort, many popular current methods such as [41, 47, 77] are single layered. Recently, *probabilistic latent semantic analysis* (pLSA) [64] has become popular (e.g.,[109]), where a hidden

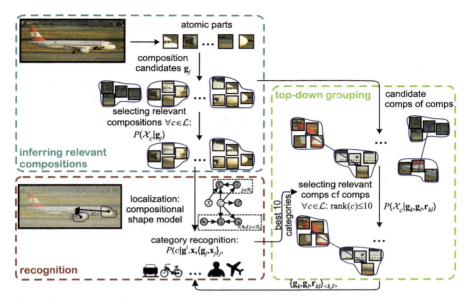

Fig. 6 Processing pipeline for automatic scene analysis. Key steps: Feature extraction, perceptual grouping to form compositions, selection of relevant compositions, object localization and recognition, and top-down grouping to form compositions of compositions which then yield an update of object hypotheses

representation layer of abstract concepts is introduced. Other examples for hierarchical approaches are the feature hierarchies of [43], the hierarchical parts and structure model of [23] or the deep compositional hierarchies of [101].

4. *Learning Paradigm:* Another modeling decision is related to the learning paradigm, i.e., pursuing a generative versus a discriminative approach. Generative models have been very popular in the vision community, e.g., [2,14,22,47,77,98, 112]. They naturally establish correspondences between model components and image features. Discriminative approaches are for instance [41] and [118]. To recognize faces in real-time Viola and Jones [118] use boosting to learn simple features which are based on local intensity differences.

5. *Degree of Supervision:* Similar to the influential paper by Fergus et al. [47] several other approaches (e.g.,[2, 41, 99]) have been proposed that only need training images (showing objects and even background clutter) and the overall category label of an image. The restriction of user assistance is desirable for scaling methods up to large numbers of categories with large training sets. A system that can be trained in an unsupervised manner is for instance that of [56], whereas Felzenszwalb and Huttenlocher have taken a supervised approach to object detection in [45]. Furthermore, Jin and Geman [67] present a compositional architecture with manually built structure for license plate reading. In their conclusion they emphasize the complexity of the future challenge of learning such a compositional model. In [97] we have addressed exactly this problem in the even less constraint case of large numbers of natural object classes.

3.2 A Compositional Approach to Object Categorization

Despite the complexity of the recognition challenge, learning of object representations from a small number of samples is possible due to the *compositional nature* of our (visual) world. As Attneave [8] points out, the visual stimulus is highly redundant in the sense that there exist significant spatial interdependencies in visual scenes. *Compositionality* (cf. Geman's work [54]) serves as a fundamental principle in cognition and especially in human vision [20] that exploits these dependencies. It refers to the prominent ability of perception to represent complex entities by means of comparably few, simple, and widely usable parts. Additional information that is missing in the individual parts is added by incorporating relations between them. The compositional approach presented in [97] automatically learns characteristic compositions of atomic parts. A visualization of relevant compositions by clustering is shown in Fig. 7. Perceptual grouping is applied to obtain candidate compositions. The statistics of relevant compositions that are both reliable and discriminative are then learned from the training data. To avoid overfitting to spurious compositions, cross-validation is performed.

3.3 Object Detection in Cluttered Scenes

Object detection in cluttered natural scenes requires matching object models to the observations in the scene. Since the objective function for matching is a highly non-convex function over scale space, the task of finding good matches is an extremely challenging optimization problem. The two leading approaches to this problem are sliding windows, e.g.,[42, 118], and voting methods, which are based on the Hough transform [66]. Sliding windows scan over possible locations and scales, evaluate a binary classifier, and use post-processing such as non-max suppression to detect objects. The computational burden of this procedure is daunting although various techniques have been proposed to deal with the complexity issue, e.g.,cascaded evaluation [118], interest point filtering, or branch-and-bound [74]. In contrast to this, Hough voting [66] parametrizes the object hypothesis (e.g.,the location of the object center) and lets each local part vote for a point in hypothesis space.

In [100] we have shown that object scale is an inherently global property, which makes local scale estimates unreliable and, thus, leads to a *scale-location-ambiguity*. An illustration of this approach is presented in Fig. 8. When the Hough transform is extended to provide hypotheses for location *and* scale, each local feature casts votes that form lines through scale space rather than just a single point as in current voting methods [48, 77, 102]. Since all points on a voting line are statistically dependent, they should agree on a single object hypothesis rather than being treated as independent votes. Ideally, all points on an object would yield lines that intersect in a single point. Due to intra-category variation, and background clutter, the points of intersection are, however, degraded into scattered clouds. Finding these clusters

Parameter Estimation in Image Processing and Computer Vision 327

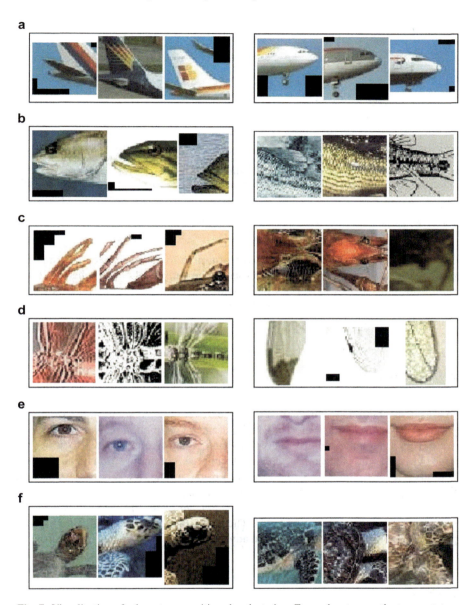

Fig. 7 Visualization of relevant compositions by clustering. For each category, the two prototypical compositions with highest relevance are illustrated by visualizing the closest compositions to that prototype. (**a**) airplanes, (**b**) bass, (**c**) crayfish, (**d**) dragonfly, (**e**) faces, and (**f**) hawksbill

becomes difficult since their number is unknown (it is the number of objects in the scene) and because the assignment of votes to objects is not provided (segmentation problem). To address these issues we frame the search for globally consistent object hypotheses as a weighted, pairwise clustering of local votes without scale

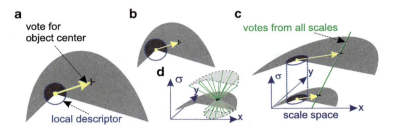

Fig. 8 Voting in scale space with scale-location-ambiguity. The *circle* indicates the spatial support of a local feature. Based on the descriptor, the difference of object scale between (**a**) and (**b**) is not detectable. Thus, a local feature casts votes for the object center on all scales, (**c**). These votes lie on a line through scale space, since the position of the center relative to a feature varies with object scale. Without noise, all voting lines from an object intersect in a single point in scale space, (**d**). For all other scales this point is blurred as indicated by the *dotted outline*

estimates. Clustering avoids a local search through hypothesis space [77] and the pairwise setting circumvents having to specify the number of objects ahead of time. Moreover, clustering voting lines deals with the large number of false positives [57] which point votes produce and that hamper the commonly used local search heuristics such as binning [77].

4 Conclusion

In this contribution, we have given a brief overview of some areas of image processing and computer vision in which parameter estimation plays a central role. We have not attempted to give a complete and exhaustive overview. However, we have outlined some of the approaches and techniques commonly used in the area and demonstrated them on typical applications. From our overview it becomes apparent that most parameter estimation problems in image processing and computer vision are highly non-linear. Inherent are often large amounts of image data and the requirement of fast computation times. Therefore, the presented applications would profit from algorithmic and conceptual advances.

References

1. Adelson EH, Bergen JR (1985) Spatiotemporal energy models for the perception of motion. Journal of the Optical Society of America A 2(2):284–299
2. Agarwal S, Awan A, Roth D (2004) Learning to detect objects in images via a sparse, part-based representation. IEEE Transactions on Pattern Analysis and Machine Intelligence 26(11):1475–1490
3. Ambrosio L, Masnou S (2003a) A direct variational approach to a problem arising in image reconstruction. Interfaces and Free Boundaries 5:63–81

Parameter Estimation in Image Processing and Computer Vision 329

4. Ambrosio L, Masnou S (2003b) A direct variational approach to a problem arising in image reconstruction. Interfaces Free Bound 5(1):63–81
5. Ambrosio L, Tortorelli VM (1992) On the approximation of free discontinuity problems. Boll Un Mat Ital B 6(7):105–123
6. Amit Y, Geman D (1998) A computational model for visual selection. Neural Computation 11(7):1691–1715
7. Anandan P (1989) A computational framework and an algorithm for the measurement of visual motion. International Journal of Computer Vision 2:283–319
8. Attneave F (1954) Some informational aspects of visual perception. Psychological Review 61(3):183–193
9. Ballester C, Bertalmio M, Caselles V, Sapiro G, Verdera J (2001a) Filling-in by joint interpolation of vector fields and gray levels. IEEE Transactions on Image Processing 10(8):1200–1211
10. Ballester C, Caselles V, Verdera J (2003b) Disocclusion by Joint Interpolation of Vector Fields and Gray Levels. Multiscale Modeling & Simulation 2(1):80, DOI 10.1137/S1540345903422458
11. Barron JL, Fleet DJ, Beauchemin S (1994) Performance of optical flow techniques. International Journal of Computer Vision 12(1):43–77
12. Beauchemin SS, Barron JL (1995) The computation of optical flow. ACM Computing Surveys 27(3):433–467
13. Bellettini G, Dal Maso G, Paolini M (1993) Semicontinuity and relaxation properties of a curvature depending functional in 2d. Ann Scuola Norm Sup Pisa Cl Sci 20:247–297
14. Berg AC, Berg TL, Malik J (2005) Shape matching and object recognition using low distortion correspondence. In: Proceedings of the IEEE Conference on Computer Vision and Pattern Recognition, pp 26–33
15. Berkels B, Kondermann C, Garbe C, Rumpf M (2009) Reconstructing optical flow fields by motion inpainting. In: Energy Minimization Methods in Computer Vision and Pattern Recognition, Springer Verlag, vol LNCS 5681, pp 388–400, DOI 10.1007/978-3-642-03641-5_29
16. Bertalmio M, Sapiro G, Caselles V, Ballester C (2000a) Image inpainting. In: Computer Graphics (SIGGRAPH '00 Proceedings), pp 417–424
17. Bertalmio M, Sapiro G, Randall G (2000b) Morphing active contours. PAMI 22(7):733–743
18. Bertalmio M, Bertozzi A, Sapiro G (2001) Navier-Stokes, fluid dynamics, and image and video inpainting. Proceedings of the International Conference on Computer Vision and Pattern Recognition, IEEE I:355–362
19. Bertalmio M, Vese L, Sapiro G, Osher S (2003) Simultaneous structure and texture image inpainting. IEEE Transactions on Image Processing 12(8):882–889, DOI 10.1109/TIP.2003.815261
20. Biederman I (1987) Recognition-by-components: A theory of human image understanding. Psychological Review 94(2):115–147
21. Bigün J (1988) Local symmetry features in image processing. PhD thesis, Linköping University, Linköping, Sweden
22. Borenstein E, Sharon E, Ullman S (2004) Combining top-down and bottom-up segmentation. In: Proceedings of the IEEE Conference on Computer Vision and Pattern Recognition, Workshop Percept Org in Comp Vision
23. Bouchard G, Triggs B (2005) Hierarchical part-based visual object categorization. In: Proceedings of the IEEE Conference on Computer Vision and Pattern Recognition, pp 710–715
24. Bredies K, Kunisch K, Pock T (2010) Total generalized variation. Siam Journal On Imaging Sciences 3(3):492, DOI 10.1137/090769521
25. Bruhn A, Weickert J, Feddern C, Kohlberger T, Schnörr C (2003) Real-time optic flow computation with variational methods. In: Petkov N, Westenberg M (eds) Computer Analysis of Images and Patterns, Lecture Notes in Computer Science, vol 2756, Springer Berlin/Heidelberg, pp 222–229
26. Bruhn A, Weickert J, Schnörr C (2005) LucasKanade meets HornSchunck: Combining local and global optic flow methods. Int J Computer Vision 61(3):211–231

27. Bruhn A, Weickert J, Kohlberger T, Schnörr C (2006) A multigrid platform for real-time motion computation with discontinuity-preserving variational methods. International Journal of Computer Vision 70(3):257–277
28. Burt PJ, Adelson EH (1983) The laplacian pyramid as a compact image code. IEEE TransCOMM 31:532–540
29. Caselles V, Morel JM, Sbert C (1998) An axiomatic approach to image interpolation. Image Processing, IEEE Transactions on 7(3):376–386, DOI 10.1109/83.661188
30. Chan T, Shen J (2001a) Mathematical models for local nontexture inpaintings. SIAM Journal on Applied Mathematics 62(3):1019–1043, DOI 10.1137/S0036139900368844
31. Chan T, Shen J (2001b) Non-texture inpainting by curvature-driven diffusions (ccd). J Visual Comm Image Rep 12:436–449
32. Chan T, Shen J (2002) Mathematical models for local non-texture inpaintings. SIAM J Appl Math 62:1019–1043
33. Chan TF, Tai XC (2003) Level set and total variation regularization for elliptic inverse problems with discontinuous coefficients. Journal of Computational Physics 193(1):40–66, DOI 10.1016/j.jcp.2003.08.003
34. Chan TF, Osher S, Shen J (2001) The digital tv filter and nonlinear denoising. IEEE Transactions on Image Processing 10(2):231–241
35. Chan TF, Kang SH, Shen J (2002) Euler's elastica and curvature-based inpainting. SIAM Appl Math 63(2):564–592
36. Cohen I (93) Nonlinear variational method for optical flow computation. In: Proc. of the Eighth Scandinavian Conference on Image Analysis, vol 1, pp 523–530
37. Cremers D, Schnörr C (2002) Motion competition: Variational integration of motion segmentation and shape regularization. In: Van Gool L (ed) Pattern Recognition - Proc. of the DAGM, Lecture Notes in Computer Science, vol 2449, pp 472–480
38. Cremers D, Schnörr C (2003) Statistical shape knowledge in variational motion segmentation. Image and Vision Computing 21:77–86
39. Cremers D, Soatto S (2005) Motion competition: A variational approach to piecewise parametric motion segmentation. International Journal of Computer Vision 62:249–265
40. Criminisi A, Pérez P, Toyama K (2004) Region filling and object removal by exemplar-based image inpainting. IEEE Transactions on Image Processing 13(9):1200–1212, DOI 10.1109/TIP.2004.833105
41. Csurka G, Dance CR, Fan L, Willamowski J, Bray C (2004) Visual categorization with bags of keypoints. In: ECCV, *Workshop on Stat Learn in CV'04*
42. Dalal N, Triggs B (2005) Histograms of oriented gradients for human detection. In: CVPR, pp 886–893
43. Epshtein B, Ullman S (2005) Feature hierarchies for object classification. In: ICCV, pp 220–227
44. Esedoglu S, Jianhong S (2002) Digital inpainting based on the Mumford-Shah-Euler image model. Euro Jnl Appl Math 13:353–370
45. Felzenszwalb PF, Huttenlocher DP (2005) Pictorial structures for object recognition. IJCV 61(1):55–79
46. Fennema C, Thompson W (1979) Velocity determination in scenes containing several moving objects. Computer Graphics and Image Processing 9:301–315
47. Fergus R, Perona P, Zisserman A (2003) Object class recognition by unsupervised scale-invariant learning. In: CVPR, pp 264–271
48. Ferrari V, Fevrier L, Jurie F, Schmid C (2008) Groups of adjacent contour segments for object detection. IEEE Transactions on Pattern Analysis and Machine Intelligence 30(1): 36–51
49. Fischler MA, Elschlager RA (1973) The representation and matching of pictorial structures. IEEE Transactions on Computers c-22(1):67–92
50. Fleet DJ (1992) Measurement of Image Velocity. Kluwer Academic Publishers, Dordrecht, The Netherlands

Parameter Estimation in Image Processing and Computer Vision 331

51. Fleet DJ, Jepson AD (1990) Computation of component image velocity from local phase information. International Journal of Computer Vision 5:77–104
52. Fukushima K (1980) Neocognitron: A self-organizing neural network model for a mechanism of pattern recognition unaffected by shift in position. Biological Cybernetics 36(4):193–202
53. Galvin B, McCane B, Novins K, Mason D, Mills S (1998) Recovering motion fields: an evaluation of eight optical flow algorithms. In: BMVC 98. Proceedings of the Ninth British Machine Vision Conference, vol 1, pp 195–204
54. Geman S, Potter DF, Chi Z (2002) *Composition Systems*. Quarterly of Applied Mathematics 60:707–736
55. Glazer F, Reynolds G, Anandan P (1983) Scene matching through hierarchical correlation. In: Proc. Conference on Computer Vision and Pattern Recognition, Washington, pp 432–441
56. Grauman K, Darrell T (2006) Pyramid match kernels: Discriminative classification with sets of image features. Tech. Rep. MIT-2006-020, MIT
57. Grimson W, Huttenlocher D (1990) On the sensitivity of the hough transform for object recognition. Pattern Analysis and Machine Intelligence, IEEE Transactions on 12(3):255 – 274, DOI 10.1109/34.49052
58. Grossauer H (2004) A combined pde and texture synthesis approach to inpainting. In: Pajdla T, Matas J (eds) Computer Vision - ECCV 2004, Lecture Notes in Computer Science, vol 3022, Springer-Verlag, pp 214–224, DOI 10.1007/978-3-540-24671-8_17
59. Guichard F (1998) A morphological, affine, and galilean invariant scale–space for movies. IEEE Transactions on Image Processing 7(3):444–456
60. Haußecker H, Spies H (1999) Motion. In: Jähne B, Haußecker H, Geißler P (eds) Handbook of Computer Vision and Applications, vol 2, Academic Press, chap 13
61. Heeger D (1987) Model for the extraction of image flow. Journal of the Optical Society of America 4(8):1455–1471
62. Heeger DJ (1988) Optical flow using spatiotemporal filters. International Journal of Computer Vision 1:279–302
63. Hinterberger W, Scherzer O, Schnörr C, Weickert J (2001) Analysis of optical flow models in the framework of calculus of variations. Tech. rep., Numerical Functional Analysis and Optimization, Revised version of Technical Report No. 8/2001. Computer Science Series, University of Mannheim, Germany
64. Hofmann T (2001) Unsupervised learning by probabilistic latent semantic analysis. Machine Learning 42(1):177–196
65. Horn B, Schunk B (1981) Determining optical flow. Artificial Intelligence 17:185–204
66. Hough P (1962) Method and means for recognizing complex patterns. *U.S. Patent 3069654*
67. Jin Y, Geman S (2006) Context and hierarchy in a probabilistic image model. In: Proceedings of the IEEE Conference on Computer Vision and Pattern Recognition, pp 2145–2152
68. Kass M, Witkin A, Terzopoulos D (1987) Snakes: Active contour models. In: ICCV87, pp 259–268
69. Kondermann C, Kondermann D, Jähne B, <u>CS Garbe</u> (2007a) An adaptive confidence measure for optical flows based on linear subspace projections. In: Hamprecht F, Schnörr C, Jähne B (eds) Pattern Recognition, Springer Verlag, vol LNCS 4713, pp 132–141, DOI 10.1007/978-3-540-74936-3_14
70. Kondermann C, Mester R, Garbe C (2008) A statistical confidence measure for optical flows. In: Forsyth D, Torr P, Zisserman A (eds) Computer Vision - ECCV 2008, Springer Verlag, vol LNCS 5304, pp 290–301, DOI 10.1007/978-3-540-88690-7_22
71. Lades M, Vorbrüggen JC, Buhmann JM, Lange J, von der Malsburg C, Würtz RP, Konen W (1993) Distortion invariant object recognition in the dynamic link architecture. IEEE Transactions on Computers 42:300–311
72. Lai S, Vemuri B (1995) Robust and efficient algorithms for optical flow computation. In: Proc. of Int. Symp. Comp. Vis., pp 455–460
73. Laine A, Fan J (1996) Frame representations for texture segmentation. IEEE Transactions on Image Processing 5(5):771–780, URL citeseer.ist.psu.edu/article/laine96frame.html

74. Lampert CH, Blaschko MB, Hofmann T (2008) Beyond sliding windows: Object localization by efficient subwindow search. In: Proceedings of the IEEE Conference on Computer Vision and Pattern Recognition
75. Lazebnik S, Schmid C, Ponce J (2006) Beyond bags of features: Spatial pyramid matching for recognizing natural scene categories. In: Proceedings of the IEEE Conference on Computer Vision and Pattern Recognition
76. Lefébure M, Cohen LD (2001) Image registration, optical flow and local rigidity. Journal of Mathematical Imaging and Vision 14:131–147
77. Leibe B, Leonardis A, Schiele B (2004) Combined object categorization and segmentation with an implicit shape model. In: ECCV, *Workshop Stat Learn'04*
78. Lenzen F, Schäfer H, CS Garbe (2011) Denoising Time-of-Flight data with adaptive total variation. In: Bebis G, Boyle R, Koracin D, Parvin B (eds) Advances in Visual Computing, ISVC 2011, vol LNCS 6938, pp 337–346, DOI 10.1007/978-3-642-24028-7_31
79. Liapis, S S, Tziritas, G (2004) Colour and texture segmentation using wavelet frame analysis, deterministic relaxation, and fast marching algorithms. Journal of Visual Communication and Image Representation 15:1–26
80. Little JJ, Verri A (1989) Analysis of differential and matching methods for optical flow. In: IEEE Workshop on Visual Motion, Irvine, CA, pp 173–180
81. Lowe DG (2004) Distinctive image features from scale-invariant keypoints. International Journal of Computer Vision 60(2):91–110
82. Lucas B, Kanade T (1981) An iterative image registration technique with an application to stereo vision. In: DARPA Image Understanding Workshop, pp 121–130
83. Lucas BD (1984) Generalized image matching by the method of differences. PhD thesis, Carnegie-Mellon University, Pittsburgh, PA
84. Luis Alvarez JS Joachim Weickert (1999) A scale-space approach to nonlocal optical flow calculations. Proceedings of the Second International Conference on Scale-Space Theories in Computer Vision pp 235–246
85. Maji S, Malik J (2009) Object detection using a max-margin hough transform. In: CVPR
86. Masnou S (2002) Disocclusion: A variational approach using level lines. IEEE Transactions on Image Processing 11(2):68–76
87. Masnou S, Morel J (1998a) Level lines based disocclusion. In: Proc. of ICIP, vol 3, pp 259–263
88. Masnou S, Morel JM (1998b) Level lines based disocclusion. In: 5th IEEE International Conference on Image Processing (ICIP), Chicago, vol 3, pp 259–263
89. Matsushita Y, Ofek E, Ge W, Tang X, Shum HY (2006) Full-frame video stabilization with motion inpainting. Pattern Analysis and Machine Intelligence, IEEE Transactions on 28(7):1150–1163, DOI 10.1109/TPAMI.2006.141
90. McCane B, Novins K, Crannitch D, Galvin B (2001) On benchmarking optical flow. Computer Vision and Image Understanding 84(1):126–143
91. Mumford D, Shah J (1989) Optimal approximation by piecewise smooth functions and associated variational problems. Communications on Pure and Applied Mathematics 42:577–685
92. Müller-Urbaniak S (1994) Eine Analyse des Zweischritt-[Theta]-Verfahrens zur Lösung der instationären Navier-Stokes-Gleichungen. Tech. rep.
93. Nagel HH (1983) Displacement vectors derived from second-order intensity variations in image sequences. Computer Graphics and Image Processing 21:85–117
94. Nagel HH, Enkelmann W (1986) An investigation of smoothness constraints for the estimation of displacement vector fields from image sequences. IEEE Trans Pattern Anal Mach Intell 8(5):565–593
95. Nicolescu,M, Medioni,G (2002) 4d voting for matching, densification and segmentation into motion layers. In: Proceedings of the International Conference on Pattern Recognition, vol 3
96. Nicolescu,M, Medioni,G (2003) Layered 4d representation and voting for grouping from motion. Pattern Analysis and Machine Intelligence 25(4):492–501
97. Ommer B, Buhmann J (2010) Learning the compositional nature of visual object categories for recognition. PAMI 32(3):501–516

Parameter Estimation in Image Processing and Computer Vision 333

98. Ommer B, Buhmann JM (2005) Object categorization by compositional graphical models. In: Energy Minimization Methods in Computer Vision and Pattern Recognition, LNCS 3757, pp 235–250
99. Ommer B, Buhmann JM (2006) Learning compositional categorization models. In: ECCV, pp 316–329
100. Ommer B, Malik J (2009) Multi-scale object detection by clustering lines. In: Computer Vision, 2009 IEEE 12th International Conference on, pp 484 –491. DOI 10.1109/ICCV.2009. 5459200
101. Ommer B, Sauter M, Buhmann JM (2006) Learning top-down grouping of compositional hierarchies for recognition. In: CVPR, *Workshop POCV*
102. Opelt A, Pinz A, Zisserman A (2006) Incremental learning of object detectors using a visual shape alphabet. In: CVPR, pp 3–10
103. Oppenheim AV, Schafer RW (2009) Discrete-Time Signal Processing, 3rd edn. Prentice Hall
104. Petrovic,A E, Vanderheynst,P (2004) Multiresolution segmentation of natural images: From linear to nonlinear scale-space representation. IEEE Transactions on Image Processing 13(8):1104–1114
105. Pietikäinen M, Rosenfeld A (1981) Image segmentation by texture using pyramid node linking. Systems, Man and Cybernetics, IEEE Transactions on 11(12):822–825, DOI 10.1109/ TSMC.1981.4308623
106. Preußer, Droske M, Garbe CS, Telea A, Rumpf M (2007) A phase field method for joint denoising, edge detection and motion estimation. SIAM Appl Math, 68(3): 599–618, DOI 10.1137/060677409
107. Proakis JG, Manolakis DK (2006) Digital Signal Processing, 4th edn. Prentice Hall
108. Schnörr C (1992) Computation of discontinuous optical flow by domain decomposition and shape optimization. International Journal Computer Vision 8(2):153–165
109. Sivic J, Russell BC, Efros AA, Zisserman A, Freeman WT (2005) Discovering objects and their localization in images. In: ICCV
110. Spies H, Garbe CS (2002) Dense parameter fields from total least squares. In: Van Gool L (ed) Pattern Recognition, Springer-Verlag, Zurich, CH, Lecture Notes in Computer Science, vol LNCS 2449, pp 379–386
111. Strehl,A, Aggarwal,JK (2000) A new Bayesian relaxation framework for the estimation and segmentation of multiple motions. In: Proceedings of the 4th IEEE Southwest Symposium on Image Analysis and Interpretation (SSIAI 2000), 2–4 April 2000. Austin, Texas, USA, IEEE, pp 21–25, URL citeseer.ist.psu.edu/strehl00new.html
112. Sudderth EB, Torralba AB, Freeman WT, Willsky AS (2005) Learning hierarchical models of scenes, objects, and parts. In: ICCV
113. Tan L (2007) Digital Signal Processing: Fundamentals and Applications. Academic Press
114. Telea A, Preusser T, Garbe C, Droske M, Rumpf M (2006) A variational approach to joint denoising, edge detection and motion estimation. In: Proc. DAGM 2006, pp 525–353
115. Tretiak O, Pastor L (1984) Velocity estimation from image sequences with second order differential operators. In: Proc. 7th International Conference on Pattern Recognition, pp 20–22
116. Unser M (1995) Texture classification and segmentation using wavelet frames. IEEE Transaction on Image Processing 11:1549–1560
117. Vidal R, Ma Y (2004) A unified algebraic approach to 2-d and 3-d motion segmentation. In: ECCV (1), pp 1–15
118. Viola PA, Jones MJ (2001) Rapid object detection using a boosted cascade of simple features. In: CVPR, pp 511–518
119. Wang JYA, Adelson EH (1993) Layered representation for motion analysis. In: Wang JYA, Adelson EH (eds) Proceedings CVPR'93, New York City, NY, Washington, DC, pp 361–366
120. Wang JYA, Adelson EH (1994) Representating moving images with layers. IEEE Transaction on Image Processing 3(5):625–638
121. Wang,JYA,Adelson,EH (1994) Representing moving images with layers. The IEEE Transactions on Image Processing Special Issue: Image Sequence Compression 3(5):625–638

122. Waxman AM, Wu J, Bergholm F (1988) Convected activation profiles and receptive fields for real time measurement of short range visual motion. In: Proc. Conf. Comput. Vis. Patt. Recog, Ann Arbor, pp 771–723
123. Weickert J, Schnörr C (2001) A theoretical framework for convex regularizers in PDE–based computation of image motion. Int J Computer Vision 45(3):245–264